MULTIBODY DYNAMICS WITH UNILATERAL CONTACTS

WILEY SERIES IN NONLINEAR SCIENCE

Abdullaev	Theory of Solitons in Inhomogeneous Media
Bolotin	Stability Problems in Fracture Mechanics
Nayfeh	Method of Normal Forms
Nayfeh and Balachandran	Applied Nonlinear Dynamics
Nayfeh and Pai	Linear and Nonlinear Structural Mechanics
Ott, Sauer, and Yorke	Coping with Chaos
Pfeiffer and Glocker	Multibody Dynamics with Unilateral Contacts
Qu	Robust Control of Nonlinear Uncertain Systems
Rozhdestvensky	Matched Asymptotics of Lifting Flows
Vakakis, et al.	Normal Modes and Localization in Nonlinear Systems

MULTIBODY DYNAMICS WITH UNILATERAL CONTACTS

FRIEDRICH PFEIFFER

CHRISTOPH GLOCKER
Technische Universität München

A Wiley-Interscience Publication
JOHN WILEY & SONS, INC.
New York • Chichester • Brisbane • Toronto • Singapore

This text is printed on acid-free paper.

Library of Congress Cataloging-in-Publication Data

Pfeiffer, Friedrich, 1935–
Multibody dynamics with unilateral contacts / Friedrich Pfeiffer, Christoph Glocker.
p. cm. — (Wiley series in nonlinear science)
Includes bibliographical references and index.
ISBN 0-471-15565-9 (cloth : alk. paper)
1. Machinery, Dynamics of. 2. Nonlinear systems. I. Glocker, Christoph. II. Title. III. Series.
TJ173.P48 1996
621.8′11—dc20 96-12451

Printed in the United States of America

10 9 8 7 6 5 4 3 2 1

Everything should be made as simple as possible but not simpler.

Albert Einstein

PREFACE

Machines and mechanisms continually become more complex and more perfect and, thus, are consistently accompanied by more mathematical modeling and simulation. Sophisticated machines require sophisticated methods, which, nevertheless, must relate to reality. Theories for rigid or elastic multibody systems and FEM/BEM-algorithms are typical examples which have considerably influenced progress in mechanical engineering.

Machines and mechanisms are systems of interconnected bodies in which the interconnections are often modeled by applying bilateral constraints. Although these models are sometimes correct, they often are not. Noise-generating mechanisms, fatigue, and wear in many cases are caused from relative motion between two bodies that is usually characterized by impulsive and stick-slip phenomena. Because machines are multibody configurations with many dependent contacts, adequate theories are required. Moreover, many technical systems rely on impulsive and stick-slip processes to perform their intended functions. The same type of theory is needed here as well. In this book we consider multibody systems with multiple dependent contacts and develop an adequate theory. In spite of the fact that the theory covers a huge, and still growing, number of applications, it was not available in a form accessible to engineers. This book tries to fill this gap. Our intention is not to give another version of multibody system theory, but focus on multibody systems with multiple, unilateral, and, often, undecoupled contacts.

The credit for establishing the mathematical foundation for nonsmooth mechanics belongs to a few European colleagues, especially Professor Moreau in Montpellier and Professor Panagiotopoulos in Thessaloniki. Their theories are mainly based on convex analysis and on accompanying fields such as linear and quadratic programming in optimization theory.

The Lehrstuhl B für Mechanik (LBM) in Munich originally began in the 1980s with a series of practical problems, but then became more involved with the mathematical foundations. This book is the result of ten years' work with many dissertations and practical contributions concerning dependent contacts in multibody systems. The theory in Part 1 has reached a state which allows the treatment of very general problems of nonsmooth dynamics. The new ideas with respect to impacts with friction have been confirmed by many experiments, although additional research is necessary to improve the model. The level of confidence in the theory is very high. The number of successful industry applications, presented in Part 2, confirms the relevancy of our modeling approach, which turns out to be quite general, including many classical methods as special cases. Its significance is increasing so quickly that we easily could fill an additional volume with sophisticated applications.

We have to thank many co-workers, associates, and friends for supporting us in writing this book. Dipl.-Ing. Markus Wösle supervised the printing and the computer generation of the figures. He also evaluated some of the examples and did some proofreading, as did Dipl.-Ing. Jürgen Braun. We are particularly indebted to Professor Ali Nayfeh of Virginia Tech, who invited us to contribute this book to his series on nonlinear dynamics. Many thanks are due to our editors at Wiley for their friendly assistance and cooperation. We apologize to those whose work was inadvertently omitted in the literature. We welcome all comments and corrections from readers.

München, im November 1995

Friedrich Pfeiffer
Christoph Glocker

CONTENTS

PART 2: Applications

MULTIBODY DYNAMICS WITH UNILATERAL CONTACTS

PART 1

THEORY

1

INTRODUCTION

1.1 Modeling Mechanical Systems

Modeling mechanical systems such as machines and mechanisms is a matter of engineering intuition and of the relevant powers of practical imagination. Models of any area of mechanics, and of all physical fundamental subjects, necessarily include assumptions, usually in an approximate form; moreover, the realization of models is often accompanied by numerical problems. In technical applications models are limited by the data situation. Establishing a complete data set for a machine might take more time than creating a model.

The first step in considering models must be a very clear and precise elaboration of the goals of the model. Is it for simulating an object, or do I want to establish a plant model for control design, a parameter model for design improvements possibly in connection with optimization processes, or a system model replacing extensive laboratory and field tests? The different requirements will result in different model approaches. In any case the chances of establishing a good model depend very strongly on a deep understanding of the physical-technical processes of the object to be modeled. A good model means a good representation of mechanical properties and therefore a good correspondence to practice and its measurements.

One word on experiments and their modeling character: With the exception of direct field tests with complete machines or transportation systems, most experiments, even in industry, are models including all the properties mentioned. Therefore, measurements are not a dogma, but researchers must know what equipment was used, what sensors were employed, how they were applied, where they were used, how signal processing was performed, and so on. Good measurements are as rare as good theoretical models. But, on the other hand, an optimized combination of experiments and theory might accelerate considerably progress in research and development with respect to a problem. This seems to be noteworthy, although it is not the topic of this book.

Good models are economical models; they include everything to achieve the goals, but not more. For an example, the size of multibody models representing vibrational systems depends on the largest frequencies of interest. These frequencies also indicate if some bodies must be modeled as elastic bodies. Grabbing of the clutch in cars, for instance, usually is observed in a frequency range of 6–15 Hz. From this it is sufficient to represent that process by a 3-4 mass configuration and a realistic stick-slip model. Anything more would be not economical.

In this book we establish a unique theory on multibody systems with multiple contacts. Mechanically we deal with arrangements of an arbitrary number of rigid or elastic bodies which possess, in addition to their continuous constraints being represented by steady constraint equations, an arbitrary number of unilateral contacts characterized by noncontinuous constraint equations. Multibody systems with impact- and friction-driven processes are a typical example. To model systems of that kind we may think about quite a number of possibilities.

First, we might leave the concept of multibody theory and model all bodies and all couplings of a machine by a FEM system, which industry really does. This results in extremely large models that can be very helpful when correctly applied. Computing times will be very large, the correctness of meshsizing is not always ensured and the numerical results need interpretation. Existing FEM codes cannot deal with unilateral problems correctly.

As a second variant we may return to our multibody system approach but model all joints, linear and nonlinear couplings, and all contacts in a more detailed way, taking into consideration, for example, local deformation effects, including local nonlinear behavior. Again we would have large computing times, and, as in all cases, we must verify our local coupling models by experiments.

In the following we start with a multibody approach including arbitrary continuous joints and couplings, the last represented by any type of steady force law. With respect to unilateral contacts we shall consider impacts and friction and a combination of both. Classical contact laws are applied throughout but with a specific adaptation to multiple–contact situations. As we shall see, this leads to a formulation allowing for an application of a linear complementarity algorithm which can be interpreted as a modified form of the well-known simplex algorithm. Bringing the equations of motion into such a form requires that, at the very instant of change from static to sliding or from sliding to static friction, the coefficients of static and sliding friction be equal. This is neither a loss of generality nor a violation of our physical understanding of technical processes, for the following reasons. In technical applications we apply friction characteristics in the form of a friction force F_F as a function of relative velocity $v_{\rm rel}$ or as a friction torque dependent on the relative angular velocity (Fig. 1.1). Typically, such curves start at zero rela-

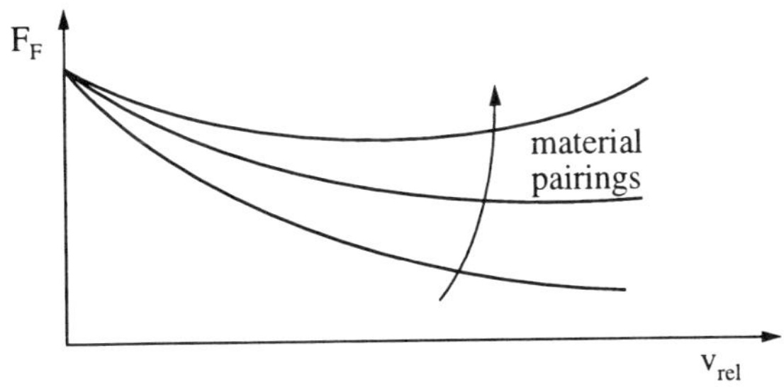

Figure 1.1: Friction Characteristics (Stribeck curves)

tive velocity with a negative slope, which, by the way, is the main reason for self-excited oscillations in frictional systems. Therefore, at the very instant of a change from stiction to sliding, or vice versa, the friction force (torque) and, thus, the friction coefficients remain approximately the same, justifying the above requirement. It excludes only a jump, which in no way is a good approximation to reality. For sliding contacts, of course, any frictional force law may be applied.

1.2 Single-Contact Dynamics

All classical textbooks on mechanics and most current research concentrate on mechanical systems with only one or two degrees of freedom and with one impulsive or frictional contact. Books and papers on chaotical properties very often use as mechanical examples impact or stick-slip systems. In the following we review the basic ideas [5, 27, 35, 43, 49, 53, 59].

Two bodies will impact if their relative distance becomes zero. This event is then a starting point for a process, which usually is assumed to have an extremely short duration. Nevertheless, deformation of the two bodies occurs, being composed of compression and expansion phases (Fig. 1.2). The forces governing this deformation depend on the initial dynamics and kinematics of the contacting bodies. The impulsive process ends when the normal force of contact vanishes and changes sign. The condition of zero relative distance cannot be used as an indicator for the end of an impact.

In the general case of impact with friction we must also consider a possible change from sliding to sticking, or vice versa, which includes frictional aspects as treated later.

In the simple case of only normal velocities we sometimes can idealize impacts according to Newton's impact laws, which relate the relative velocity after an impact with that before an impact. Such an idealization can only be performed if the force budget allows it. In the case of impacts by hard

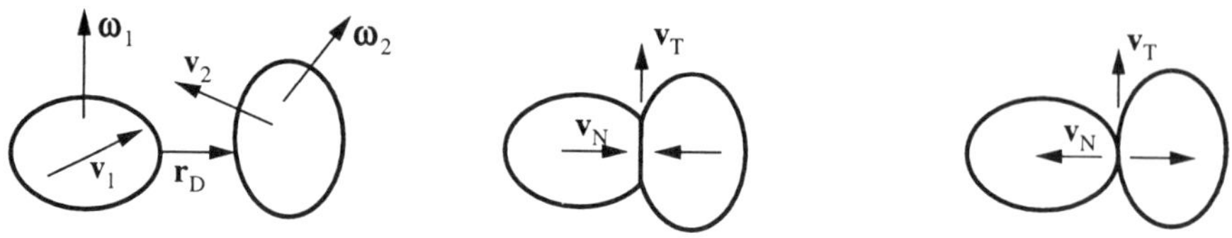

Figure 1.2: Details of an Impact

loaded bodies we must analyze the deformation in detail. Gear hammering taking place under heavy loads and gear rattling taking place under no load are typical examples.

As in all other contact dynamical problems, impacts possess complementarity properties. For ideal classical inelastic impacts either the relative velocity is zero and the accompanying normal constraint impulse is not zero, or vice versa. The scalar product of relative velocity and normal impulse is thus always zero. For the more complicated case of an impact with friction we shall find such a complementarity in each phase of the impact (Chapter 8). Friction in one contact only is characterized by a contact condition of vanishing relative distance and by two frictional conditions, either sliding or sticking (Fig. 1.3).

From the contact constraint $\boldsymbol{r}_D = \mathbf{0}$ we get a normal constraint force $\boldsymbol{F}_N$ which, according to Coulomb's laws, is proportional to the friction forces. For sliding $\boldsymbol{F}_{TS} = -\mu \boldsymbol{F}_N \operatorname{sgn}(\boldsymbol{v}_{\text{rel}})$, and for stiction $\boldsymbol{F}_{T0} = \mu_0 \boldsymbol{F}_N$, where μ and μ_0 are the coefficient of sliding and static friction, respectively. Stiction is indicated by $\boldsymbol{v}_{\text{rel}} = 0$ and by a surplus of the static friction force over the constraint force; i.e., $\mu_0|\boldsymbol{F}_N| - |\boldsymbol{F}_{TC}| \geq 0$. If this friction saturation becomes zero the stiction situation will end and sliding will start again with a nonzero relative acceleration $\boldsymbol{a}_{\text{rel}}$. Again we find here complementary behavior: Either the relative velocity (acceleration) is zero and the friction saturation is not zero, or vice versa. The product of relative acceleration and friction saturation is always zero.

1.3 Multiple-Contact Dynamics

We consider a multibody system with n bodies and f degrees of freedom. In addition we have n_G unilateral contacts where impacts and friction may occur. Each contact event is indicated by some indicator function — for example, the beginning by a relative distance or a relative velocity and the end by a relevant constraint force condition. The constraint equation itself is always a kinematical relationship. If a constraint is active it generates a constraint force; if it is passive no constraint force appears.

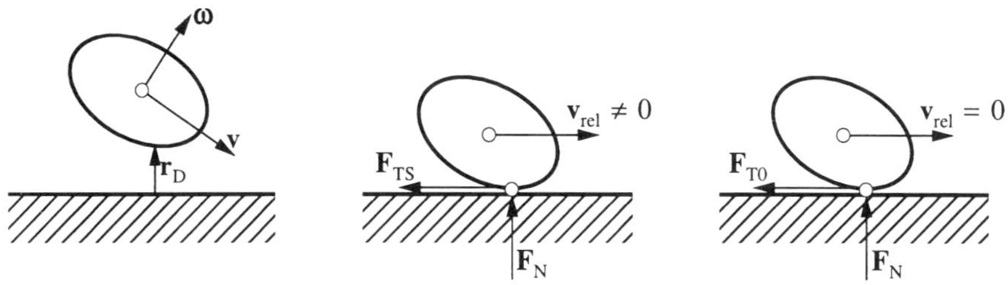

Figure 1.3: Sliding and Static Friction

In multibody systems with multiple contacts these contacts may be decoupled by springs or any other force law, or they may not. In the last case a change of the contact situation in only one contact results in a modified contact situation in the other contacts. If we characterize these situations by the combination of all active and passive constraint equations in all existing contacts, we get a combinatorial problem of considerable extent by any change in the unilateral and coupled contacts. Let us consider this problem in more detail.

Figure 1.4 shows ten masses which may stick or slide on each other. The little mass tower is excited by a periodically vibrating table. Gravity forces and friction forces act on each mass,and each mass can move to the left with v^-, to the right with v^+, or not move at all. Each type of motion is connected with some passive or active constraint situation. Combining all ten masses, each of which has three possibilities of motion, results in $3^{10} = 59,049$ possible combinations of constraints. But only one is the correct constraint configuration. To find this one configuration is a crucial task of combinatorial search or an elegant way of applying the complementarity idea. We shall focus on this way.

As pointed out all contact dynamical problems possess complementarity properties [34, 40, 41, 50, 60, 61, 67, 68]. For any unilateral contact the relative kinematics is zero and some constraint forces are not zero, or vice versa. The scalar product of magnitudes representing relative kinematics and constraint forces is always zero. This property possesses the character of a basic law in unilateral dynamics, the application of which makes multiple-contact problems solvable. Introducing these considerations into the equations of motion and into the active set of constraint equations allows a reduction of these equations to a standard complementarity problem, which is closely related to linear programming problems. The basic idea consists of the property that the complementary behavior of unilateral contact problems reduces the solution space for the constraint magnitudes drastically. Usually a unique solution can be found, and the combinatorial problem is solved.

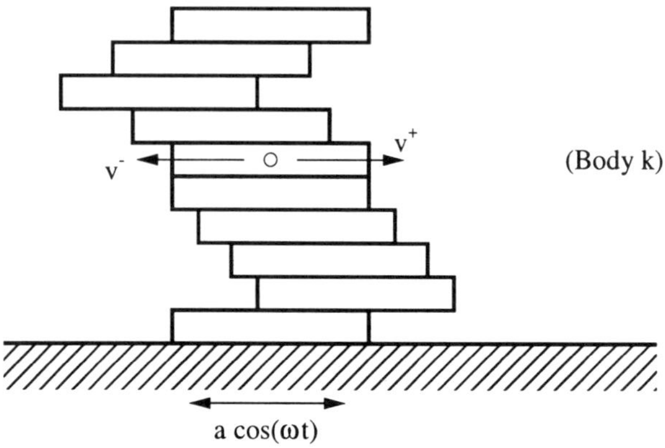

Figure 1.4: A Combinatorial Problem

Changes of the contact situation, and thus the constraint configuration, depend on the evolution of the state and, therefore, on the motion itself. They generate a discontinuously varying structure of the equations of motion. Such systems are often called systems with time-variant structure or with time-variant topology. It is a typical property of all mechanical systems with impacts and friction in unilateral contacts.

2

MULTIBODY KINEMATICS

Kinematics is geometry of motion. Applied to multibody systems it describes the linear and angular positions of all bodies within the system and provides methods for calculating their velocities and accelerations. It also takes into account the directions of unconstrained and constrained motion which might occur when bodies are linked together by certain joints.

2.1 Geometry and Definitions

Multibody kinematics requires a precise and unique definition of coordinate frames and the transformations between them [11, 75]. In the following we shall use the inertial base I, the body-fixed base B or B_i and some arbitrary reference frame R or R_i for convenience (Fig. 2.1). We say that a vector $\boldsymbol{v} \in V$ is a component of vector space V. It can be represented in any of the mentioned coordinate frames.

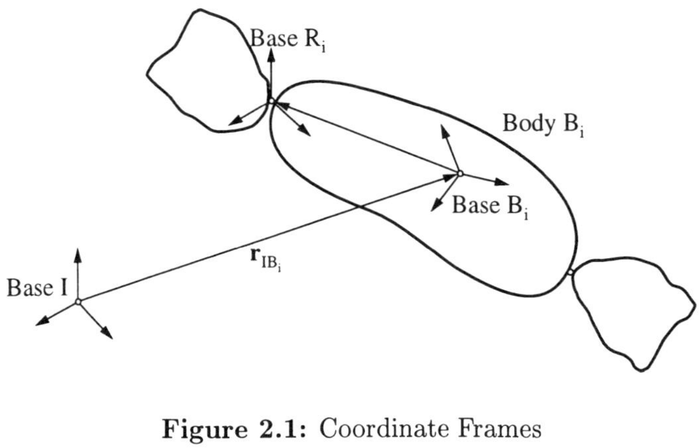

Figure 2.1: Coordinate Frames

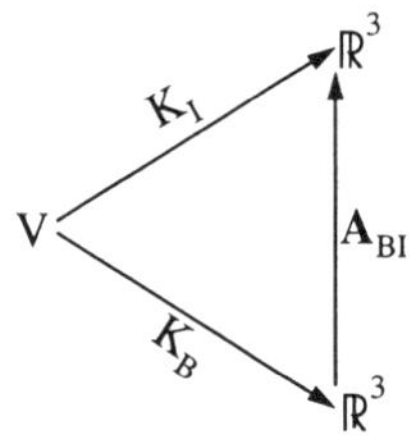

Figure 2.2: Transformation Triangle

For such a representation we apply the convention

$$\begin{aligned} K_B(\boldsymbol{v}) &:= {}_B\boldsymbol{v} \in \mathbb{R}^3 \,, \\ K_I(\boldsymbol{v}) &:= {}_I\boldsymbol{v} \in \mathbb{R}^3 \,, \\ K_R(\boldsymbol{v}) &:= {}_R\boldsymbol{v} \in \mathbb{R}^3 \,, \end{aligned} \tag{2.1}$$

which says that the components of the vector $\boldsymbol{v}$ are written in the coordinate frames B, I, R, respectively. Furthermore, we define the composition

$$K_I = \boldsymbol{A}_{IB} \circ K_B \,, \tag{2.2}$$

which has to be applied to any of such transformations in an adequate form (Fig. 2.2) [83]. Figure 2.2 nicely gives a direct geometrical interpretation of $\boldsymbol{A}_{BI}\boldsymbol{A}_{IB} = \boldsymbol{E}$ resulting from eq. (2.2) and $K_B = \boldsymbol{A}_{BI} \circ K_I$. We may derive this important result following another route.

Figure 2.3 shows two reference frames B and R and an arbitrary vector $\boldsymbol{v}$ with given coordinates with respect to frame R, ${}_R\boldsymbol{v} = ({}_Rv_x,\, {}_Rv_y,\, {}_Rv_z)^T$. In order to get its coordinates ${}_B\boldsymbol{v} = ({}_Bv_x,\, {}_Bv_y,\, {}_Bv_z)^T$ we only have to write it as a linear combination of the basis vectors of R, but using frame B instead. Hence,

$$\begin{aligned} {}_B\boldsymbol{v} &= {}_B\boldsymbol{e}_x^R\, {}_Rv_x + {}_B\boldsymbol{e}_y^R\, {}_Rv_y + {}_B\boldsymbol{e}_z^R\, {}_Rv_z \\ &= \left({}_B\boldsymbol{e}_x^R,\, {}_B\boldsymbol{e}_y^R,\, {}_B\boldsymbol{e}_z^R\right) \begin{pmatrix} {}_Rv_x \\ {}_Rv_y \\ {}_Rv_z \end{pmatrix} \\ &= \boldsymbol{A}_{BR} \cdot {}_R\boldsymbol{v} \,, \quad \boldsymbol{A}_{BR} = \left({}_B\boldsymbol{e}_x^R,\, {}_B\boldsymbol{e}_y^R,\, {}_B\boldsymbol{e}_z^R\right) \,, \end{aligned} \tag{2.3}$$

which expresses the well-known fact that the transformation matrix $\boldsymbol{A}_{BR}$ from R to B is composed of the unit basis vectors of frame R written down in frame B. From eq. (2.3) we get immediately

$$\boldsymbol{A}_{BR}^T \boldsymbol{A}_{BR} = \boldsymbol{E} \,, \quad \boldsymbol{A}_{BR}^T = \boldsymbol{A}_{BR}^{-1} = \boldsymbol{A}_{RB} \,. \tag{2.4}$$

The evaluation of the transformation matrices $\boldsymbol{A}_{IB}$ or $\boldsymbol{A}_{BI}$ follows well-known standard methods of rigid body kinematics. To rotate a coordinate frame B

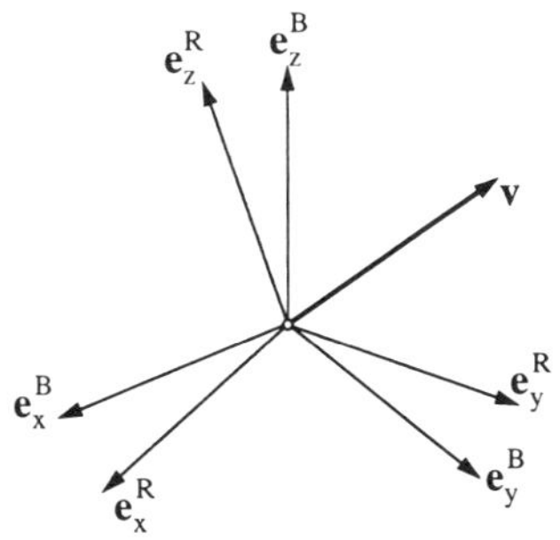

Figure 2.3: Transformation of Vectors

into a frame R we may use Euler or Kardan angles or any other set of angles which is convenient for our problem (Fig. 2.4).

Every mapping can be composed of elementary rotations around some known axes. In the example of Kardan angles the three elementary rotations are given by

$$\boldsymbol{A}_\alpha = \begin{pmatrix} 1 & 0 & 0 \\ 0 & \cos\alpha & -\sin\alpha \\ 0 & \sin\alpha & \cos\alpha \end{pmatrix} ;$$

$$\boldsymbol{A}_\beta = \begin{pmatrix} \cos\beta & 0 & \sin\beta \\ 0 & 1 & 0 \\ -\sin\beta & 0 & \cos\beta \end{pmatrix} ; \tag{2.5}$$

$$\boldsymbol{A}_\gamma = \begin{pmatrix} \cos\gamma & -\sin\gamma & 0 \\ \sin\gamma & \cos\gamma & 0 \\ 0 & 0 & 1 \end{pmatrix} .$$

From this the complete transformation from B to R is simply

$$\boldsymbol{A}_{RB} = \boldsymbol{A}_\gamma \boldsymbol{A}_\beta \boldsymbol{A}_\alpha ,$$

where the transformation sequence

$$B \xrightarrow{\boldsymbol{A}_\alpha} B_1 \xrightarrow{\boldsymbol{A}_\beta} B_2 \xrightarrow{\boldsymbol{A}_\gamma} R$$

consisting of only elementary rotations has been used. The frames B_1 and B_2 are intermediate systems which result from the first two elementary rotations. In the case of Euler angles the evaluation is similar and leads, with respect to Fig. 2.4, to an overall transformation matrix $\boldsymbol{A}_{RB} = \boldsymbol{A}_\gamma(\psi)\boldsymbol{A}_\alpha(\vartheta)\boldsymbol{A}_\gamma(\varphi)$.

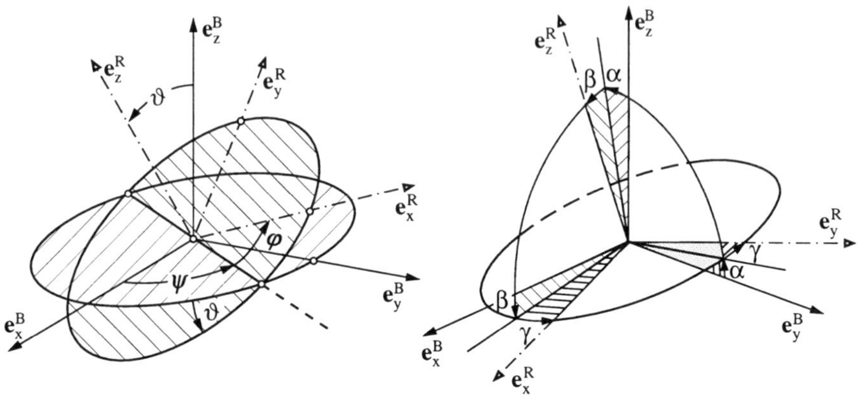

Figure 2.4: Euler and Kardan Angles [55]

From the equality $\boldsymbol{A}_{RB} = \boldsymbol{A}_\gamma \boldsymbol{A}_\beta \boldsymbol{A}_\alpha$ and the structure of the elementary rotations we immediately get $\det(\boldsymbol{A}_{RB}) = 1$, which is a general characteristic feature of any rotational mapping. For practical calculations these are helpful formulas, where in many cases the structure of existing machines and mechanisms allows a simple connection of the components by one degree of freedom only and thus by only one elementary rotation.

An important process in considering multibody kinematics consists of the evaluation of many successive coordinate frames (Fig. 2.1), which has been used in the composition of the transformation matrix $\boldsymbol{A}_{RB}$.
With respect to Fig. 2.5 we get, for example,

$$\begin{aligned} {}_I\boldsymbol{v} &= \boldsymbol{A}_{IB} \cdot {}_B\boldsymbol{v}\,, \\ {}_R\boldsymbol{v} &= \boldsymbol{A}_{RB} \cdot {}_B\boldsymbol{v}\,, \\ {}_I\boldsymbol{v} &= \boldsymbol{A}_{IR} \cdot {}_R\boldsymbol{v} \;=\; \boldsymbol{A}_{IR}\boldsymbol{A}_{RB}\, {}_B\boldsymbol{v}\,. \end{aligned}$$

Comparing the first and last equations results in the important relationship

$$\boldsymbol{A}_{IB} = \boldsymbol{A}_{IR} \cdot \boldsymbol{A}_{RB}\,, \tag{2.6}$$

which says that the transformation matrix from B to I can be composed by the transformation matrix from B to R and by the one from R to I.

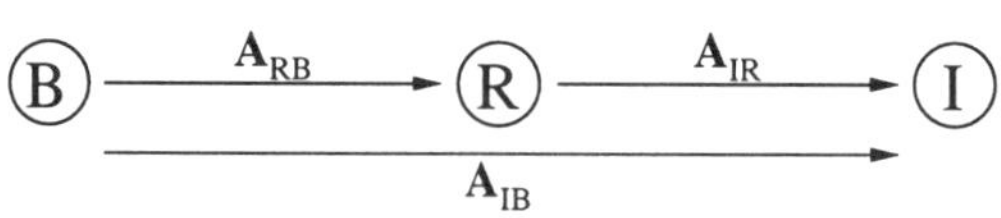

Figure 2.5: Successive Coordinate Frames

2.2 Time Derivations

One tedious task in multibody kinematics is the evaluation of the above transformations. The other tedious task consists of determining all velocities and accelerations in any of the chosen coordinate frames, mostly in the inertial and body-fixed frames. Therefore, a crucial factor in establishing multibody kinematics is time derivatives with respect to moving coordinate systems.

Let us start again with some vector $\boldsymbol{r} \in V$ as a component of a vector space V. Applying strictly the definitions of eq. (2.1), we remember that $K_B(\dot{\boldsymbol{r}}) := {}_B(\dot{\boldsymbol{r}})$ means that the components of the time derivative $\dot{\boldsymbol{r}}$ are given in frame B. In contrast, $({}_B\boldsymbol{r})^\bullet$ denotes the time derivatives of the components of a vector $\boldsymbol{r}$ given in frame B which we abbreviate ${}_B\dot{\boldsymbol{r}} := ({}_B\boldsymbol{r})^\bullet$. This is a necessary formal definition with respect to the reference for the $(\dot{\boldsymbol{r}})$-components. Realizing a time derivation needs some additional considerations. We know from basic mechanics that a vector given in an inertial coordinate frame can be derived with respect to time directly. An inertial system is the only one where mapping and time differentiation can be interchanged. Taken in our form

$$K_I(\dot{\boldsymbol{r}}) := {}_I(\dot{\boldsymbol{r}}) = ({}_I\boldsymbol{r})^\bullet = {}_I\dot{\boldsymbol{r}} \,. \tag{2.7}$$

Consequently, and wherever we want to perform time derivation, we have to go back to an inertial form and transform the result to the desired frame. Let us apply this idea to the time derivative of some vector with respect to a moving reference B:

$$\begin{aligned} {}_I\boldsymbol{r} &= \boldsymbol{A}_{IB} \cdot {}_B\boldsymbol{r} \,, \\ {}_I(\dot{\boldsymbol{r}}) &= \dot{\boldsymbol{A}}_{IB} \cdot {}_B\boldsymbol{r} + \boldsymbol{A}_{IB} \cdot {}_B\dot{\boldsymbol{r}} \,. \end{aligned}$$

Multiplying the last equation from the left by $\boldsymbol{A}_{BI}$ results in

$$\boldsymbol{A}_{BI} \cdot {}_I(\dot{\boldsymbol{r}}) = \boldsymbol{A}_{BI}\dot{\boldsymbol{A}}_{IB} \cdot {}_B\boldsymbol{r} + \boldsymbol{A}_{BI}\boldsymbol{A}_{IB} \cdot {}_B\dot{\boldsymbol{r}} \,. \tag{2.8}$$

We then derive the important formula of all relative kinematics (sometimes called the Coriolis-equation)

$${}_B(\dot{\boldsymbol{r}}) = {}_B\dot{\boldsymbol{r}} + {}_B\tilde{\boldsymbol{\omega}}_{IB} \cdot {}_B\boldsymbol{r} \,. \tag{2.9}$$

In words: The time derivative of $\boldsymbol{r}$ represented in the coordinates of the moving frame B is equal to the time derivative of the $\boldsymbol{r}$-components as given in B and the vector product of the angular velocity between B and I (written in B) and the vector ${}_B\boldsymbol{r}$ with its components in B. We now have to explain the last term of eq. (2.9).

We first show that $\boldsymbol{A}_{BI}\dot{\boldsymbol{A}}_{IB} = {}_B\tilde{\boldsymbol{\omega}}_{IB}$. For this purpose we consider the rotation of a body with respect to I (Fig. 2.6), where B is a body-fixed frame.

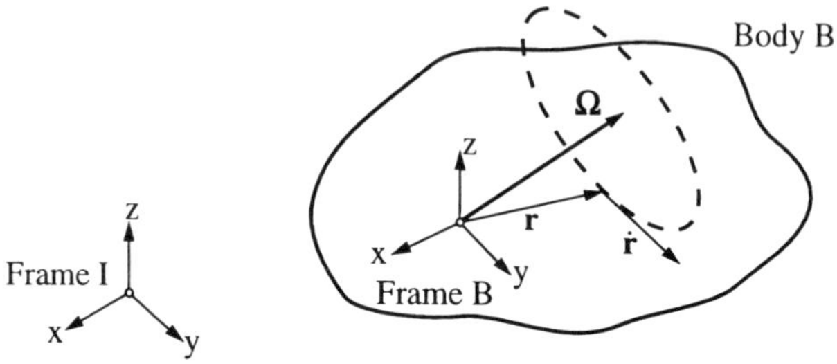

Figure 2.6: Rotation of a Body B

Thus, the angular velocity $\boldsymbol{\Omega}$ of the body is the same as the angular velocity between the frames B and I, $\boldsymbol{\omega}_{IB}$. Next we connect the origin of frame B and an arbitrary point of the rigid body by a vector $\boldsymbol{r}$. Then we can derive in one step [83]

$$\begin{aligned} K_B(\dot{\boldsymbol{r}}) &= K_B(\boldsymbol{\Omega} \times \boldsymbol{r}) \\ &= K_B(\boldsymbol{\omega}_{IB} \times \boldsymbol{r}) \\ &= {}_B\boldsymbol{\omega}_{IB} \times {}_B\boldsymbol{r} \\ &= {}_B\tilde{\boldsymbol{\omega}}_{IB}\, {}_B\boldsymbol{r}\,. \end{aligned} \tag{2.10}$$

In a second step we argue in the following way (eq. 2.2):

$$\begin{aligned} K_B(\dot{\boldsymbol{r}}) &= \boldsymbol{A}_{BI}\, K_I(\dot{\boldsymbol{r}}) \\ &= \boldsymbol{A}_{BI}\, ({}_I\boldsymbol{r})^{\bullet} \\ &= \boldsymbol{A}_{BI}\, (\boldsymbol{A}_{IB}\, {}_B\boldsymbol{r})^{\bullet} \\ &= \boldsymbol{A}_{BI}\, (\dot{\boldsymbol{A}}_{IB}\, {}_B\boldsymbol{r} + \boldsymbol{A}_{IB}\, {}_B\dot{\boldsymbol{r}}) \\ &= (\boldsymbol{A}_{BI}\dot{\boldsymbol{A}}_{IB})\, {}_B\boldsymbol{r}\,, \end{aligned} \tag{2.11}$$

where ${}_B\dot{\boldsymbol{r}} = 0$ in the body-fixed frame. A comparison of eqs. (2.10) and (2.11) yields

$${}_B\tilde{\boldsymbol{\omega}}_{IB} = \boldsymbol{A}_{BI}\dot{\boldsymbol{A}}_{IB}\,, \tag{2.12}$$

which is the first term of eq. (2.8). Transforming this expression into the I-frame and noting the transformation necessities of a tensor give

$${}_I\tilde{\boldsymbol{\omega}}_{IB} = \boldsymbol{A}_{IB}(\boldsymbol{A}_{BI}\dot{\boldsymbol{A}}_{IB})\boldsymbol{A}_{BI} = \dot{\boldsymbol{A}}_{IB}\boldsymbol{A}_{BI}\,. \tag{2.13}$$

The skew-symmetry of $\tilde{\boldsymbol{\omega}}$ follows from eq. (2.4) with $R \equiv I$, which we differentiate with respect to time to get

$$\dot{\boldsymbol{A}}_{IB}\boldsymbol{A}_{BI} + \boldsymbol{A}_{IB}\dot{\boldsymbol{A}}_{BI} = 0\,.$$

We can then write (eq. 2.4)

$$\left(\dot{\boldsymbol{A}}_{IB}\boldsymbol{A}_{BI}\right) = -\left(\dot{\boldsymbol{A}}_{BI}^{T}\boldsymbol{A}_{IB}^{T}\right)^{T} = -\left(\dot{\boldsymbol{A}}_{IB}\boldsymbol{A}_{BI}\right)^{T}, \tag{2.14}$$

because we may easily prove that $\dot{\boldsymbol{A}}_{BI}^{T} = \left(\boldsymbol{A}_{BI}^{T}\right)^{\bullet} = (\boldsymbol{A}_{IB})^{\bullet} = \dot{\boldsymbol{A}}_{IB}$. Thus the tensor $\tilde{\boldsymbol{\omega}}$ is skew-symmetric and possesses the principal form

$$\tilde{\boldsymbol{\omega}} = \begin{pmatrix} 0 & -\omega_z & \omega_y \\ \omega_z & 0 & -\omega_x \\ -\omega_y & \omega_x & 0 \end{pmatrix}, \tag{2.15}$$

from which we derive the correspondence

$$\tilde{\boldsymbol{\omega}} \cdot \boldsymbol{r} \equiv \boldsymbol{\omega} \times \boldsymbol{r}\,. \tag{2.16}$$

With the above properties we can easily derive a well-known formula with respect to Fig. 2.3 and the unit vectors used in eq. (2.3). Multiplying eq. (2.13) from the right by $\boldsymbol{A}_{IB}$ gives

$$\dot{\boldsymbol{A}}_{IB} = {}_I\tilde{\boldsymbol{\omega}}_{IB}\boldsymbol{A}_{IB} \qquad \text{with } \boldsymbol{A}_{IB} = \left({}_I\boldsymbol{e}_x^B, {}_I\boldsymbol{e}_y^B, {}_I\boldsymbol{e}_z^B\right). \tag{2.17}$$

Differentiating every component results in

$${}_I\dot{\boldsymbol{e}}_i^B = {}_I\boldsymbol{\omega}_{IB} \times {}_I\boldsymbol{e}_i^B\,,$$

which expresses the physical fact that the time derivative of a unit vector can only result in a change in its direction but not in a change in its magnitude.

2.3 Velocities and Accelerations

With the preceding chapters we have a sound basis for the evaluation of velocities and accelerations in various coordinate frames.

A typical situation of relative kinematics is shown in Fig. 2.7. We frequently apply a number of reference points (e.g., P, R, O) and various coordinate systems (e.g., I, C, B), where C might be some frame convenient for the problem under consideration. Point O has velocity $\boldsymbol{v}_O$, and point P has velocity $\boldsymbol{v}_P$. The body is exhibited to some angular velocity $\boldsymbol{\Omega}$ which we define later in detail. Our convenience reference C might have an angular velocity $\boldsymbol{\omega}_{IC}$ between C and I. Point R is fixed in the inertial frame. As seen from the point R and according to eq. (2.9), the absolute velocity of point P represented in the moving frame C can be written as

$${}_C\boldsymbol{v}_P = {}_C(\dot{\boldsymbol{r}}_{RP}) = {}_C\dot{\boldsymbol{r}}_{RP} + {}_C\boldsymbol{\omega}_{IC} \times {}_C\boldsymbol{r}_{RP}\,. \tag{2.18}$$

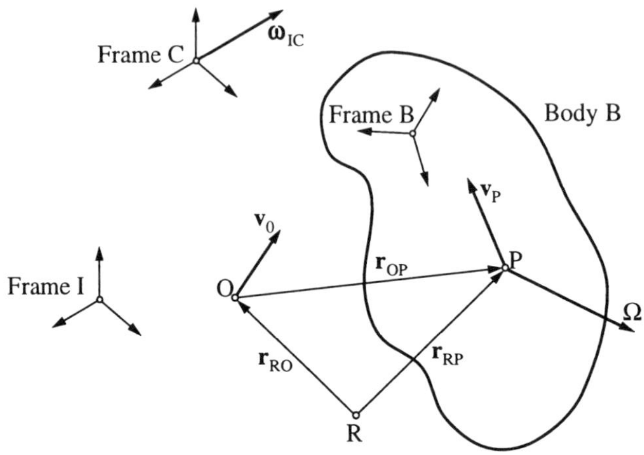

Figure 2.7: A Typical Configuration of Relative Kinematics

With the geometrical relationship $\boldsymbol{r}_{RP} = \boldsymbol{r}_{RO} + \boldsymbol{r}_{OP}$ (Fig. 2.7) we obtain the well-known relation

$$\begin{aligned} {}_C\boldsymbol{v}_P &= {}_C\boldsymbol{v}_O + {}_C\dot{\boldsymbol{r}}_{OP} + {}_C\boldsymbol{\omega}_{IC} \times {}_C\boldsymbol{r}_{OP} \\ &\quad \text{with} \quad {}_C\boldsymbol{v}_O = {}_C\dot{\boldsymbol{r}}_{RO} + {}_C\boldsymbol{\omega}_{IC} \times {}_C\boldsymbol{r}_{RO} \,. \end{aligned} \tag{2.19}$$

The physical interpretation is straightforward and easy.

In many cases, though, our moving coordinate frame will be some body-fixed system B. If, for example, in eq. (2.19) we replace point O by a new body-fixed point P and the old point P by a second body-fixed point Q (Fig. 2.8), we obtain

$$\begin{aligned} {}_B\boldsymbol{v}_Q &= {}_B\boldsymbol{v}_P + {}_B\dot{\boldsymbol{r}}_{PQ} + {}_B\boldsymbol{\omega}_{IB} \times {}_B\boldsymbol{r}_{PQ} \\ &\quad \text{with} \quad {}_B\boldsymbol{v}_P = {}_B\dot{\boldsymbol{r}}_{RP} + {}_B\boldsymbol{\omega}_{IB} \times {}_B\boldsymbol{r}_{RP} \,. \end{aligned} \tag{2.20}$$

In frame B we have ${}_B\dot{\boldsymbol{r}}_{PQ} = 0$, which is the rigid body definition. Therefore,

$${}_B\boldsymbol{v}_Q = {}_B\boldsymbol{v}_P + {}_B\boldsymbol{\Omega} \times {}_B\boldsymbol{r}_{PQ} \tag{2.21}$$

where ${}_B\boldsymbol{v}_P$ is given by the second equation in (2.20). Equation (2.21) can be easily written down in a body-fixed frame and then, if needed, transformed in any base C:

$$\begin{aligned} \boldsymbol{A}_{CB}\,{}_B\boldsymbol{v}_Q &= \boldsymbol{A}_{CB}\,{}_B\boldsymbol{v}_P + (\boldsymbol{A}_{CB}\,{}_B\tilde{\boldsymbol{\Omega}}\boldsymbol{A}_{BC})(\boldsymbol{A}_{CB}\,{}_B\boldsymbol{r}_{PQ}) \,, \\ {}_C\boldsymbol{v}_Q &= {}_C\boldsymbol{v}_P + {}_C\tilde{\boldsymbol{\Omega}}\,{}_C\boldsymbol{r}_{PQ} \,. \end{aligned} \tag{2.22}$$

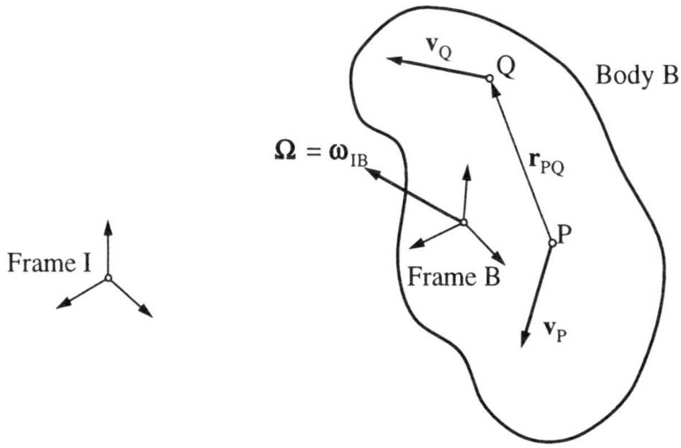

Figure 2.8: Body-Related Kinematics

Proceeding to the accelerations we follow the same kind of argument as for the velocities. We imagine the velocities in Figs. 2.7 and 2.8 being replaced by accelerations and start again with the basic equation (2.9). Then we get for an arbitrarily moving coordinate system C (Fig. 2.7) the acceleration of point P in the form

$$ {}_C\boldsymbol{a}_P = {}_C(\dot{\boldsymbol{v}}_P) = {}_C\dot{\boldsymbol{v}}_P + {}_C\boldsymbol{\omega}_{IC} \times {}_C\boldsymbol{v}_P \,, \tag{2.23} $$

with ${}_C\boldsymbol{v}_P$ is given by eq. (2.19). Performing all derivations with respect to time, we get the well-known formula

$$ \begin{aligned} {}_C\boldsymbol{a}_P &= {}_C\boldsymbol{a}_O + {}_C\ddot{\boldsymbol{r}}_{OP} + ({}_C\dot{\tilde{\boldsymbol{\omega}}}_{IC} + {}_C\tilde{\boldsymbol{\omega}}_{IC}\,{}_C\tilde{\boldsymbol{\omega}}_{IC})\,{}_C\boldsymbol{r}_{OP} + 2\,{}_C\tilde{\boldsymbol{\omega}}_{IC}\,{}_C\dot{\boldsymbol{r}}_{OP} \\ &\text{with} \quad {}_C\boldsymbol{a}_O = {}_C\dot{\boldsymbol{v}}_O + {}_C\tilde{\boldsymbol{\omega}}_{IC}\,{}_C\boldsymbol{v}_O \,. \end{aligned} \tag{2.24} $$

If we again pass from the points O, P to the new body-fixed points P, Q we receive for a body-fixed coordinate base

$$ {}_B\boldsymbol{a}_Q = {}_B\boldsymbol{a}_P + ({}_B\dot{\tilde{\boldsymbol{\Omega}}} + {}_B\tilde{\boldsymbol{\Omega}}\,{}_B\tilde{\boldsymbol{\Omega}})\,{}_B\boldsymbol{r}_{PQ} \tag{2.25} $$

which can be transformed to an arbitrarily moving base by applying the procedure of eq. (2.22). Note particularly that in an arbitrary frame the angular acceleration becomes by eq. (2.9)

$$ {}_C(\dot{\boldsymbol{\Omega}}) = {}_C\dot{\boldsymbol{\Omega}} + {}_C\boldsymbol{\omega}_{IC} \times {}_C\boldsymbol{\Omega} \,, \tag{2.26} $$

which for body-fixed coordinates reduces to (${}_B\boldsymbol{\omega}_{IB} = {}_B\boldsymbol{\Omega}$!)

$$ {}_B(\dot{\boldsymbol{\Omega}}) = {}_B\dot{\boldsymbol{\Omega}} \,. \tag{2.27} $$

2.4 Recursive Methods

Multibody systems are characterized by chains. Even when these chains are partially closed they can be cut to add additional constraints. Therefore a treelike structure can always be achieved, which from the kinematical standpoint of view requires recursive algorithms.

Figure 2.9 shows two bodies of such a chain where body i is the predecessor of body j. We assume that all kinematical magnitudes of body i are known in its body-fixed frame i and which are, in particular,

- the rotational orientation of body i with respect to the initial frame I, given by the transformation matrix $\boldsymbol{A}_i$,
- the absolute angular velocity ${}_i\dot{\boldsymbol{\Omega}}_i$,
- the absolute angular acceleration ${}_i(\dot{\boldsymbol{\Omega}}_i)$,
- the displacement ${}_i\boldsymbol{r}_i$ of the bodies center of mass S_i,
- the absolute velocity ${}_i\boldsymbol{v}_i$ of S_i, and
- the absolute acceleration ${}_i\boldsymbol{a}_i$ of S_i.

The bodies are connected by a joint which allows some rotational and translational relative motion in certain directions. These directions can be accounted for by a transformation matrix $\boldsymbol{A}_{ji}$, which describes the rotational displacements of frame j with respect to frame i, and a vector ${}_i\boldsymbol{r}_{ij}$, which connects both centers of masses and is written in frame i for convenience. Both $\boldsymbol{A}_{ji}$ and ${}_i\boldsymbol{r}_{ij}$ have to be formulated in dependence on the special structure of the joint and cannot be stated generally. They must take into account the directions of unconstrained motion, which requires the introduction of

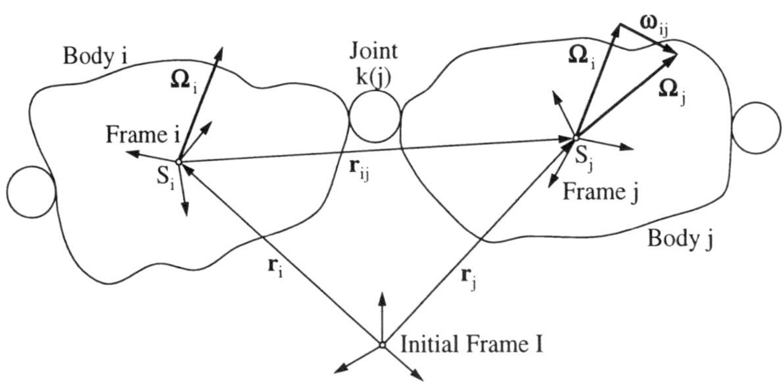

Figure 2.9: Successive Bodies in a Tree Structure

certain joint coordinates. Let $\boldsymbol{q}_k \in \mathbb{R}^k$, $k \leq 6$, be such a set of coordinates which uniquely determines the k unconstrained directions between body i and body j by the expressions $\boldsymbol{A}_{ji}(\boldsymbol{q}_k, t)$ and ${}_i\boldsymbol{r}_{ij}(\boldsymbol{q}_k, t)$. They might in addition explicitly depend on time t in order to take into account external kinematic excitations. Both terms can be formulated independently from the remaining multibody system and are therefore assumed to be known. With the help of eq. (2.13) the relative angular velocity between frame i and j results in ${}_i\tilde{\boldsymbol{\omega}}_{ij} = \dot{\boldsymbol{a}}_{ij}\boldsymbol{A}_{ji}$. With that in mind and differentiating the components of ${}_i\boldsymbol{r}_{ij}$ and ${}_i\boldsymbol{\omega}_{ij}$ with respect to time, we obtain six expressions which completely describe the joint and are independent of the remaining dynamical system:

$$\begin{aligned} \boldsymbol{A}_{ji} &= \boldsymbol{A}_{ji}(\boldsymbol{q}_k, t)\,, & {}_i\boldsymbol{r}_{ij} &= {}_i\boldsymbol{r}_{ij}(\boldsymbol{q}_k, t)\,, \\ {}_i\boldsymbol{\omega}_{ij} &= {}_i\boldsymbol{\omega}_{ij}(\boldsymbol{q}_k, \dot{\boldsymbol{q}}_k, t)\,, & {}_i\dot{\boldsymbol{r}}_{ij} &= {}_i\dot{\boldsymbol{r}}_{ij}(\boldsymbol{q}_k, \dot{\boldsymbol{q}}_k, t)\,, \\ {}_i\dot{\boldsymbol{\omega}}_{ij} &= {}_i\dot{\boldsymbol{\omega}}_{ij}(\boldsymbol{q}_k, \dot{\boldsymbol{q}}_k, \ddot{\boldsymbol{q}}_k, t)\,, & {}_i\ddot{\boldsymbol{r}}_{ij} &= {}_i\ddot{\boldsymbol{r}}_{ij}(\boldsymbol{q}_k, \dot{\boldsymbol{q}}_k, \ddot{\boldsymbol{q}}_k, t)\,. \end{aligned} \tag{2.28}$$

Note that the terms in the second line are always linear with respect to $\dot{\boldsymbol{q}}_k$, whereas the third line depends linearly on $\ddot{\boldsymbol{q}}_k$. This fact results from the time differentiation procedure.

Our aim is now to state all the kinematical magnitudes of body j in dependence on the known terms of its predecessor and the known joint properties of eq. (2.28). When we have performed this step, the entire kinematics of our treelike structure is solved in dependence on the union of all joint coordinates and their derivatives. The easiest step consists of the evaluating the angular and linear displacements. With respect to Fig. 2.5 and Fig. 2.9 they simply are

$$\boldsymbol{A}_j = \boldsymbol{A}_{ji}\,\boldsymbol{A}_i\,; \quad {}_j\boldsymbol{r}_j = \boldsymbol{A}_{ji}({}_i\boldsymbol{r}_i + {}_i\boldsymbol{r}_{ij})\,. \tag{2.29}$$

The angular velocity of body j may be obtained either by applying (2.13) to the left equation in (2.29) or directly from Fig. 2.9,

$${}_i\boldsymbol{\Omega}_j = {}_i\boldsymbol{\Omega}_i + {}_i\boldsymbol{\omega}_{ij}\,. \tag{2.30}$$

The centers of mass S_i and S_j are not points of the same rigid body. Thus, the general equation of relative kinematics (2.19) has to be used in order to obtain the velocity of S_j. We evaluate (2.19) using the body-fixed frame i, which yields

$${}_i\boldsymbol{v}_j = {}_i\boldsymbol{v}_i + {}_i\dot{\boldsymbol{r}}_{ij} + {}_i\tilde{\boldsymbol{\Omega}}_i\,{}_i\boldsymbol{r}_{ij}\,, \tag{2.31}$$

and we state both equations in matrix notation:

$$\begin{aligned} \begin{pmatrix} {}_i\boldsymbol{v}_j \\ {}_i\boldsymbol{\Omega}_j \end{pmatrix} &= \begin{pmatrix} \boldsymbol{E} & -{}_i\tilde{\boldsymbol{r}}_{ij} \\ 0 & \boldsymbol{E} \end{pmatrix} \begin{pmatrix} {}_i\boldsymbol{v}_i \\ {}_i\boldsymbol{\Omega}_i \end{pmatrix} + \begin{pmatrix} {}_i\dot{\boldsymbol{r}}_{ij} \\ {}_i\boldsymbol{\omega}_{ij} \end{pmatrix} \\ \begin{pmatrix} {}_j\boldsymbol{v}_j \\ {}_j\boldsymbol{\Omega}_j \end{pmatrix} &= \begin{pmatrix} \boldsymbol{A}_{ji}\,{}_i\boldsymbol{v}_j \\ \boldsymbol{A}_{ji}\,{}_i\boldsymbol{\Omega}_j \end{pmatrix} \end{aligned} \tag{2.32}$$

which is already the desired recursive relationship of the velocities. We proceed in the same manner with the accelerations: The absolute time derivatives of eq. (2.30) are obtained with the help of eq. (2.26) as

$$
\begin{aligned}
{}_i(\dot{\boldsymbol{\Omega}}_j) &= {}_i(\dot{\boldsymbol{\Omega}}_i) + {}_i(\dot{\boldsymbol{\omega}}_{ij}) \\
&= {}_i(\dot{\boldsymbol{\Omega}}_i) + {}_i\dot{\boldsymbol{\omega}}_{ij} + {}_i\tilde{\boldsymbol{\Omega}}_i\, {}_i\boldsymbol{\omega}_{ij}\,.
\end{aligned}
\tag{2.33}
$$

Finally we state the accelerations of S_j by using eq. (2.24):

$$
{}_i\boldsymbol{a}_j = {}_i\boldsymbol{a}_i + {}_i\ddot{\boldsymbol{r}}_{ij} + {}_i\dot{\tilde{\boldsymbol{\Omega}}}_i\, {}_i\boldsymbol{r}_{ij} + {}_i\tilde{\boldsymbol{\Omega}}_i\, {}_i\tilde{\boldsymbol{\Omega}}_i\, {}_i\boldsymbol{r}_{ij} + 2\, {}_i\tilde{\boldsymbol{\Omega}}_i\, {}_i\dot{\boldsymbol{r}}_{ij} \tag{2.34}
$$

and we write both equations in matrix notation by considering the equivalence (2.27), ${}_i\dot{\boldsymbol{\Omega}}_i = {}_i(\dot{\boldsymbol{\Omega}}_i)$:

$$
\begin{aligned}
\begin{pmatrix} {}_i\boldsymbol{a}_j \\ {}_i(\dot{\boldsymbol{\Omega}}_j) \end{pmatrix} &= \begin{pmatrix} \boldsymbol{E} & -{}_i\tilde{\boldsymbol{r}}_{ij} \\ 0 & \boldsymbol{E} \end{pmatrix} \begin{pmatrix} {}_i\boldsymbol{a}_i \\ {}_i(\dot{\boldsymbol{\Omega}}_i) \end{pmatrix} + \begin{pmatrix} {}_i\ddot{\boldsymbol{r}}_{ij} \\ {}_i\dot{\boldsymbol{\omega}}_{ij} \end{pmatrix} \\
&\quad + \begin{pmatrix} {}_i\tilde{\boldsymbol{\Omega}}_i\, {}_i\tilde{\boldsymbol{\Omega}}_i\, {}_i\boldsymbol{r}_{ij} + 2\, {}_i\tilde{\boldsymbol{\Omega}}_i\, {}_i\dot{\boldsymbol{r}}_{ij} \\ {}_i\tilde{\boldsymbol{\Omega}}_i\, {}_i\boldsymbol{\omega}_{ij} \end{pmatrix}; \\
\begin{pmatrix} {}_j\boldsymbol{a}_j \\ {}_j(\dot{\boldsymbol{\Omega}}_j) \end{pmatrix} &= \begin{pmatrix} \boldsymbol{A}_{ji}\, {}_i\boldsymbol{a}_j \\ \boldsymbol{A}_{ji}\, {}_i(\dot{\boldsymbol{\Omega}}_j) \end{pmatrix}.
\end{aligned}
\tag{2.35}
$$

With eqs. (2.29), (2.32) and (2.35) a complete recursive description of the kinematics of the multibody system is given in dependence on the joint coordinates. With respect to the evaluation we have to start with a body nearest to the environment, which is a body without a predecessor, and then pass all other bodies occurring along the chains in the tree structure.

3

DYNAMICS OF RIGID BODY SYSTEMS

The investigation of generally constrained dynamic systems starts with an appropriate formulation of the equations of motion. We consider systems of rigid or elastic bodies under the influence of active forces which may be represented by compliance or damping elements, and we restrict their movement by including constraint conditions. Some of these constraints may act permanently on the system in a bilateral manner, which results in a time-independent reduction of the degrees of freedom. The remaining constraints are unilateral and influence the system in dependence of the contact conditions, which leads to a state-dependent and varying number of degrees of freedom. Such contact conditions arise from collisions, contact or separation, and stick-slip transitions with respect to Coulomb friction.

3.1 Equations of Motion

After having established a kinematical foundation we are now able to derive the equations of motion for systems without unilateral constraints [10, 11]. A typical body in such a system is shown in Fig. 3.1. For each body i of our n-body system we can write the equations of the linear and angular momenta, $\dot{\boldsymbol{p}} \in \mathbb{R}^3$ and $\dot{\boldsymbol{L}} \in \mathbb{R}^3$ respectively, as

$$\left\{ \begin{pmatrix} \dot{\boldsymbol{p}} \\ \dot{\boldsymbol{L}} \end{pmatrix} - \begin{pmatrix} \boldsymbol{E} & \boldsymbol{0} \\ \tilde{\boldsymbol{r}}_{SA} & \boldsymbol{E} \end{pmatrix} \begin{pmatrix} \boldsymbol{F}_A \\ \boldsymbol{M}_A \end{pmatrix} - \begin{pmatrix} \boldsymbol{E} & \boldsymbol{0} \\ \tilde{\boldsymbol{r}}_{SP} & \boldsymbol{E} \end{pmatrix} \begin{pmatrix} \boldsymbol{F}_P \\ \boldsymbol{M}_P \end{pmatrix} \right\}_i = \boldsymbol{0}; \qquad i = 1, \ldots, n, \tag{3.1}$$

with

$$\begin{pmatrix} \dot{\boldsymbol{p}} \\ \dot{\boldsymbol{L}} \end{pmatrix} = \begin{pmatrix} m\boldsymbol{E} & \boldsymbol{0} \\ \boldsymbol{0} & \boldsymbol{I}_S \end{pmatrix}_i \begin{pmatrix} \dot{\boldsymbol{v}}_S \\ \dot{\boldsymbol{\Omega}} \end{pmatrix}_i + \begin{pmatrix} \boldsymbol{0} \\ \tilde{\boldsymbol{\Omega}} \boldsymbol{I}_S \boldsymbol{\Omega} \end{pmatrix}_i . \tag{3.2}$$

In these equations, which sometimes are called Newton-Euler equations, and in Fig. 3.1 S denotes the center of mass of the body under consideration with

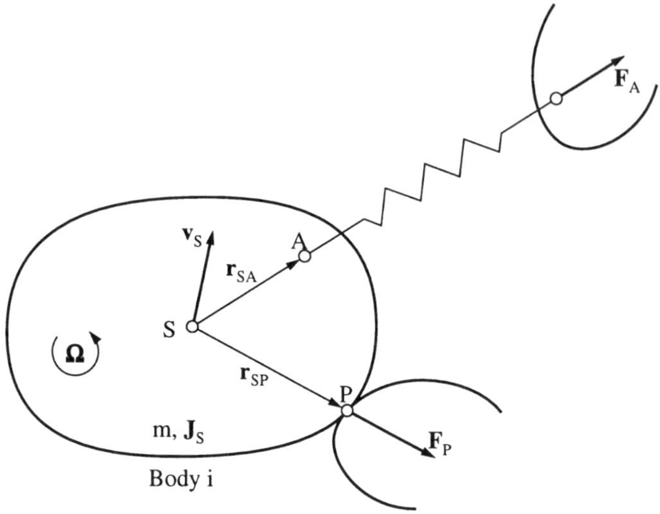

Figure 3.1: Constrained Multibody System

mass m and inertia matrix $\boldsymbol{I}_S \in \mathbb{R}^{3,3}$ about S, $\boldsymbol{v}_S \in \mathbb{R}^3$ is the absolute linear velocity of point S, and $\boldsymbol{\Omega} \in \mathbb{R}^3$ is the absolute angular velocity of the body. The forces $\boldsymbol{F} \in \mathbb{R}^3$ and moments $\boldsymbol{M} \in \mathbb{R}^3$ which act on the body are divided into two sets by the indices A for "active" and P for "passive." For example, $\boldsymbol{F}_A$ means an active force acting on point A, and $\boldsymbol{M}_P$ is a passive moment (constraint moment) not related to any point on the body.

The introduction of active and passive forces originally comes from Lagrange [47]. Cutting adjacent bodies of a multibody system, we generate a pair of cutting forces which either can be moved against each other in an active way or else they cannot. In the first case they are active and produce work or power, in the second case they are passive and do not produce work or power. Thus, the words "active" and "passive" are good illustrations of d'Alembert's or Jourdain's principles.

Finally the terms $\tilde{\boldsymbol{r}}_{SA}$ and $\tilde{\boldsymbol{r}}_{SP}$ are skew symmetric-matrices $\in \mathbb{R}^{3,3}$ which express the cross product $\boldsymbol{a} \times \boldsymbol{b}$ by a matrix-vector multiplication $\tilde{\boldsymbol{a}}\boldsymbol{b}$, and $\boldsymbol{E}$ is the identity matrix in $\mathbb{R}^{3,3}$. In the following we need an additional kinematical relation, namely the dependency of the velocities $\boldsymbol{v}_A$ and $\boldsymbol{v}_P$ on $\boldsymbol{v}_S$:

$$\begin{aligned} \begin{pmatrix} \boldsymbol{v}_A \\ \boldsymbol{\Omega} \end{pmatrix}_i &= \begin{pmatrix} \boldsymbol{E} & -\tilde{\boldsymbol{r}}_{SA} \\ \boldsymbol{0} & \boldsymbol{E} \end{pmatrix}_i \begin{pmatrix} \boldsymbol{v}_S \\ \boldsymbol{\Omega} \end{pmatrix}_i ; \\ \begin{pmatrix} \boldsymbol{v}_P \\ \boldsymbol{\Omega} \end{pmatrix}_i &= \begin{pmatrix} \boldsymbol{E} & -\tilde{\boldsymbol{r}}_{SP} \\ \boldsymbol{0} & \boldsymbol{E} \end{pmatrix}_i \begin{pmatrix} \boldsymbol{v}_S \\ \boldsymbol{\Omega} \end{pmatrix}_i . \end{aligned} \tag{3.3}$$

We also assume that the system has only f degrees of freedom due to $6n - f$ bilateral holonomic (or nonholonomic) constraints. Thus, we can choose a set of f independent (velocity)coordinates $\dot{\boldsymbol{q}}$ which describe the system uniquely and fulfill each of the constraints. This set we call the *generalized coordinates* of our system, and we choose it, for example, as the union of all joint coordinates which have been introduced in Chapter 2.

From eq. (2.28) and the remarks below, every velocity, acceleration or variation can be obviously expressed in a linear manner by the corresponding terms of the generalized coordinates; for example we have

$$\begin{aligned}
\begin{pmatrix} \boldsymbol{v}_S \\ \boldsymbol{\Omega} \end{pmatrix}_i &= \begin{pmatrix} \boldsymbol{J}_S \\ \boldsymbol{J}_R \end{pmatrix}_i \dot{\boldsymbol{q}} + \begin{pmatrix} \tilde{\boldsymbol{\jmath}}_S \\ \tilde{\boldsymbol{\jmath}}_R \end{pmatrix}_i ; \\
\begin{pmatrix} \dot{\boldsymbol{v}}_S \\ \dot{\boldsymbol{\Omega}} \end{pmatrix}_i &= \begin{pmatrix} \boldsymbol{J}_S \\ \boldsymbol{J}_R \end{pmatrix}_i \ddot{\boldsymbol{q}} + \begin{pmatrix} \bar{\boldsymbol{\jmath}}_S \\ \bar{\boldsymbol{\jmath}}_R \end{pmatrix}_i ; \\
\begin{pmatrix} \delta\boldsymbol{v}_S \\ \delta\boldsymbol{\Omega} \end{pmatrix}_i &= \begin{pmatrix} \boldsymbol{J}_S \\ \boldsymbol{J}_R \end{pmatrix}_i \delta\dot{\boldsymbol{q}} .
\end{aligned} \tag{3.4}$$

It may be noticed that in the second equation of (3.4) the additional terms, when expressing the absolute accelerations $(\dot{\boldsymbol{v}}_S, \dot{\boldsymbol{\Omega}})$, are collected in $(\bar{\boldsymbol{\jmath}}_S, \bar{\boldsymbol{\jmath}}_R)$ and are not further specified. However, each term in (3.4) can be derived in a recursive manner by using the results from eqs. (2.32) and (2.35). The matrices $(\boldsymbol{J}_B, \boldsymbol{J}_R)$ which result from the operation

$$\begin{pmatrix} \partial\boldsymbol{v}_B/\partial\dot{\boldsymbol{q}} \\ \partial\boldsymbol{\Omega}/\partial\dot{\boldsymbol{q}} \end{pmatrix}_i = \begin{pmatrix} \boldsymbol{J}_B \\ \boldsymbol{J}_R \end{pmatrix}_i ; \quad \begin{pmatrix} \boldsymbol{J}_B \in \mathbb{R}^{3,f} \\ \boldsymbol{J}_R \in \mathbb{R}^{3,f} \end{pmatrix}_i \tag{3.5}$$

are called the "Jacobian of translation of point B," where B is any of (S, A, P) and the "Jacobian of rotation" (index R), respectively. When we apply the operation (3.5) to the right- and left-hand sides of (3.3), we get the transformation rule of the Jacobians,

$$\begin{aligned}
\begin{pmatrix} \boldsymbol{J}_A \\ \boldsymbol{J}_R \end{pmatrix} &= \begin{pmatrix} \boldsymbol{E} & -\tilde{\boldsymbol{r}}_{SA} \\ \boldsymbol{0} & \boldsymbol{E} \end{pmatrix} \begin{pmatrix} \boldsymbol{J}_S \\ \boldsymbol{J}_R \end{pmatrix} ; \\
\begin{pmatrix} \boldsymbol{J}_P \\ \boldsymbol{J}_R \end{pmatrix} &= \begin{pmatrix} \boldsymbol{E} & -\tilde{\boldsymbol{r}}_{SP} \\ \boldsymbol{0} & \boldsymbol{E} \end{pmatrix} \begin{pmatrix} \boldsymbol{J}_S \\ \boldsymbol{J}_R \end{pmatrix} .
\end{aligned} \tag{3.6}$$

Now the virtual power of the system (3.1) can be formulated as

$$\sum_{i=1}^{n} \left\{ \begin{pmatrix} \delta\boldsymbol{v}_S \\ \delta\boldsymbol{\Omega} \end{pmatrix}^T \left[\begin{pmatrix} \dot{\boldsymbol{p}} \\ \dot{\boldsymbol{L}} \end{pmatrix} - \begin{pmatrix} \boldsymbol{E} & \boldsymbol{0} \\ \tilde{\boldsymbol{r}}_{SA} & \boldsymbol{E} \end{pmatrix} \begin{pmatrix} \boldsymbol{F}_A \\ \boldsymbol{M}_A \end{pmatrix} \right.\right.$$
$$\left.\left. - \begin{pmatrix} \boldsymbol{E} & \boldsymbol{0} \\ \tilde{\boldsymbol{r}}_{SP} & \boldsymbol{E} \end{pmatrix} \begin{pmatrix} \boldsymbol{F}_P \\ \boldsymbol{M}_P \end{pmatrix} \right] \right\}_i = \boldsymbol{0} ,$$

which, with help of the third equation of (3.4), leads to

$$\delta\dot{q}^T \sum_{i=1}^{n} \left\{ \begin{pmatrix} J_S \\ J_R \end{pmatrix}^T \left[\begin{pmatrix} \dot{p} \\ \dot{L} \end{pmatrix} - \begin{pmatrix} E & 0 \\ \tilde{r}_{SA} & E \end{pmatrix} \begin{pmatrix} F_A \\ M_A \end{pmatrix} - \begin{pmatrix} E & 0 \\ \tilde{r}_{SP} & E \end{pmatrix} \begin{pmatrix} F_P \\ M_P \end{pmatrix} \right] \right\}_i = \mathbf{0}\,. \tag{3.7}$$

At this point we have to remember that the generalized coordinates have been chosen in such a manner that the constraints are fulfilled for any arbitrary $\dot{q}$. Thus, eq. (3.7) must hold for every $\delta\dot{q}$, which is only possible if the sum is equal to zero. Using the transformation rule for the Jacobians (3.6), eq. (3.7) results in

$$\sum_{i=1}^{n} \left\{ \begin{pmatrix} J_S \\ J_R \end{pmatrix}^T \begin{pmatrix} \dot{p} \\ \dot{L} \end{pmatrix} - \begin{pmatrix} J_A \\ J_R \end{pmatrix}^T \begin{pmatrix} F_A \\ M_A \end{pmatrix} - \begin{pmatrix} J_P \\ J_R \end{pmatrix}^T \begin{pmatrix} F_P \\ M_P \end{pmatrix} \right\}_i = \mathbf{0}\,. \tag{3.8}$$

The sets of forces and moments with index A and P in (3.8) should be discussed further. If we have chosen the set with index P in such a manner that

$$\sum_{i=1}^{n} \left\{ \begin{pmatrix} J_P \\ J_R \end{pmatrix}^T \begin{pmatrix} F_P \\ M_P \end{pmatrix} \right\}_i \equiv \mathbf{0}\,, \tag{3.9}$$

which corresponds to the principle of lost power (Jourdain), then the dynamics of the system is only influenced by the terms with index A:

$$\sum_{i=1}^{n} \left\{ \begin{pmatrix} J_S \\ J_R \end{pmatrix}^T \begin{pmatrix} \dot{p} \\ \dot{L} \end{pmatrix} - \begin{pmatrix} J_A \\ J_R \end{pmatrix}^T \begin{pmatrix} F_A \\ M_A \end{pmatrix} \right\}_i = \mathbf{0}\,. \tag{3.10}$$

Thus the forces and moments in (3.9) are not needed to determine the dynamics of our system, and therefore we can use (3.9) as a definition for what we call passive forces (or constraint forces): Each force or moment $\in \mathbb{R}^3$ (or pairs, triples, etc., of forces or moments or combinations of them) which fulfills (3.9) is called passive and does not influence the dynamics of the system. If we use a set of generalized coordinates these terms are not needed in the equations of motion (3.10). Geometrically, the $6n-f$ bilateral and holonomic constraints define an f-dimensional manifold in $\mathbb{R}^{6n}$ on which the trajectories of our system must lie. The tangent plane to this manifold then is given by a set of f linearly independent vectors of $\mathbb{R}^{6n}$ which are nothing else than the f columns of the overall Jacobian $J = (J_{P1}^T, J_{R1}^T, \dots, J_{Pn}^T, J_{Rn}^T)^T \in \mathbb{R}^{6n,f}$. A force vector $\in \mathbb{R}^{6n}$ consisting of the passive forces $\in \mathbb{R}^3$ and moments

$\in \mathbb{R}^3$ defined above is then, by eq. (3.9) perpendicular to the tangent plane and, hence, to the manifold.

Finally, a more condensed representation of eq. (3.10) can be written. Substituting (3.2) into (3.10) and expressing the resulting accelerations using the second equation of (3.4) yield

$$\sum_{i=1}^{n}\left\{\begin{pmatrix} J_S \\ J_R \end{pmatrix}^T \begin{pmatrix} mE & 0 \\ 0 & I_S \end{pmatrix} \begin{pmatrix} J_S \\ J_R \end{pmatrix} \ddot{q} \right.$$
$$+ \begin{pmatrix} J_S \\ J_R \end{pmatrix}^T \left[\begin{pmatrix} 0 \\ \tilde{\Omega} I_S \Omega \end{pmatrix} + \begin{pmatrix} mE & 0 \\ 0 & I_S \end{pmatrix} \begin{pmatrix} \bar{j}_S \\ \bar{j}_R \end{pmatrix} \right]$$
$$\left. - \begin{pmatrix} J_A \\ J_R \end{pmatrix}^T \begin{pmatrix} F_A \\ M_A \end{pmatrix} \right\}_i = \mathbf{0}\,.$$

This results in an expression of the form [30]

$$M(q,t) \cdot \ddot{q} - h(q,\dot{q},t) = \mathbf{0} \in \mathbb{R}^f \tag{3.11}$$

with a symmetric positive-definite mass matrix $M \in \mathbb{R}^{f,f}$ and a vector $h \in \mathbb{R}^f$ which consists of all gyroscopical accelerations together with the sum of all active moments and forces. In the following sections eq. (3.11) is mainly used for further investigations; however, we will occcasionally also need the representation of eq. (3.10).

3.2 Nonlinear Applied Forces

3.2.1 Some Remarks

Before discussing the real discontinuous behavior connected with impacts and friction it makes sense to consider the characteristics of nonlinear forces and torques representing such phenomena. Modeling impacts by Newton's and Poisson's laws stands for an idealization which reasonably can be applied only if the impacting bodies are not exposed to large loads. The bouncing ball on a table or gear rattling in change over transmissions are well-known examples. On the other hand, having such large loads on impacting bodies requires another model. We must analyze the structure of the impact in detail by taking into account a compression and an expansion phase and by examining at each numerical step the indicators of relative kinematics and of the impact forces. This approach affords the knowledge of the nonlinear force law resulting from the local stiffness behavior of the contact zones, whereas the impulsive approach requires the coefficients of Newton's and, in the case of frictional impacts, Poisson's laws. Both types of data are not always available or, if available, can be very difficult to evaluate. We shall come back to this problem in the application chapter.

Sometimes stick-slip phenomena are approximated by sigmoid-shaped characteristics which avoid kinks or jumps resulting from a direct application of the Stribek curve (Fig. 1.1). This might be a good approximation if there are no real sticking events in the contacts under consideration. In addition to the fact that such an approach will generate "stiff differential equations" we must see the practical relevancy. Contacts as realized in machines or mechanisms comprise either dry or viscous friction. In the first case, and naturally depending on the system dynamics, we must expect sliding and stiction and transitions between them. This excludes the above approach. In the second case the laws of friction forces are determined by the tribological process in the contact. They are usually very much different in type and structure from the dry friction laws. For these reasons any dry friction approximation applying sigmoid characteristics must be considered with special care.

3.2.2 Couplings by Force Laws

Couplings by force laws do not restrict the freedom of movement of the participating bodies. They generate forces which depend only on the state of the system and on time. Such forces enter the equations of motion (eqs. 3.10) on the right-hand side; they are active forces.

We usually idealize force couplings in such a way that we assume single discrete forces acting in one direction. This line of action is defined by the connection of two body points (see Fig. 3.2) and formally given by the unit vectors $\boldsymbol{e}_i$ or $\boldsymbol{e}_j$ with respect to body i or body j. Obviously, we have $\boldsymbol{e}_i = -\boldsymbol{e}_j$ ($\in \mathbb{R}^3$). Considering several force or torque laws between the same body points C_{ki} and C_{kj}, we may assign several force elements to these points. From Fig. 3.2 we get generally

$$\begin{aligned} \boldsymbol{F}_i &= \boldsymbol{e}_i \zeta_k \in \mathbb{R}^3, \qquad |\boldsymbol{F}_i| = |\boldsymbol{F}_j|, \\ \boldsymbol{F}_j &= \boldsymbol{e}_j \zeta_k \in \mathbb{R}^3, \qquad \boldsymbol{e}_i = -\boldsymbol{e}_j. \end{aligned} \tag{3.12}$$

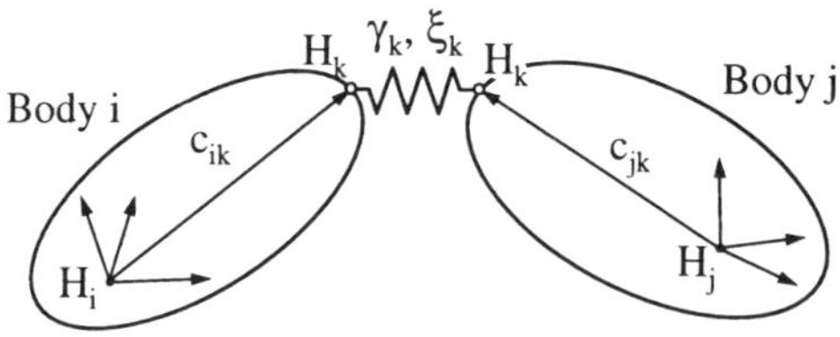

Figure 3.2: Coupling by a Force Law between Body i and Body j

These are applied forces (or torques) in three-dimensional space ($\mathbb{R}^3$). Passing to configuration space ($\mathbb{R}^f$) we have to apply the transformation $\boldsymbol{J}_A \in \mathbb{R}^{3,f}$ to the forces and $\boldsymbol{J}_R \in \mathbb{R}^{3,f}$ to the torques and then to add it to eqs. (3.11). We obtain

$$\boldsymbol{M}\ddot{\boldsymbol{q}} - \boldsymbol{h} - \boldsymbol{J}_{Ai}^T \boldsymbol{F}_i - \boldsymbol{J}_{Aj}^T \boldsymbol{F}_j - \boldsymbol{J}_{Ri}^T \boldsymbol{M}_i - \boldsymbol{J}_{Rj}^T \boldsymbol{M}_j = 0 , \tag{3.13}$$

where $\boldsymbol{M}$ is given by a similar law as $\boldsymbol{F}$:

$$\boldsymbol{M}_i = \boldsymbol{e}_i \eta_k ; \quad \boldsymbol{M}_j = \boldsymbol{e}_j \eta_k ; \quad |\boldsymbol{M}_i| = |\boldsymbol{M}_j| . \tag{3.14}$$

Inserting eqs. (3.12) and (3.14) into (3.13) yields

$$\boldsymbol{M}\ddot{\boldsymbol{q}} - \boldsymbol{h} - \sum_i (\boldsymbol{w}_A \zeta_k)_i + \sum_i (\boldsymbol{w}_R \eta_k)_i = 0 \tag{3.15}$$

with

$$\boldsymbol{w}_{Ai} = \boldsymbol{J}_{Ai}^T \boldsymbol{e}_i + \boldsymbol{J}_{Aj}^T \boldsymbol{e}_j , \quad \boldsymbol{w}_{Ri} = \boldsymbol{J}_{Ri} \boldsymbol{e}_i + \boldsymbol{J}_{Rj} \boldsymbol{e}_j .$$

Obviously, the vectors $\boldsymbol{w}_{Ai}, \boldsymbol{w}_{Ri}$ are unit vectors in configuration space in the direction of the corresponding force or torque laws. The subscript j always designates a neighbor body to body i. From machine dynamics some typical examples for such force laws and their lines of action are given in Chapter 11.

3.2.3 Some Examples

To illustrate the foregoing discussion let us start with the force laws connected with impacts under load which might be typical for machine components with backlashes. Neglecting friction in the first step, we may expect a force law according to Fig. 3.3.

For a positive relative distance γ the force $\zeta = F_A = 0$ and for a negative distance $\zeta > 0$, which means [19]

$$F_A(\gamma) = \zeta = \begin{cases} -c\gamma & \text{for } \gamma < 0 , \\ 0 & \text{for } \gamma \geq 0 . \end{cases} \tag{3.16}$$

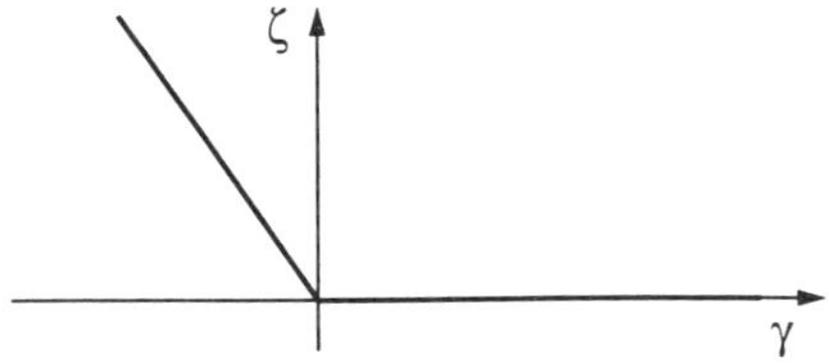

Figure 3.3: Characteristic Force Law for a System with Backlash [19]

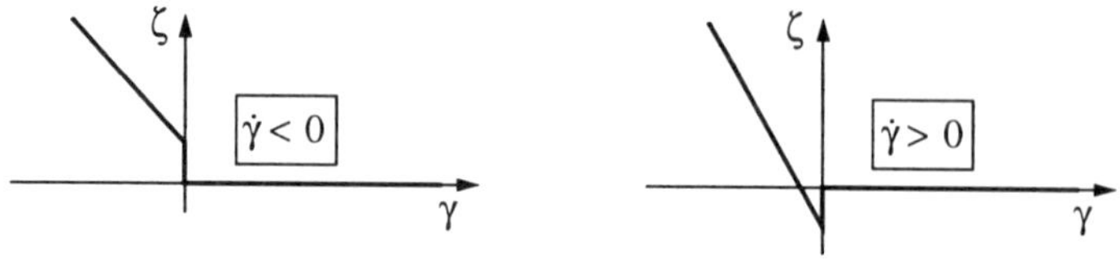

Figure 3.4: Characteristic Force Law with Backlash and Damping [19]

If we extend this law by some viscous friction term

$$F_A(\gamma, \dot{\gamma}) = \zeta = \begin{cases} -c\gamma - d\dot{\gamma} & \text{for } \gamma < 0\,, \\ 0 & \text{for } \gamma \geq 0\,, \end{cases} \tag{3.17}$$

we get the characteristics of Fig. 3.4.

Applying such a contact force law to a simple one-DOF oscillator will reveal some typical problems of contact dynamics. Combining the equation of motion ($m\ddot{\gamma} = F_A(\gamma, \dot{\gamma}) - mg$) for the model of Fig. 3.5 with the force law equation (3.17) results in a contact force possessing a jump at the beginning and a negative contact force at the end. But a negative contact force is not possible because the bodies would separate. If instead of eq. (3.17) we try the law

$$F_A(\gamma, \dot{\gamma}) = \begin{cases} F_N & \text{for } F_N > 0\,, \\ 0 & \text{for } F_N \leq 0\,, \end{cases} \tag{3.18}$$

we avoid the pulling contact force at the end of the contact, but we get a force before the mass m starts to hit the table. From these arguments we see that we have to treat the beginning and end of the contact, in general of any contact, in different ways.

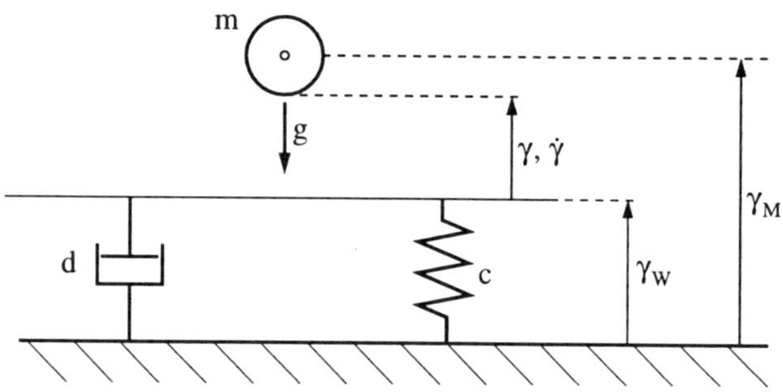

Figure 3.5: Impact on a Table [19]

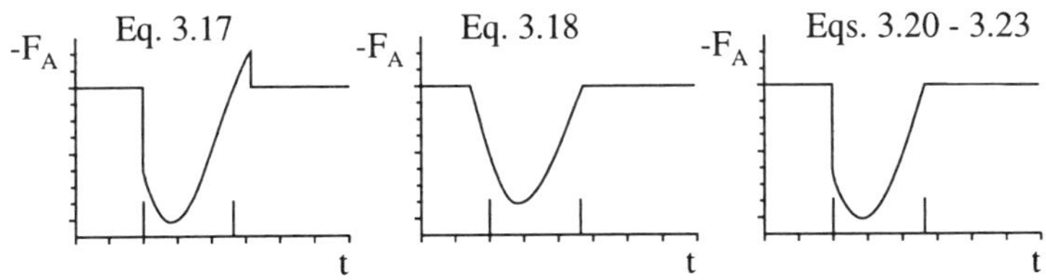

Figure 3.6: Contact Forces for Different Force Laws [19]

The contact force jumps impulsively to a positive value (pressure force!) as soon as the relative distance γ becomes zero. The contact ends whenever the normal force F_N becomes zero, the relative distance γ still nonzero ($\gamma < 0$). Therefore the two bodies separate before they are in a nondeformed state. In our simple example of Fig. 3.5 the motion of the table could be described by

$$d\dot{\gamma}_w + c\gamma_w = 0\,, \tag{3.19}$$

where we have neglected the inertia influence. The next impact might take place before $\gamma_w = 0$. A complete and physically consistent description therefore contains the following set:

- Free Flight

$$\left.\begin{aligned} m\ddot{\gamma}_M &= -mg \\ d\dot{\gamma}_w &= -c\gamma_w \end{aligned}\right\} \text{ for } \gamma_M \geq \gamma_w\,. \tag{3.20}$$

- Transition Free Flight → Contact

$$\gamma = \gamma_M - \gamma_w = 0\,. \tag{3.21}$$

- Contact

$$\left.\begin{aligned} m\ddot{\gamma}_M + d\dot{\gamma}_M + c\gamma_M &= -mg \\ \gamma_w &= \gamma_M \end{aligned}\right\} \text{ for } -d\dot{\gamma}_M - c\gamma_M > 0\,. \tag{3.22}$$

- Transition Contact → Free Flight

$$F_N = -d\dot{\gamma}_M - c\gamma_M = 0\,. \tag{3.23}$$

If we apply the three types of force laws (eqs. 3.16, 3.17 and the set 3.20–3.23) to the model of Fig. 3.5, we obtain the results in Fig. 3.6.

Two important conclusions can be drawn from these considerations. First, the beginning and the end of a contact must be treated in different ways, which means, secondly, that the beginning of a contact is indicated by relative

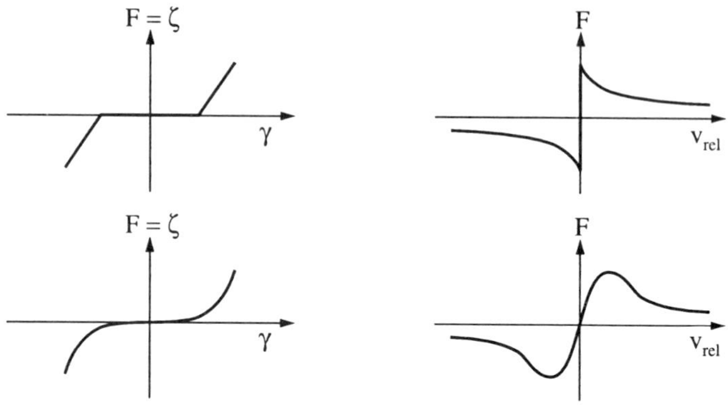

Figure 3.7: Smooth Approximations to Discontinuous Force Laws

kinematical magnitudes — in our case by the relative distance — and the end by some contact forces — in our example the normal force. This is true for any approach to a contact mechanism. In the case of classical impact theory and its assumption of infinitely small impact time the beginning and end close up, and the process is governed by the relative velocities due to Newton's law or by the momenta as given by Poisson's law. As a result forces are usually not regarded in that idealized model.

Another approximation sometimes applied to impulsive and frictional problems tries to avoid discontinuities and, thus, the tedious task of evaluating the discontinuous points as indicated above. We may approximate the force functions of Figs. 3.3 or 3.4 by continuous curves, and we may perform the same type of approach for the frictional characteristics of Fig. 1.1. The result might look like Fig. 3.7. There exist applications where such an approach leads to reasonable results even in comparison with measurements. Considering the expenses to be paid for the different methods, we must regard the switching point evaluations and the additional complementarity algorithms on one side and the stiff differential equations with their numerical problems on the other side. From many practical applications we know that the discontinuous model obviously is nearer to reality and includes fewer numerical problems, although computing times are large also.

4

CONTACT KINEMATICS

Equation (3.11) describes the dynamics of a bilaterally constrained system with f degrees of freedom. In order to take into account additional unilateral constraints like contact or friction constraints, we have to modify the dynamical equation (3.11). Before doing so we must derive some kinematic contact conditions. These are, for example, the distances between the bodies which may come into contact, their relative velocities and accelerations. In the following we shall focus on plane contacts, which give us some capability of establishing the concise geometrical contact equations. The evaluation of such equations for spatial contacts of arbitrarily shaped bodies is still a matter of ongoing research.

4.1 Contour Geometry

In order to derive the kinematic and kinetic contact equations of our system we first look at the geometry of a single body [22, 24]. Figure 4.1 shows such a body, which may have a rotational velocity $\boldsymbol{\Omega}$ and a rotational acceleration $\dot{\boldsymbol{\Omega}}$. Its body-fixed point P may move by a velocity $\boldsymbol{v}_P$ and may be accelerated by $\boldsymbol{a}_P$. The smooth and planar contour Σ of the body is assumed to be strictly convex and can be described in parametric form by the vector ${}_B\boldsymbol{r}_{P\Sigma}(s)$, using the body-fixed frame B. The parameter s corresponds to the arc length of the body. Thus, the moving trihedral $(\boldsymbol{t}, \boldsymbol{n}, \boldsymbol{b})$ of our curve can be stated in frame B by the well-known relations

$$ {}_B\boldsymbol{t} = {}_B\boldsymbol{r}'_{P\Sigma}\,; \qquad \kappa\,{}_B\boldsymbol{n} = {}_B\boldsymbol{r}''_{P\Sigma}\,; \qquad (\,\cdot\,)' = \frac{d}{ds}\,; \tag{4.1} $$

$$ {}_B\boldsymbol{n} = {}_B\boldsymbol{b} \times {}_B\boldsymbol{t}\,; \qquad {}_B\boldsymbol{b} = {}_B\boldsymbol{t} \times {}_B\boldsymbol{n}\,; \qquad {}_B\boldsymbol{t} = {}_B\boldsymbol{n} \times {}_B\boldsymbol{b}\,, \tag{4.2} $$

where κ denotes the curvature of the contour at point s, and the normal $\boldsymbol{n}$ always points inward and is directed to the center of the circle of curvature. Planar curves have the property that the binormal ${}_B\boldsymbol{b}$ is constant, and thus

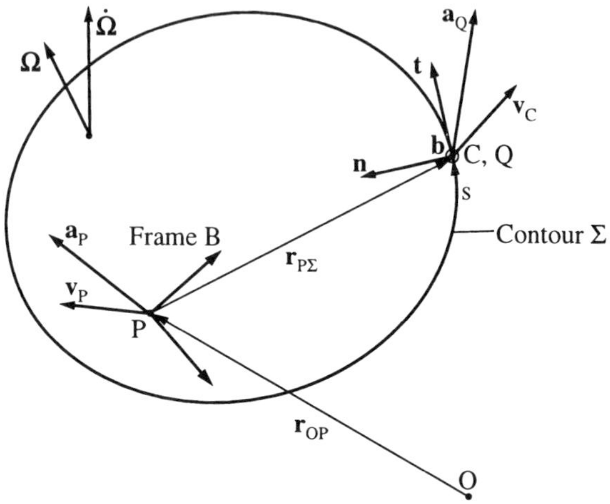

Figure 4.1: Geometry of a Single Body

independent of s. With it differentiation of (4.2) with respect to s yields

$$\begin{aligned} {}_B\boldsymbol{n}' &= {}_B\boldsymbol{b} \times {}_B\boldsymbol{t}' = {}_B\boldsymbol{b} \times {}_B\boldsymbol{n}\kappa = -\kappa_B\boldsymbol{t} \\ {}_B\boldsymbol{t}' &= \kappa_B\boldsymbol{n}\,. \end{aligned} \tag{4.3}$$

Now we imagine a point Σ moving along the contour with a given velocity $\dot{s}$. Due to that motion the normal and tangent vectors change direction with respect to the body-fixed frame B, which we can express by using eq. (4.3):

$$\begin{aligned} {}_B\dot{\boldsymbol{n}} &= {}_B\boldsymbol{n}'\dot{s} = -\kappa\dot{s}\,{}_B\boldsymbol{t}\,, \\ {}_B\dot{\boldsymbol{t}} &= {}_B\boldsymbol{t}'\dot{s} = +\kappa\dot{s}\,{}_B\boldsymbol{n}\,. \end{aligned} \tag{4.4}$$

On the other hand, the absolute changes of $\boldsymbol{n}$ and $\boldsymbol{t}$ are given by the Coriolis equation (2.9),

$$\begin{aligned} {}_B(\dot{\boldsymbol{n}}) &= {}_B\dot{\boldsymbol{n}} + {}_B\tilde{\boldsymbol{\Omega}}_B\boldsymbol{n}\,, \\ {}_B(\dot{\boldsymbol{t}}) &= {}_B\dot{\boldsymbol{t}} + {}_B\tilde{\boldsymbol{\Omega}}_B\boldsymbol{t}\,, \end{aligned} \tag{4.5}$$

where we must keep in mind that ${}_B\boldsymbol{\omega}_{IB} = {}_B\boldsymbol{\Omega}$ for body-fixed frames B. Putting (4.4) into (4.5) we get a coordinate-free representation of the overall changes $\dot{\boldsymbol{n}}, \dot{\boldsymbol{t}}$,

$$\begin{aligned} \dot{\boldsymbol{n}} &= \tilde{\boldsymbol{\Omega}}\boldsymbol{n} - \kappa\dot{s}\boldsymbol{t}\,, \\ \dot{\boldsymbol{t}} &= \tilde{\boldsymbol{\Omega}}\boldsymbol{t} + \kappa\dot{s}\boldsymbol{n}\,, \end{aligned} \tag{4.6}$$

which we can evaluate in any basis. The main advantage of (4.6) consists of the eliminated, frame-dependent differentiations ${}_B\dot{\boldsymbol{n}}$ and ${}_B\dot{\boldsymbol{t}}$. Only algebraic manipulations have to be done in the evaluation of (4.6). In the same manner we proceed with the contour vector $\boldsymbol{r}_{P\Sigma}$. According to (4.4), (4.5) we write

$$\begin{aligned} {}_B\dot{\boldsymbol{r}}_{P\Sigma} &= {}_B\boldsymbol{r}'_{P\Sigma}\,\dot{s} = \dot{s}\,{}_B\boldsymbol{t}\,, \\ {}_B(\dot{\boldsymbol{r}}_{P\Sigma}) &= {}_B\dot{\boldsymbol{r}}_{P\Sigma} + {}_B\tilde{\boldsymbol{\Omega}}\,{}_B\boldsymbol{r}_{P\Sigma}\,, \end{aligned} \tag{4.7}$$

and eliminate ${}_B\dot{\boldsymbol{r}}_{P\Sigma}$. Then we get the absolute changes of $\boldsymbol{r}_{P\Sigma}$,

$$\dot{\boldsymbol{r}}_{P\Sigma} = \tilde{\boldsymbol{\Omega}}\boldsymbol{r}_{P\Sigma} + \dot{s}\boldsymbol{t}\,. \tag{4.8}$$

Due to $\boldsymbol{v}_\Sigma = \boldsymbol{v}_P + \dot{\boldsymbol{r}}_{P\Sigma}$, the absolute velocity of the moving contour point is given by

$$\boldsymbol{v}_\Sigma = \boldsymbol{v}_P + \tilde{\boldsymbol{\Omega}}\boldsymbol{r}_{P\Sigma} + \dot{s}\boldsymbol{t}\,, \tag{4.9}$$

where

$$\boldsymbol{v}_C := \boldsymbol{v}_P + \tilde{\boldsymbol{\Omega}}\boldsymbol{r}_{P\Sigma}\,. \tag{4.10}$$

The velocity $\boldsymbol{v}_C$ in (4.10) results from the rigid body kinematical equation (2.21) and corresponds to the velocity of a body-fixed point at the contour. From (4.9) and (4.10) we see that

$$\boldsymbol{v}_\Sigma = \boldsymbol{v}_C + \dot{s}\boldsymbol{t}\,. \tag{4.11}$$

Next, we want to derive the absolute acceleration of C by differentiating (4.10) with respect to time:

$$\dot{\boldsymbol{v}}_C = \dot{\boldsymbol{v}}_P + \dot{\tilde{\boldsymbol{\Omega}}}\boldsymbol{r}_{P\Sigma} + \tilde{\boldsymbol{\Omega}}\dot{\boldsymbol{r}}_{P\Sigma}\,. \tag{4.12}$$

With $\dot{\boldsymbol{v}}_C = \boldsymbol{a}_C$, $\dot{\boldsymbol{v}}_P = \boldsymbol{a}_P$ and $\dot{\boldsymbol{r}}_{P\Sigma}$ from eq. (4.8) we get

$$\boldsymbol{a}_C = \boldsymbol{a}_P + \dot{\tilde{\boldsymbol{\Omega}}}\boldsymbol{r}_{P\Sigma} + \tilde{\boldsymbol{\Omega}}\tilde{\boldsymbol{\Omega}}\boldsymbol{r}_{P\Sigma} + \tilde{\boldsymbol{\Omega}}\boldsymbol{t}\dot{s} \tag{4.13}$$

which is *not* the acceleration of a body-fixed point on the contour. Only the part

$$\boldsymbol{a}_Q := \boldsymbol{a}_P + \dot{\tilde{\boldsymbol{\Omega}}}\boldsymbol{r}_{P\Sigma} + \tilde{\boldsymbol{\Omega}}\tilde{\boldsymbol{\Omega}}\boldsymbol{r}_{P\Sigma} \tag{4.14}$$

corresponds to such an acceleration with respect to eq. (2.25), so we can write

$$\boldsymbol{a}_C = \boldsymbol{a}_Q + \tilde{\boldsymbol{\Omega}}\boldsymbol{t}\dot{s}\,. \tag{4.15}$$

Later we have to determine the relative velocities of contact points in the normal and tangential directions and their time derivatives. For that purpose we introduce the scalars

$$v_n = \boldsymbol{n}^T\boldsymbol{v}_C\,; \quad v_t = \boldsymbol{t}^T\boldsymbol{v}_C \tag{4.16}$$

and state their derivatives as

$$\begin{aligned} \dot{v}_n &= \dot{\boldsymbol{n}}^T \boldsymbol{v}_C + \boldsymbol{n}^T \dot{\boldsymbol{v}}_C \,, \\ \dot{v}_t &= \dot{\boldsymbol{t}}^T \boldsymbol{v}_C + \boldsymbol{t}^T \dot{\boldsymbol{v}}_C \,. \end{aligned} \tag{4.17}$$

With $\dot{\boldsymbol{n}}, \dot{\boldsymbol{t}}$ from (4.6), $\dot{\boldsymbol{v}}_C = \boldsymbol{a}_C$ from (4.15), and noting $\boldsymbol{n}^T \tilde{\boldsymbol{\Omega}} \boldsymbol{t} = \boldsymbol{b}^T \boldsymbol{\Omega}$, $\boldsymbol{t}^T \tilde{\boldsymbol{\Omega}} \boldsymbol{t} = 0$ we derive

$$\begin{aligned} \dot{v}_n &= \boldsymbol{n}^T (\boldsymbol{a}_Q - \tilde{\boldsymbol{\Omega}} \boldsymbol{v}_C) - \kappa \dot{s} \boldsymbol{t}^T \boldsymbol{v}_C + \dot{s} \boldsymbol{b}^T \boldsymbol{\Omega} \\ \dot{v}_t &= \boldsymbol{t}^T (\boldsymbol{a}_Q - \tilde{\boldsymbol{\Omega}} \boldsymbol{v}_C) + \kappa \dot{s} \boldsymbol{n}^T \boldsymbol{v}_C \end{aligned} \tag{4.18}$$

which will be used in the next sections.

4.2 The Distance between Bodies

In this section we show how the distance between two bodies like the body in Fig. 4.1 is computed and how the contact points of our problem are defined. Then we derive the relative velocities of the contact points and state their time derivatives [22, 24].

Figure 4.2 shows two bodies with the nomenclature used in Fig. 4.1. The sense of rotation of the curve parameters s_1, s_2 is chosen in such a manner that the binormals of both moving trihedrals are the same, $\boldsymbol{b}_1 = \boldsymbol{b}_2$. The origins of the trihedrals are connected by a distance vector $\boldsymbol{r}_D$.

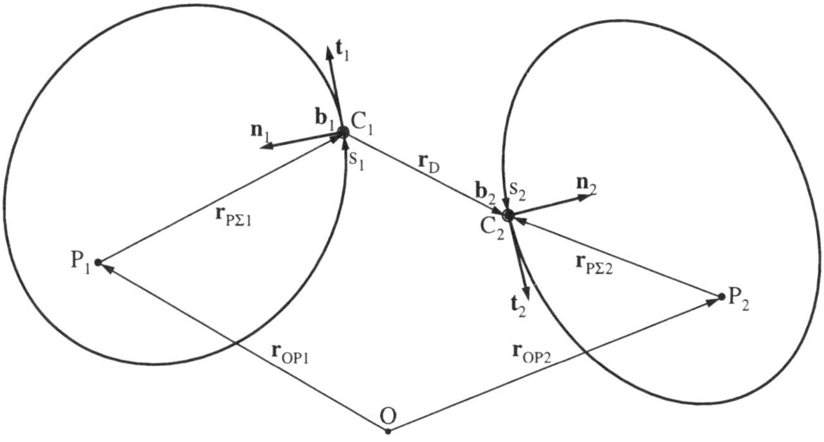

Figure 4.2: General Orientation of Two Bodies

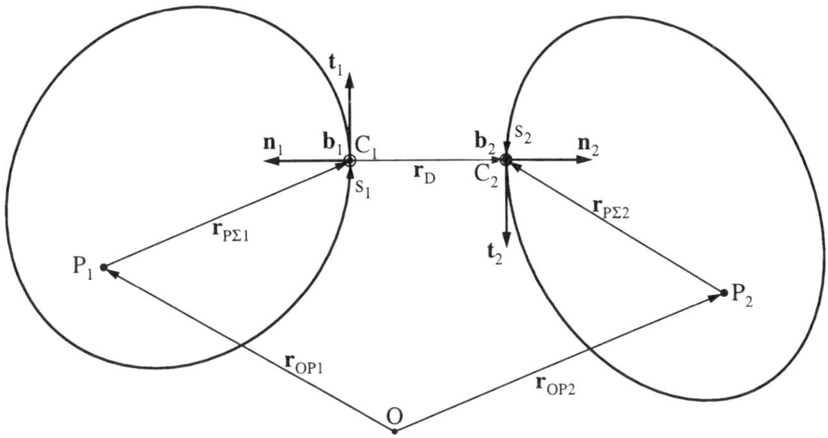

Figure 4.3: Oriented Trihedrals

In order to derive the distance of the bodies we proceed by two steps. First, we orient the axes of the trihedrals to be perpendicular to each other by

$$\boldsymbol{n}_1^T(s_1) \cdot \boldsymbol{t}_2(s_2) = 0 \quad \Leftrightarrow \quad \boldsymbol{n}_2^T(s_2) \cdot \boldsymbol{t}_1(s_1) = 0\,. \tag{4.19}$$

Now both normals are parallel as well as both tangents. Note that only one of the equations in (4.19) has to be fulfilled, because the statements in (4.19) are equivalent.

In the second step we demand the distance vector $\boldsymbol{r}_D$ to point in the same direction as the already oriented normals, which means perpendicular to one of the tangents:

$$\boldsymbol{r}_D^T(s_1, s_2) \cdot \boldsymbol{t}_1(s_1) = 0\,, \quad \boldsymbol{r}_D^T(s_1, s_2) \cdot \boldsymbol{t}_2(s_2) = 0\,. \tag{4.20}$$

Even these equations are equivalent if (4.19) is already fulfilled. Altogether we need one of the equations in (4.19) and one in (4.20), for example,

$$\boldsymbol{n}_1^T(s_1) \cdot \boldsymbol{t}_2(s_2) = 0\,, \quad \boldsymbol{r}_D^T(s_1, s_2) \cdot \boldsymbol{t}_1(s_1) = 0\,. \tag{4.21}$$

The solution (s_1, s_2) of the nonlinear problem (4.21) generates a configuration which is shown in Fig. 4.3, where the normal and tangential vectors of both bodies are antiparallel to each other, and the distance vector $\boldsymbol{r}_D$ is perpendicular to both surfaces.

The values (s_1, s_2) are called the "contact parameters" of our problem and the corresponding points (C_1, C_2) the "contact points." For this configuration the relations between the axes of both trihedrals are given by

$$\boldsymbol{n}_1 = -\boldsymbol{n}_2 \,; \quad \boldsymbol{t}_1 = -\boldsymbol{t}_2 \,; \quad \boldsymbol{b}_{12} := \boldsymbol{b}_1 = \boldsymbol{b}_2 \,. \tag{4.22}$$

With these results we easily write the distance g_N between the bodies:

$$g_N(\boldsymbol{q}, t) = \boldsymbol{r}_D^T \boldsymbol{n}_2 = -\boldsymbol{r}_D^T \boldsymbol{n}_1 \,. \tag{4.23}$$

Since the normal vectors always point inward, g_N is positive for separation and negative for overlapping. Therefore, a changing sign of g_N from positive to negative indicates a transition from initially separated bodies to contact.

4.3 The Relative Velocities of the Contact Points

Since the displacements of the contact points are known, we can derive their relative velocities. These velocities refer to body-fixed points on the contours and are used later in the impact equations. For a vanishing distance g_N the bodies in Fig. 4.3 come into contact by a collision which leads to instantaneous jumps in the relative velocities. Their changes are determined by certain impact laws which take into account the dissipative behavior in the normal direction by using a restitution coefficient ε, as well as the friction-affected tangential shock due to the law of Coulomb. The relative velocities are also used as indicators for potentially sticking contacts. We look at a situation where two bodies slide on each other, keeping continual contact. Due to some external forces or by the sliding friction force, the sliding velocity may decrease until zero, which indicates a possible transition to sliding. Thus, the tangential relative velocity in the case of contact may be used as an indicator for supervising a necessary condition of sticking, which is the zero velocity of the contact points.

In order to derive these relative velocities we start our discussion with the general situation in Fig. 4.2 and remember that $\boldsymbol{r}_D$ connects two points moving along the contours. With respect to (4.9) the absolute changes of $\boldsymbol{r}_D$ are

$$\dot{\boldsymbol{r}}_D = \boldsymbol{v}_{\Sigma 2} - \boldsymbol{v}_{\Sigma 1} \,. \tag{4.24}$$

Next, we differentiate the three scalar products of eqs. (4.21) and (4.23) with respect to time:

$$\begin{aligned} \left(\boldsymbol{r}_D^T \boldsymbol{t}_1\right)^{\bullet} &= \dot{\boldsymbol{r}}_D^T \boldsymbol{t}_1 + \boldsymbol{r}_D^T \dot{\boldsymbol{t}}_1 \\ \left(\boldsymbol{r}_D^T \boldsymbol{n}_1\right)^{\bullet} &= \dot{\boldsymbol{r}}_D^T \boldsymbol{n}_1 + \boldsymbol{r}_D^T \dot{\boldsymbol{n}}_1 \\ \left(\boldsymbol{n}_1^T \boldsymbol{t}_2\right)^{\bullet} &= \dot{\boldsymbol{n}}_1^T \boldsymbol{t}_2 + \boldsymbol{n}_1^T \dot{\boldsymbol{t}}_2 \end{aligned} \tag{4.25}$$

where $\dot{\boldsymbol{r}}_D$ is given by (4.24), and $\dot{\boldsymbol{n}}, \dot{\boldsymbol{t}}$ can be taken from eq. (4.6). Using eq. (4.11), we rewrite eqs. (4.25) as

$$\begin{aligned}
\left(\boldsymbol{r}_D^T\boldsymbol{t}_1\right)^\bullet &= \boldsymbol{t}_1^T\left(\boldsymbol{v}_{C2}-\boldsymbol{v}_{C1}-\boldsymbol{\Omega}_1\times\boldsymbol{r}_D\right)\\
&\quad+\boldsymbol{t}_1^T\boldsymbol{t}_2\dot{s}_2-\boldsymbol{t}_1^T\boldsymbol{t}_1\dot{s}_1+\boldsymbol{r}_D^T\boldsymbol{n}_1\kappa_1\dot{s}_1\\
\left(\boldsymbol{r}_D^T\boldsymbol{n}_1\right)^\bullet &= \boldsymbol{n}_1^T\left(\boldsymbol{v}_{C2}-\boldsymbol{v}_{C1}-\boldsymbol{\Omega}_1\times\boldsymbol{r}_D\right)\\
&\quad+\boldsymbol{n}_1^T\boldsymbol{t}_2\dot{s}_2-\boldsymbol{n}_1^T\boldsymbol{t}_1\dot{s}_1\boldsymbol{r}_D^T\boldsymbol{t}_1\kappa_1\dot{s}_1\\
\left(\boldsymbol{n}_1^T\boldsymbol{t}_2\right)^\bullet &= -\boldsymbol{t}_1^T\boldsymbol{t}_2\kappa_1\dot{s}_1+\boldsymbol{n}_1^T\boldsymbol{n}_2\kappa_2\dot{s}_2\\
&\quad+\left(\boldsymbol{t}_2\times\boldsymbol{n}_1\right)^T\left(\boldsymbol{\Omega}_2-\boldsymbol{\Omega}_1\right)
\end{aligned} \tag{4.26}$$

These time derivatives belong to the *general* situation of Fig. 4.2 where the trihedrals have not yet been oriented. For the oriented configuration in Fig. 4.3 the additional eqs. (4.22) and (4.23) hold; thus (4.26) can be simplified by using the properties

$$\begin{aligned}
&\boldsymbol{t}_1^T\boldsymbol{t}_1=1\,; \quad \boldsymbol{t}_1^T\boldsymbol{t}_2=-1\,; \quad \boldsymbol{r}_D^T\boldsymbol{t}_1=0\,,\\
&\boldsymbol{n}_1^T\boldsymbol{t}_1=0\,; \quad \boldsymbol{n}_1^T\boldsymbol{t}_2=0\,; \quad \boldsymbol{r}_D^T\boldsymbol{n}_1=-g_N\,,\\
&\boldsymbol{t}_2\times\boldsymbol{n}_1=-\boldsymbol{b}_{12}\,.
\end{aligned} \tag{4.27}$$

We now recall that the orientation of the trihedrals was done by the two conditions $\boldsymbol{n}_1^T\boldsymbol{t}_2 = 0$, $\boldsymbol{r}_D^T\boldsymbol{t}_1 = 0$ in eq. (4.21). Once oriented, we demand this configuration to continue for all times; in other words, changes in $\left(\boldsymbol{n}_1^T\boldsymbol{t}_2=0;\ \boldsymbol{r}_D^T\boldsymbol{t}_1=0\right)$ are forbidden. These additional conditions,

$$\left(\boldsymbol{r}_D^T\boldsymbol{t}_1\right)^\bullet=0\,; \quad \left(\boldsymbol{n}_1^T\boldsymbol{t}_2\right)^\bullet=0\,, \tag{4.28}$$

ensure that the collinearity of $(\boldsymbol{n}_1, \boldsymbol{n}_2, \boldsymbol{r}_D)$ in Fig. 4.3 is always guaranteed. Putting (4.27) and (4.28) into (4.26) finally yields

$$\begin{aligned}
0 &= \boldsymbol{t}_1^T\left(\boldsymbol{v}_{C2}-\boldsymbol{v}_{C1}\right)-g_N\boldsymbol{b}_{12}^T\boldsymbol{\Omega}_1-\dot{s}_1-\dot{s}_2-g_N\kappa_1\dot{s}_1\\
-\dot{g}_N &= \boldsymbol{n}_1^T\left(\boldsymbol{v}_{C2}-\boldsymbol{v}_{C1}\right)\\
0 &= \kappa_1\dot{s}_1-\kappa_2\dot{s}_2-\boldsymbol{b}_{12}^T\left(\boldsymbol{\Omega}_2-\boldsymbol{\Omega}_1\right)\,.
\end{aligned} \tag{4.29}$$

After some basic calculations we obtain from the first and third equations in (4.29) the changes of the arc lengths:

$$\begin{aligned}
\dot{s}_1 &= \frac{\kappa_2\boldsymbol{t}_1^T(\boldsymbol{v}_{C2}-\boldsymbol{v}_{C1})-\kappa_2 g_N\boldsymbol{b}_{12}^T\boldsymbol{\Omega}_1+\boldsymbol{b}_{12}^T(\boldsymbol{\Omega}_2-\boldsymbol{\Omega}_1)}{\kappa_1+\kappa_2+g_N\kappa_1\kappa_2}\\
\dot{s}_2 &= \frac{\kappa_1\boldsymbol{t}_1^T(\boldsymbol{v}_{C2}-\boldsymbol{v}_{C1})-\kappa_1 g_N\boldsymbol{b}_{12}^T\boldsymbol{\Omega}_2-\boldsymbol{b}_{12}^T(\boldsymbol{\Omega}_2-\boldsymbol{\Omega}_1)}{\kappa_1+\kappa_2+g_N\kappa_1\kappa_2}
\end{aligned} \tag{4.30}$$

which can be geometrically interpreted as follows:

We connect the surfaces of the two bodies in Fig. 4.3 by a rubber band in such a manner that its ends can move without friction along the contours. Due to the tension in the band, its position will coincide with the distance vector $\boldsymbol{r}_D$ in Fig. 4.3. Now we start moving around both bodies arbitrarily. Under this motion the rubber band always tries using the shortest length while passing its ends along the contours with certain velocities. These velocities are exactly the magnitudes $\dot{s}_1$ and $\dot{s}_2$ in (4.30). Clearly, the tangential relative velocity of the body-fixed contour points which pass the ends of the rubber band are given by

$$-\dot{s}_1 - \dot{s}_2 = \frac{\boldsymbol{t}_1^T(\boldsymbol{v}_{C1} - \boldsymbol{v}_{C2})(\kappa_1 + \kappa_2) + g_N \boldsymbol{b}_{12}^T(\kappa_2 \boldsymbol{\Omega}_1 + \kappa_1 \boldsymbol{\Omega}_2)}{\kappa_1 + \kappa_2 + g_N \kappa_1 \kappa_2} \tag{4.31}$$

In particular, for a vanishing distance ($g_N = 0$), eq. (4.31) becomes

$$(-\dot{s}_1 - \dot{s}_2)|_{g_N=0} = \boldsymbol{t}_1^T (\boldsymbol{v}_{C1} - \boldsymbol{v}_{C2}) \ , \tag{4.32}$$

which is the relative velocity of the body-fixed contact points (see eq. 4.10) projected in the tangential direction. This expression will be used continuously in the following sections and denoted by $\dot{g}_T$:

$$\dot{g}_T = \boldsymbol{t}_1^T (\boldsymbol{v}_{C1} - \boldsymbol{v}_{C2}) \ . \tag{4.33}$$

The relative velocity in the normal direction has been derived by the second equation of (4.29):

$$\dot{g}_N = \boldsymbol{n}_1^T (\boldsymbol{v}_{C1} - \boldsymbol{v}_{C2}) \ , \tag{4.34}$$

and corresponds to the first time derivative of the distance g_N from (4.23).

At this point note that the rubber-band concept fails under certain conditions, because the denominators of eqs. (4.30) and (4.31) may become zero. This happens if

$$-g_N = \frac{\kappa_1 + \kappa_2}{\kappa_1 \kappa_2} = \frac{1}{\kappa_1} + \frac{1}{\kappa_2} = R_1 + R_2 \ , \tag{4.35}$$

where R_i are the radii corresponding to the curvatures κ_i. Since $R_i > 0$, eq. (4.35) can only be fulfilled for negative values of g_N. In that case the bodies penetrate each other in such a manner that the circles of curvature lie concentrically (see Fig. 4.4).

Since overlapping is a forbidden configuration, this situation will never occur. Even at numerical evaluation the impenetrability condition is violated only by some tolerances, which generally does not lead to the above-mentioned problems. One should keep in mind, however, that this singularity point may arise at separation when the rubber-band concept is applied to convex-concave contact partners.

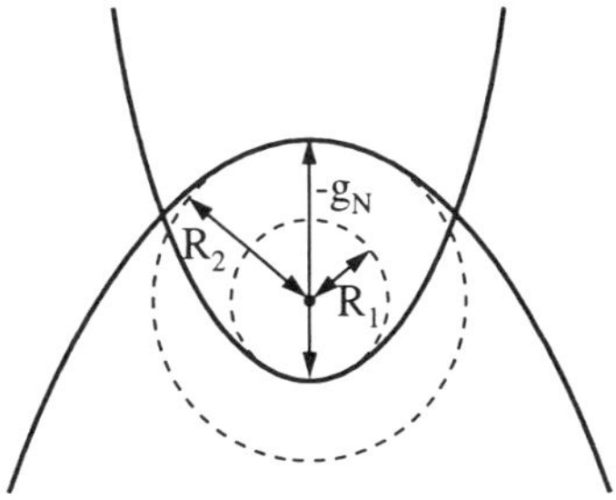

Figure 4.4: Singularity at Penetration

Finally, we state the relative velocities in terms of the generalized coordinates. For that purpose, (4.33) and (4.34) are rewritten, using (4.22), as

$$\dot{g}_N = \boldsymbol{n}_1^T \boldsymbol{v}_{C1} + \boldsymbol{n}_2^T \boldsymbol{v}_{C2} ; \quad \dot{g}_T = \boldsymbol{t}_1^T \boldsymbol{v}_{C1} + \boldsymbol{t}_2^T \boldsymbol{v}_{C2} , \tag{4.36}$$

and the velocities $\boldsymbol{v}_{C1}, \boldsymbol{v}_{C2}$ are expressed in terms of the Jacobians (3.4) and (3.5):

$$\boldsymbol{v}_{C1} = \boldsymbol{J}_{C1}\dot{\boldsymbol{q}} + \tilde{\boldsymbol{j}}_{C1} ; \quad \boldsymbol{v}_{C2} = \boldsymbol{J}_{C2}\dot{\boldsymbol{q}} + \tilde{\boldsymbol{j}}_{C2} . \tag{4.37}$$

Putting (4.37) into (4.36) yields

$$\dot{g}_N = \boldsymbol{w}_N^T \dot{\boldsymbol{q}} + \tilde{w}_N ; \quad \dot{g}_T = \boldsymbol{w}_T^T \dot{\boldsymbol{q}} + \tilde{w}_T \tag{4.38}$$

with

$$\begin{aligned} \boldsymbol{w}_N &= \boldsymbol{J}_{C1}^T \boldsymbol{n}_1 + \boldsymbol{J}_{C2}^T \boldsymbol{n}_2 ; \quad & \boldsymbol{w}_T &= \boldsymbol{J}_{C1}^T \boldsymbol{t}_1 + \boldsymbol{J}_{C2}^T \boldsymbol{t}_2 , \\ \tilde{w}_N &= \tilde{\boldsymbol{j}}_{C1}^T \boldsymbol{n}_1 + \tilde{\boldsymbol{j}}_{C2}^T \boldsymbol{n}_2 ; \quad & \tilde{w}_T &= \tilde{\boldsymbol{j}}_{C1}^T \boldsymbol{t}_1 + \tilde{\boldsymbol{j}}_{C2}^T \boldsymbol{t}_2 , \end{aligned} \tag{4.39}$$

which we use in the following as a representation of the relative velocities. It may be noticed here that a negative value of $\dot{g}_N$ corresponds to an approaching process of the bodies and coincides at vanishing distance $g_N = 0$ with the relative velocity in the normal direction before an impact. In the case of a continual contact ($g_N = \dot{g}_N = 0$) the term $\dot{g}_T$ shows the relative sliding velocity of the bodies, which we can use to determine the time points of transitions from sliding ($\dot{g}_T \neq 0$) to sticking or rolling ($\dot{g}_T = 0$).

4.4 Changes of the Relative Velocities

Contact between two bodies is only maintained if the distance g_N and the normal relative velocity $\dot{g}_N$ are equal to zero for all times t. In particular $\dot{g}_N = 0$ must not change, which can be accounted for by the additional condition

$\ddot{g}_N = 0$. On the other hand, a transition from continual contact ($g_N = \dot{g}_N = 0$) to separation demands velocities greater than zero, $\dot{g}_N > 0$, which are only achieved by $\ddot{g}_N > 0$. Similarly, this behavior can be transferred to the tangential constraint. A sticking or rolling contact fulfills $g_N = \dot{g}_N = \dot{g}_T = 0$ and will continue to stick only if, in addition, $\ddot{g}_N = \ddot{g}_T = 0$ holds. Transitions to sliding ($\dot{g}_T \neq 0$) are only possible if the tangential relative acceleration $\ddot{g}_T$ becomes unequal to zero. Therefore, the magnitudes ($\ddot{g}_N, \ddot{g}_T$) determine whether states like sticking or contact are maintained, or transitions from sticking to sliding or contact to separation occur. All these state transitions arise only at distances $g_N = 0$, which can be used for simplifying the following equations. We start our discussion using the relative velocities of eq. (4.36) together with the abbreviations (4.16):

$$\dot{g}_N = v_{n1} + v_{n2}\,; \quad \dot{g}_T = v_{t1} + v_{t2}\,; \tag{4.40}$$

we differentiate them with respect to time:

$$\ddot{g}_N = \dot{v}_{n1} + \dot{v}_{n2}\,; \quad \ddot{g}_T = \dot{v}_{t1} + \dot{v}_{t2}\,. \tag{4.41}$$

The terms $\dot{v}_n, \dot{v}_t$ are already known by eq. (4.18). Inserting them, we get

$$\begin{aligned} \ddot{g}_N &= \boldsymbol{n}_1^T(\boldsymbol{a}_{Q1} - \tilde{\boldsymbol{\Omega}}_1\boldsymbol{v}_{C1}) - \kappa_1\dot{s}_1\boldsymbol{t}_1^T\boldsymbol{v}_{C1} + \dot{s}_1\boldsymbol{b}_{12}^T\boldsymbol{\Omega}_1 \\ &\quad +\boldsymbol{n}_2^T(\boldsymbol{a}_{Q2} - \tilde{\boldsymbol{\Omega}}_2\boldsymbol{v}_{C2}) - \kappa_2\dot{s}_2\boldsymbol{t}_2^T\boldsymbol{v}_{C2} + \dot{s}_2\boldsymbol{b}_{12}^T\boldsymbol{\Omega}_2 \\ \ddot{g}_T &= \boldsymbol{t}_1^T(\boldsymbol{a}_{Q1} - \tilde{\boldsymbol{\Omega}}_1\boldsymbol{v}_{C1}) + \kappa_1\dot{s}_1\boldsymbol{n}_1^T\boldsymbol{v}_{C1} \\ &\quad +\boldsymbol{t}_2^T(\boldsymbol{a}_{Q2} - \tilde{\boldsymbol{\Omega}}_2\boldsymbol{v}_{C2}) + \kappa_2\dot{s}_2\boldsymbol{n}_2^T\boldsymbol{v}_{C2} \end{aligned} \tag{4.42}$$

Finally, we state the rigid body accelerations $\boldsymbol{a}_Q$ in terms of the generalized coordinates by means of (3.4):

$$\boldsymbol{a}_{Q1} = \boldsymbol{J}_{Q1}\ddot{\boldsymbol{q}} + \overline{\boldsymbol{j}}_{Q1}\,; \quad \boldsymbol{a}_{Q2} = \boldsymbol{J}_{Q2}\ddot{\boldsymbol{q}} + \overline{\boldsymbol{j}}_{Q2}\,, \tag{4.43}$$

and remember that the velocity of point C in (4.10) and the acceleration of point Q in (4.14) result from rigid body kinematics. Because of the same position of both points (see Fig. 4.1) the Jacobians of C and Q are the same:

$$\boldsymbol{J}_{Q1} = \boldsymbol{J}_{C1}\,; \quad \boldsymbol{J}_{Q2} = \boldsymbol{J}_{C2}\,. \tag{4.44}$$

Putting (4.43) and (4.44) into (4.42) results in

$$\ddot{g}_N = \boldsymbol{w}_N^T\ddot{\boldsymbol{q}} + \overline{w}_N\,; \quad \ddot{g}_T = \boldsymbol{w}_T^T\ddot{\boldsymbol{q}} + \overline{w}_T \tag{4.45}$$

where $\boldsymbol{w}_N, \boldsymbol{w}_T$ are given by (4.39) and $\overline{w}_N, \overline{w}_T$ are

$$\begin{aligned} \overline{w}_N &= \boldsymbol{n}_1^T(\overline{\boldsymbol{j}}_{Q1} - \tilde{\boldsymbol{\Omega}}_1\boldsymbol{v}_{C1}) - \kappa_1\dot{s}_1\boldsymbol{t}_1^T\boldsymbol{v}_{C1} + \dot{s}_1\boldsymbol{b}_{12}^T\boldsymbol{\Omega}_1 \\ &\quad +\boldsymbol{n}_2^T(\overline{\boldsymbol{j}}_{Q2} - \tilde{\boldsymbol{\Omega}}_2\boldsymbol{v}_{C2}) - \kappa_2\dot{s}_2\boldsymbol{t}_2^T\boldsymbol{v}_{C2} + \dot{s}_2\boldsymbol{b}_{12}^T\boldsymbol{\Omega}_2\,, \\ \overline{w}_T &= \boldsymbol{t}_1^T(\overline{\boldsymbol{j}}_{Q1} - \tilde{\boldsymbol{\Omega}}_1\boldsymbol{v}_{C1}) + \kappa_1\dot{s}_1\boldsymbol{n}_1^T\boldsymbol{v}_{C1} \\ &\quad +\boldsymbol{t}_2^T(\overline{\boldsymbol{j}}_{Q2} - \tilde{\boldsymbol{\Omega}}_2\boldsymbol{v}_{C2}) + \kappa_2\dot{s}_2\boldsymbol{n}_2^T\boldsymbol{v}_{C2}\,. \end{aligned} \tag{4.46}$$

The unknowns $(\dot{s}_1, \dot{s}_2)$ in eq. (4.46) can be taken from eq. (4.30), where the distance g_N may be set to zero, because the relative accelerations (4.45) need only be evaluated for the contact case:

$$\begin{aligned} \dot{s}_1 &= \frac{1}{\kappa_1 + \kappa_2}\left[\kappa_2 \boldsymbol{t}_1^T(\boldsymbol{v}_{C2} - \boldsymbol{v}_{C1}) + \boldsymbol{b}_{12}^T(\boldsymbol{\Omega}_2 - \boldsymbol{\Omega}_1)\right] \\ \dot{s}_2 &= \frac{1}{\kappa_1 + \kappa_2}\left[\kappa_1 \boldsymbol{t}_1^T(\boldsymbol{v}_{C2} - \boldsymbol{v}_{C1}) - \boldsymbol{b}_{12}^T(\boldsymbol{\Omega}_2 - \boldsymbol{\Omega}_1)\right] \end{aligned} \qquad (4.47)$$

With (4.46), (4.47) and (4.39) the relative accelerations $(\ddot{g}_N, \ddot{g}_T)$ in (4.45) are completely determined. It is remarkable that all these relations can be built without any additional differentiations when the curvatures, the normal and tangential vectors, and the velocities and accelerations of the bodies are known in terms of the contact parameters.

4.5 Evaluation of the Contact Kinematics

All relations which have to be used by solving the contact problem are available by the results of the previous sections. In detail, the following steps have to be performed:

1. At any time t the displacements $\boldsymbol{q}$ and velocities $\dot{\boldsymbol{q}}$ of our system are known. The contour of the body is described by a parameter vector ${}_B\boldsymbol{r}_{P\Sigma}$ which depends only on the arc length s if it is stated in the body-fixed frame B. The Jacobian terms ${}_B\boldsymbol{J}_C$, ${}_B\boldsymbol{J}_Q$, ${}_B\bar{\boldsymbol{j}}_C$, ${}_B\bar{\boldsymbol{j}}_Q$ in eqs. (4.37), (4.43) are known from recursive relative kinematics and can be regarded as given functions of s.

2. The solution of the nonlinear system (4.21) provides the contact parameters (s_1, s_2) and orients the trihedrals with respect to Fig. 4.3. With it the contour vector $\boldsymbol{r}_{P\Sigma}$, the tangent $\boldsymbol{t}$, the normal $\boldsymbol{n}$, the binormal $\boldsymbol{b}$ and the curvature κ can be computed with respect to (4.1), (4.2), as well as the Jacobian terms mentioned.

3. The distance of the bodies as a function of $\boldsymbol{q}$ and t is given by eqs. (4.23):

$$g_N(\boldsymbol{q}, t) = -\boldsymbol{r}_D^T \boldsymbol{n}_1 = +\boldsymbol{r}_D^T \boldsymbol{n}_2 \,.$$

4. After calculating the magnitudes $\boldsymbol{w}_N, \boldsymbol{w}_T, \tilde{w}_N, \tilde{w}_T$ in (4.39), the relative velocities of the contact points in the normal and tangential direction are obtained by (4.38):

$$\dot{g}_N(\boldsymbol{q}, \dot{\boldsymbol{q}}, t) = \boldsymbol{w}_N^T \dot{\boldsymbol{q}} + \tilde{w}_N \,; \quad \dot{g}_T(\boldsymbol{q}, \dot{\boldsymbol{q}}, t) = \boldsymbol{w}_T^T \dot{\boldsymbol{q}} + \tilde{w}_T \,.$$

5. Using $\dot{s}_1, \dot{s}_2$ from (4.47) by computing the terms $\overline{w}_N, \overline{w}_T$ in (4.46), we are now able to state the changes of the relative velocities in dependency on the still unknown accelerations $\ddot{\boldsymbol{q}}$:

$$\ddot{g}_N(\boldsymbol{q}, \dot{\boldsymbol{q}}, \ddot{\boldsymbol{q}}, t) = \boldsymbol{w}_N^T \ddot{\boldsymbol{q}} + \overline{w}_N ; \quad \ddot{g}_T(\boldsymbol{q}, \dot{\boldsymbol{q}}, \ddot{\boldsymbol{q}}, t) = \boldsymbol{w}_T^T \ddot{\boldsymbol{q}} + \overline{w}_T .$$

These five steps have to be performed at each time t during numerical integration. It is important that the nonlinear system (4.21) which provides the contact parameters must be solved after each integration step. A useful choice of the initial values for computing these zeros is the contact parameters of the previous step. Another possibility of treating the contact problem is given by the following approach. We solve the nonlinear equations only once at the beginning of the numerical integration. These contact parameters are then used as the initial values of the differential equations (4.30) which we integrate together with the dynamical system. As mentioned eqs. (4.30) determine the temporal changes of (s_1, s_2) in such a manner that the orientation of the trihedrals never change. Due to the numerical integration, however, some errors may occur in (4.21) which have to be corrected, for example, by projection methods.

Up to now we have assumed that the contour vector ${}_B\boldsymbol{r}_{P\Sigma}(s)$ depends on the arc lengths s. In most cases, however, the contours are described much more easily by any other parameter σ. Thus we have to express any function of s as dependency on σ. These are particularly the tangents, normals, curvatures, and time derivatives $\dot{s}$. Using the theory of planar curves and assuming a given contour vector ${}_B\boldsymbol{r}_{P\Sigma}(\sigma)$ the relations we are looking for are

$${}_B\boldsymbol{r}_{P\Sigma} = {}_B\boldsymbol{r}_{P\Sigma}(\sigma) ; \quad (\,\cdot\,)' = \frac{d}{d\sigma} ;$$

$${}_B\boldsymbol{t} = \frac{{}_B\boldsymbol{r}'_{P\Sigma}}{|{}_B\boldsymbol{r}'_{P\Sigma}|} ; \quad {}_B\boldsymbol{n} = \frac{{}_B\boldsymbol{r}''_{P\Sigma}\, {}_B\boldsymbol{r}'^T_{P\Sigma} - {}_B\boldsymbol{r}'_{P\Sigma}\, {}_B\boldsymbol{r}''^T_{P\Sigma}}{|{}_B\boldsymbol{r}'_{P\Sigma} \times {}_B\boldsymbol{r}''_{P\Sigma}|} \frac{{}_B\boldsymbol{r}'_{P\Sigma}}{|{}_B\boldsymbol{r}'_{P\Sigma}|} ;$$

$$\kappa = \frac{|{}_B\boldsymbol{r}'_{P\Sigma} \times {}_B\boldsymbol{r}''_{P\Sigma}|}{|{}_B\boldsymbol{r}'_{P\Sigma}|^3} ; \quad \dot{s} = |{}_B\boldsymbol{r}'_{P\Sigma}|\, \dot{\sigma} .$$

Note that all these equations are written in the body-fixed frame B, and therefore the only variable contained in them is σ. After these operations, the five steps for solving the contact problem can be performed analogously with the only exception that the nonlinear contact equations are now stated in terms of σ_1, σ_2.

Recall that the contact kinematics of Chapter 4 is based on the assumption that the contact event under consideration takes place in planar sections of the two contacting bodies (Figs. 4.1, 4.2). Considering out-of-plane contact kinematics results in more complicated relationships which is a matter of ongoing research. On the other hand, most applications in the dynamics of machines and mechanisms include only such planar contacts where the positions and orientations of contact planes are known in advance. Spatial contact situations may be found in contact dynamical problems of manufacturing especially in connection with assembly processes.

4.6 Example: Contact Problem of a Parabola and a Straight Line

In this example we want to solve the contact problem of a parabola and a straight line. The system under consideration is depicted in Fig. 4.5 and corresponds to an application problem of a parabolic friction element which is mounted between the platforms of neighboring blades in airborne gas turbines. The dynamic behavior of the complete system will be presented in Section 13.1 [6, 28, 29, 80].

In order to describe the displacements of the platform and the damper, we use as coordinates the tuple $\boldsymbol{q} = (z\ \varphi\ x\ y)^T$. Hereby, z denotes the horizontal location of the platform's body-fixed frame B, and (x, y, φ) determines the translational and rotational displacements of the damper. The contours under investigation are Σ_1 and Σ_2. They are described by the functions $y = ax$ and $y = -p_1x^2 + p_2$ related to the body-fixed frames B and K, respectively. In

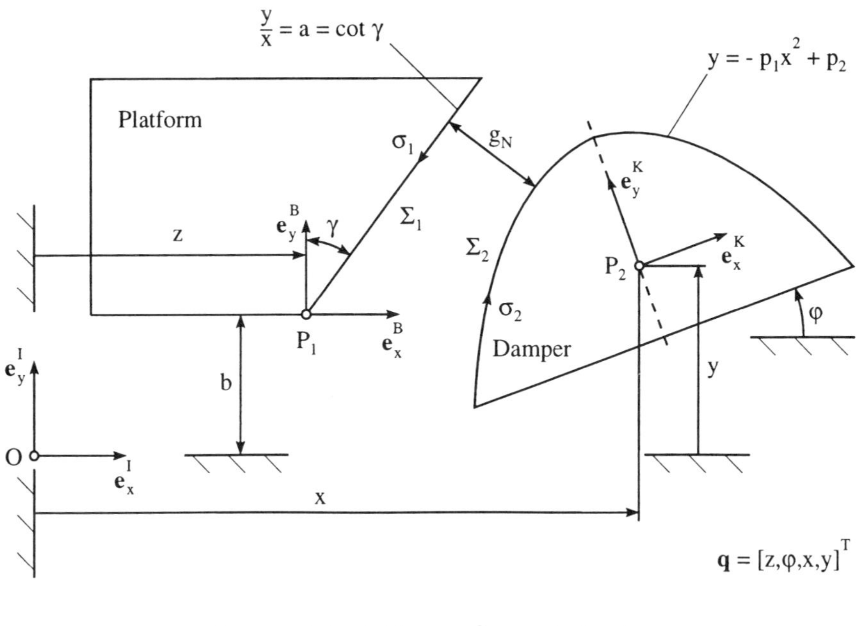

Figure 4.5: A Planar Contact Problem

the following the contact problem will be solved analytically by several steps, particulary the calculation of the contact parameters σ_i, the distance g_N, and the terms related to the normal and tangential relative velocities $\dot{g}_{N,T}$ and accelerations $\ddot{g}_{N,T}$.

1. Geometry and Displacement Vectors (see Fig. 4.2)

- Transformation matrix (2.3), (2.5)

$$\boldsymbol{A}_{KB} = \begin{pmatrix} c\varphi & s\varphi & 0 \\ -s\varphi & c\varphi & 0 \\ 0 & 0 & 1 \end{pmatrix} \qquad \begin{array}{l} c = \cos\,, \\ s = \sin\,. \end{array}$$

- Displacements of platform

$$_B\boldsymbol{r}_{OP1} = \begin{pmatrix} z \\ b \\ 0 \end{pmatrix}; \quad _B\boldsymbol{r}_{P\Sigma 1} = \begin{pmatrix} -\sigma_1 \\ -a\sigma_1 \\ 0 \end{pmatrix}; \quad a = \cot\gamma > 0\,.$$

- Displacements of damper

$$_B\boldsymbol{r}_{OP2} = \begin{pmatrix} x \\ y \\ 0 \end{pmatrix}; \quad _K\boldsymbol{r}_{P\Sigma 2} = \begin{pmatrix} \sigma_2 \\ -p_1\sigma_2^2 + p_2 \\ 0 \end{pmatrix}; \quad p_1, p_2 > 0\,.$$

2. Velocities, Angular Velocities and Jacobians

- Platform

$$\boldsymbol{\Omega}_1 = 0\,; \quad _B\boldsymbol{v}_{P1} = \begin{pmatrix} \dot{z} \\ 0 \\ 0 \end{pmatrix}; \quad _B\boldsymbol{v}_{C1} \overset{(4.10)}{=} \begin{pmatrix} \dot{z} \\ 0 \\ 0 \end{pmatrix};$$

$$_B\boldsymbol{J}_{C1} \overset{(4.37)}{=} \begin{pmatrix} 1 & 0 & 0 & 0 \\ 0 & 0 & 0 & 0 \\ 0 & 0 & 0 & 0 \end{pmatrix}; \quad _B\tilde{\boldsymbol{j}}_{C1} \overset{(4.37)}{=} \begin{pmatrix} 0 \\ 0 \\ 0 \end{pmatrix}.$$

- Damper

$$_B\boldsymbol{\Omega}_2 = \begin{pmatrix} 0 \\ 0 \\ \dot{\varphi} \end{pmatrix}; \quad _B\boldsymbol{v}_{P2} = \begin{pmatrix} \dot{x} \\ \dot{y} \\ 0 \end{pmatrix};$$

$$
\begin{aligned}
{}_B\boldsymbol{v}_{C2} &\overset{(4.10)}{=} {}_B\boldsymbol{v}_{P2} + {}_B\boldsymbol{\Omega}_2 \times \boldsymbol{A}_{BK}\, {}_K\boldsymbol{r}_{P\Sigma 2} \\
&= \begin{pmatrix} \dot{x} \\ \dot{y} \\ 0 \end{pmatrix} \\
&\quad + \begin{pmatrix} 0 \\ 0 \\ \dot{\varphi} \end{pmatrix} \times \begin{pmatrix} c\varphi & -s\varphi & 0 \\ s\varphi & c\varphi & 0 \\ 0 & 0 & 1 \end{pmatrix} \begin{pmatrix} \sigma_2 \\ -p_1\sigma_2^2 + p_2 \\ 0 \end{pmatrix} \\
&= \begin{pmatrix} \dot{x} - \sigma_2\dot{\varphi}s\varphi - \dot{\varphi}(-p_1\sigma_2^2 + p_2)c\varphi \\ \dot{y} + \sigma_2\dot{\varphi}c\varphi - \dot{\varphi}(-p_1\sigma_2^2 + p_2)s\varphi \\ 0 \end{pmatrix} ;
\end{aligned}
$$

$$
{}_B\boldsymbol{J}_{C2} \overset{(4.37)}{=} \begin{pmatrix} 0 & -\sigma_2 s\varphi - (-p_1\sigma_2^2 + p_2)c\varphi & 1 & 0 \\ 0 & +\sigma_2 c\varphi - (-p_1\sigma_2^2 + p_2)s\varphi & 0 & 1 \\ 0 & 0 & 0 & 0 \end{pmatrix} ;
$$

$$
{}_B\tilde{\boldsymbol{j}}_{C2} \overset{(4.37)}{=} \begin{pmatrix} 0 \\ 0 \\ 0 \end{pmatrix} .
$$

3. Accelerations, Angular Accelerations and Jacobians

- Platform

$$
\boldsymbol{\Omega}_1 = 0 ; \quad {}_B\boldsymbol{a}_{P1} = \begin{pmatrix} \ddot{z} \\ 0 \\ 0 \end{pmatrix} ; \quad {}_B\boldsymbol{a}_{Q1} \overset{(4.14)}{=} \begin{pmatrix} \ddot{z} \\ 0 \\ 0 \end{pmatrix} ;
$$

$$
{}_B\boldsymbol{J}_{Q1} \overset{(4.44)}{=} {}_B\boldsymbol{J}_{C1} ; \quad {}_B\bar{\boldsymbol{j}}_{Q1} \overset{(4.43)}{=} \begin{pmatrix} 0 \\ 0 \\ 0 \end{pmatrix} .
$$

- Damper

$$
{}_B(\dot{\boldsymbol{\Omega}}_2) = {}_B\,\dot{\boldsymbol{\Omega}}_2 = \begin{pmatrix} 0 \\ 0 \\ \ddot{\varphi} \end{pmatrix} ; \quad {}_B\boldsymbol{a}_{P2} = \begin{pmatrix} \ddot{x} \\ \ddot{y} \\ 0 \end{pmatrix} ;
$$

$$
\begin{aligned}
{}_B\boldsymbol{a}_{Q2} &\overset{(4.14)}{=} {}_B\boldsymbol{a}_{P2} + {}_B\dot{\boldsymbol{\Omega}}_2 \times \boldsymbol{A}_{BK}\, {}_K\boldsymbol{r}_{P\Sigma 2} \\
&\quad + {}_B\boldsymbol{\Omega}_2 \times ({}_B\boldsymbol{\Omega}_2 \times \boldsymbol{A}_{BK}\, {}_K\boldsymbol{r}_{P\Sigma 2}) \\
&= \begin{pmatrix} \ddot{x} \\ \ddot{y} \\ 0 \end{pmatrix} \\
&\quad + \begin{pmatrix} 0 \\ 0 \\ \ddot{\varphi} \end{pmatrix} \times \begin{pmatrix} c\varphi & -s\varphi & 0 \\ s\varphi & c\varphi & 0 \\ 0 & 0 & 1 \end{pmatrix} \begin{pmatrix} \sigma_2 \\ -p_1\sigma_2^2 + p_2 \\ 0 \end{pmatrix} \\
&\quad + \begin{pmatrix} 0 \\ 0 \\ \dot{\varphi} \end{pmatrix} \times \left[\begin{pmatrix} 0 \\ 0 \\ \dot{\varphi} \end{pmatrix} \times \begin{pmatrix} c\varphi & -s\varphi & 0 \\ s\varphi & c\varphi & 0 \\ 0 & 0 & 1 \end{pmatrix} \begin{pmatrix} \sigma_2 \\ -p_1\sigma_2^2 + p_2 \\ 0 \end{pmatrix} \right] \\
&= \begin{pmatrix} \ddot{x} - \sigma_2\ddot{\varphi}s\varphi - \ddot{\varphi}(-p_1\sigma_2^2 + p_2)c\varphi \\ \ddot{y} + \sigma_2\ddot{\varphi}c\varphi - \ddot{\varphi}(-p_1\sigma_2^2 + p_2)s\varphi \\ 0 \end{pmatrix} \\
&\quad + \begin{pmatrix} -\sigma_2\dot{\varphi}^2 c\varphi + \dot{\varphi}^2(-p_1\sigma_2^2 + p_2)s\varphi \\ -\sigma_2\dot{\varphi}^2 s\varphi - \dot{\varphi}^2(-p_1\sigma_2^2 + p_2)c\varphi \\ 0 \end{pmatrix} ;
\end{aligned}
$$

$$
{}_B\boldsymbol{J}_{Q2} \overset{(4.44)}{=} {}_B\boldsymbol{J}_{C2} ;
$$

$$
{}_B\bar{\boldsymbol{j}}_{Q2} \overset{(4.43)}{=} \begin{pmatrix} -\sigma_2\dot{\varphi}^2 c\varphi + \dot{\varphi}^2(-p_1\sigma_2^2 + p_2)s\varphi \\ -\sigma_2\dot{\varphi}^2 s\varphi - \dot{\varphi}^2(-p_1\sigma_2^2 + p_2)c\varphi \\ 0 \end{pmatrix} .
$$

4. Moving Trihedrals, Curvatures and Distance Vector (see Section 4.5)

- Platform

$$
{}_B\boldsymbol{r}'_{P\Sigma 1} = \begin{pmatrix} -1 \\ -a \\ 0 \end{pmatrix} ; \quad {}_B\boldsymbol{r}''_{P\Sigma 1} = \begin{pmatrix} 0 \\ 0 \\ 0 \end{pmatrix} ; \quad \kappa_1 = 0 ;
$$

$$
{}_B\boldsymbol{t}_1 = \frac{1}{\sqrt{1+a^2}} \begin{pmatrix} -1 \\ -a \\ 0 \end{pmatrix} ; \quad {}_B\boldsymbol{n}_1 = \frac{1}{\sqrt{1+a^2}} \begin{pmatrix} -a \\ 1 \\ 0 \end{pmatrix} ;
$$

$$
{}_B\boldsymbol{b}_1 = \begin{pmatrix} 0 \\ 0 \\ -1 \end{pmatrix} .
$$

- Damper

$$ {}_K\boldsymbol{r}'_{P\Sigma 2} = \begin{pmatrix} 1 \\ -2p_1\sigma_2 \\ 0 \end{pmatrix} ; \quad {}_K\boldsymbol{r}''_{P\Sigma 2} = \begin{pmatrix} 0 \\ -2p_1 \\ 0 \end{pmatrix} ; $$

$$ \kappa_2 = \frac{|{}_K\boldsymbol{r}'_{P\Sigma 2} \times {}_K\boldsymbol{r}''_{P\Sigma 2}|}{|{}_K\boldsymbol{r}'_{P\Sigma 2}|^3} = \frac{2p_1}{(1+4p_1^2\sigma_2^2)^{3/2}} ; $$

$$ {}_K\boldsymbol{t}_2 = \frac{1}{\sqrt{1+4p_1^2\sigma_2^2}} \begin{pmatrix} 1 \\ -2p_1\sigma_2 \\ 0 \end{pmatrix} ; $$

$$ {}_K\boldsymbol{n}_2 = \frac{1}{\sqrt{1+4p_1^2\sigma_2^2}} \begin{pmatrix} -2p_1\sigma_2 \\ -1 \\ 0 \end{pmatrix} ; \quad {}_K\boldsymbol{b}_2 = \begin{pmatrix} 0 \\ 0 \\ -1 \end{pmatrix} . $$

- Distance vector $\boldsymbol{r}_D$ (see Fig. 4.2)

$$ \begin{aligned} {}_B\boldsymbol{r}_D = &\ {}_B\boldsymbol{r}_{OP2} + \boldsymbol{A}_{BK}\ {}_K\boldsymbol{r}_{P\Sigma 2} - {}_B\boldsymbol{r}_{OP1} - {}_B\boldsymbol{r}_{P\Sigma 1} \\ = &\ \begin{pmatrix} x \\ y \\ 0 \end{pmatrix} + \begin{pmatrix} c\varphi & -s\varphi & 0 \\ s\varphi & c\varphi & 0 \\ 0 & 0 & 1 \end{pmatrix} \begin{pmatrix} \sigma_2 \\ -p_1\sigma_2^2 + p_2 \\ 0 \end{pmatrix} \\ & - \begin{pmatrix} z \\ b \\ 0 \end{pmatrix} - \begin{pmatrix} -\sigma_1 \\ -a\sigma_1 \\ 0 \end{pmatrix} \\ = &\ \begin{pmatrix} x - z + \sigma_1 + \sigma_2 c\varphi - (-p_1\sigma_2^2 + p_2)s\varphi \\ y - b + a\sigma_1 + \sigma_2 s\varphi + (-p_1\sigma_2^2 + p_2)c\varphi \\ 0 \end{pmatrix} . \end{aligned} $$

5. Contact Parameters

- Parabola parameter σ_2 from eq. (4.19)

$$ \boldsymbol{n}_1^T\boldsymbol{t}_2 = 0 \Rightarrow {}_B\boldsymbol{n}_1^T \boldsymbol{A}_{BK}\ {}_K\boldsymbol{t}_2 = 0 $$

$$ \Rightarrow \begin{pmatrix} -a \\ 1 \\ 0 \end{pmatrix}^T \begin{pmatrix} c\varphi & -s\varphi & 0 \\ s\varphi & c\varphi & 0 \\ 0 & 0 & 1 \end{pmatrix} \begin{pmatrix} 1 \\ -2p_1\sigma_2 \\ 0 \end{pmatrix} = 0 $$

$$ \Rightarrow \sigma_2 = \frac{-1}{2p_1} \cdot \frac{ac\varphi - s\varphi}{as\varphi + c\varphi} . \tag{4.48} $$

- Contact parameter σ_1 of the straight line from eq. (4.20)

$$\boldsymbol{r}_D^T \boldsymbol{t}_1 = 0 \;\Rightarrow\; {}_B\boldsymbol{r}_D^T \, {}_B\boldsymbol{t}_1 = 0 \,.$$

(σ_1 is not needed in the following.)

6. Distance in the Normal Direction

$$\begin{aligned} g_N &\overset{(4.23)}{=} -\boldsymbol{r}_D^T \boldsymbol{n}_1 = -{}_B\boldsymbol{r}_D^T \, {}_B\boldsymbol{n}_1 \\ &= \frac{-1}{\sqrt{1+a^2}} \begin{pmatrix} -a \\ 1 \\ 0 \end{pmatrix}^T \begin{pmatrix} x - z + \sigma_1 + \sigma_2 c\varphi - (-p_1\sigma_2^2 + p_2)s\varphi \\ y - b + a\sigma_1 + \sigma_2 s\varphi + (-p_1\sigma_2^2 + p_2)c\varphi \\ 0 \end{pmatrix} \\ &= \frac{1}{\sqrt{1+a^2}} \left[b - y + ax - az - (p_1\sigma_2^2 + p_2)(as\varphi + c\varphi) \right] \,. \end{aligned} \tag{4.49}$$

(simplified by using eq. 4.48)

7. Relative Velocities in Normal and Tangential Directions

- Remark: With σ_2 from (4.48) the relations $\boldsymbol{n}_1 = -\boldsymbol{n}_2$ and $\boldsymbol{t}_1 = -\boldsymbol{t}_2$ hold.
- Relative velocity in normal direction

$$\dot{g}_N \overset{(4.38)}{=} \boldsymbol{w}_N^T \dot{\boldsymbol{q}} + \tilde{w}_N \,;$$

$$\begin{aligned} \boldsymbol{w}_N^T &\overset{(4.39)}{=} \boldsymbol{n}_1^T (\boldsymbol{J}_{C1} - \boldsymbol{J}_{C2}) = {}_B\boldsymbol{n}_1^T ({}_B\boldsymbol{J}_{C1} - {}_B\boldsymbol{J}_{C2}) \\ &= \frac{1}{\sqrt{1+a^2}} \begin{pmatrix} -a \\ 1 \\ 0 \end{pmatrix}^T \begin{pmatrix} 1 & +\sigma_2 s\varphi + (-p_1\sigma_2^2 + p_2)c\varphi & -1 & 0 \\ 0 & -\sigma_2 c\varphi + (-p_1\sigma_2^2 + p_2)s\varphi & 0 & -1 \\ 0 & 0 & 0 & 0 \end{pmatrix} \\ &= \frac{1}{\sqrt{1+a^2}} \left[-a \;\; (\tfrac{1}{2p_1} + p_1\sigma_2^2 - p_2)(ac\varphi - s\varphi) \;\; a \;\; -1 \right] ; \end{aligned} \tag{4.50}$$

(simplified by using eq. 4.48)

$$\tilde{w}_N \overset{(4.39)}{=} \boldsymbol{n}_1^T \left(\tilde{\boldsymbol{j}}_{C1} - \tilde{\boldsymbol{j}}_{C2} \right) = 0 \,. \tag{4.51}$$

- Relative velocity in tangential direction

$$\dot{g}_T \overset{(4.38)}{=} \boldsymbol{w}_T^T \dot{\boldsymbol{q}} + \tilde{w}_T ;$$

$$\begin{aligned} \boldsymbol{w}_T^T &\overset{(4.39)}{=} \boldsymbol{t}_1^T(\boldsymbol{J}_{C1} - \boldsymbol{J}_{C2}) = {}_B\boldsymbol{t}_1^T({}_B\boldsymbol{J}_{C1} - {}_B\boldsymbol{J}_{C2}) \\ &= \frac{1}{\sqrt{1+a^2}} \begin{pmatrix} -1 \\ -a \\ 0 \end{pmatrix}^T \begin{pmatrix} 1 & +\sigma_2 s\varphi + (-p_1\sigma_2^2 + p_2)c\varphi & -1 & 0 \\ 0 & -\sigma_2 c\varphi + (-p_1\sigma_2^2 + p_2)s\varphi & 0 & -1 \\ 0 & 0 & 0 & 0 \end{pmatrix} \\ &= \frac{1}{\sqrt{1+a^2}} \left[-1 \quad -(p_1\sigma_2^2 + p_2)(as\varphi + c\varphi) \quad 1 \quad a \right] ; \end{aligned} \tag{4.52}$$

(simplified by using eq. 4.48)

$$\tilde{w}_T \overset{(4.39)}{=} \boldsymbol{t}_1^T \left(\tilde{\boldsymbol{j}}_{C1} - \tilde{\boldsymbol{j}}_{C2} \right) = 0 . \tag{4.53}$$

8. Relative Accelerations in Normal and Tangential Directions

- Remark: With σ_2 from (4.48) the relations $\boldsymbol{n}_1 = -\boldsymbol{n}_2$, $\boldsymbol{t}_1 = -\boldsymbol{t}_2$ and $\boldsymbol{b}_1 = \boldsymbol{b}_2$ hold. Furthermore, $\kappa_1 = 0$, $\boldsymbol{\Omega}_1 = 0$, $\overline{\boldsymbol{j}}_{Q1} = 0$.
- Changes of the arc length $\dot{s}_2$

$$\dot{s}_2 \overset{(4.47)}{=} -\frac{1}{\kappa_2} \boldsymbol{b}_1^T \boldsymbol{\Omega}_2 .$$

- Relative acceleration in normal direction

$$\ddot{g}_N \overset{(4.45)}{=} \boldsymbol{w}_N^T \ddot{\boldsymbol{q}} + \overline{w}_N ;$$

$$\begin{aligned} \overline{w}_N &\overset{(4.46)}{=} \boldsymbol{n}_2^T \left(\overline{\boldsymbol{j}}_{Q2} - \boldsymbol{\Omega}_2 \times \boldsymbol{v}_{C2} \right) - \kappa_2 \dot{s}_2 \boldsymbol{t}_2^T \boldsymbol{v}_{C2} + \dot{s}_2 \boldsymbol{b}_{12}^T \boldsymbol{\Omega}_2 \\ &\overset{(4.47)}{=} -\boldsymbol{n}_1^T \left(\overline{\boldsymbol{j}}_{Q2} - \boldsymbol{\Omega}_2 \times \boldsymbol{v}_{C2} \right) - \boldsymbol{b}_1^T \boldsymbol{\Omega}_2 \boldsymbol{t}_1^T \boldsymbol{v}_{C2} - \frac{1}{\kappa_2} \left(\boldsymbol{b}_1^T \boldsymbol{\Omega}_2 \right)^2 \\ &= \frac{\dot{\varphi}^2}{\sqrt{1+a^2}} (as\varphi + c\varphi) \left[p_2 - \frac{1}{2p_1} - p_1\sigma_2^2 \left(3 + 8p_1^2\sigma_2^2 \right) \right] . \end{aligned} \tag{4.54}$$

(simplyfied by using eq. 4.48)

- Relative acceleration in tangential direction

$$\ddot{g}_T \overset{(4.45)}{=} \boldsymbol{w}_T^T \ddot{\boldsymbol{q}} + \overline{w}_T ;$$

$$\begin{aligned}
\overline{w}_T &\overset{(4.46)}{=} \boldsymbol{t}_2^T \left(\overline{\boldsymbol{j}}_{Q2} - \boldsymbol{\Omega}_2 \times \boldsymbol{v}_{C2} \right) + \kappa_2 \dot{s}_2 \boldsymbol{n}_2^T \boldsymbol{v}_{C2} \\
&\overset{(4.47)}{=} -\boldsymbol{t}_1^T \left(\overline{\boldsymbol{j}}_{Q2} - \boldsymbol{\Omega}_2 \times \boldsymbol{v}_{C2} \right) - \boldsymbol{b}_1^T \boldsymbol{\Omega}_2 \boldsymbol{n}_1^T \boldsymbol{v}_{C2} \\
&= \frac{\dot{\varphi}^2}{\sqrt{1+a^2}} (as\varphi + c\varphi) \left[2p_1\sigma_2(p_2 - p_1\sigma_2^2) - \sigma_2 \right] . \qquad (4.55)
\end{aligned}$$

(simplyfied by using eq. 4.48)

5

MULTIPLE-CONTACT CONFIGURATIONS

The equations of motion of a bilaterally constrained system are already known from eq. (3.11). In this chapter we discuss several methods of taking into account additional constraints, especially friction-affected contact constraints. For this purpose, we have to include the arising contact forces into the equations of motion, and have to regard the constrained directions by the kinematical contact equations of Chapter 4. At sliding contacts the directions of the contact forces and the constrained displacements are *not* collinear, which is an important property of friction which leads to additional difficulties by using minimal coordinates. Moreover, overconstrained systems, which can be easily handled in the frictionless case, provide uniqueness problems when connected by Coulomb's law. Before discussing these phenomena we first consider the equations of the superimposed constraints.

5.1 Superimposed Constraints

Two bodies in the state of separation ($g_N > 0$) are depicted in Fig. 4.3. A completely inelastic impact or a smooth touchdown, which are subjects of Chapter 8, result in both bodies remaining in contact with each other. This leads to at least one additional constraint and, hence, to the occurrence of contact forces in the normal and tangential directions which are shown in Fig. 5.1.

As a consequence the coordinates $\boldsymbol{q}$ are no longer generalized coordinates but only a set of descriptor variables due to the new constraints. This situation can be handled by two different approaches: First, we could choose a new set of generalized coordinates $\boldsymbol{p}$, which is smaller then $\boldsymbol{q}$, in order to further reduce the system similarly to the method used in Section 2. The constraint forces of the contact would then not be needed in eq. (3.10). This is, however, unacceptable for several reasons: Systems with more than one possible contact point would force us to choose for any imaginable contact configuration a certain set of generalized coordinates with varying dimension.

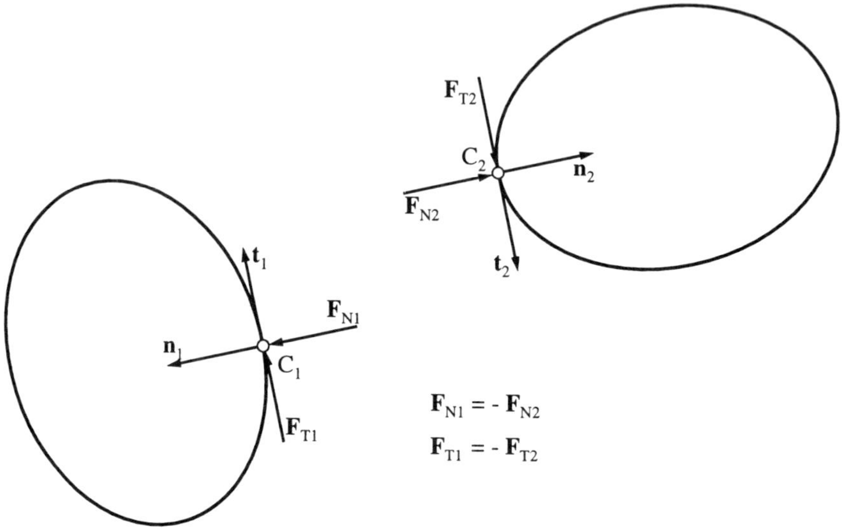

Figure 5.1: Contact Forces

A set of n possible contact points in the frictionless case, for example, would produce

$$\sum_{k=1}^{n} \frac{n!}{k!(n-k)!} = 2^n - 1$$

different sets of generalized coordinates $\boldsymbol{p} \in \mathbb{R}^{f-k}$. This situation is additionally complicated if the constraints are not independent or Coulomb friction is considered, where the active tangential friction force depends on the passive normal force in the case of sliding and becomes passive in the case of sticking. Thus we suggest a more direct method where the contact forces are included in eq. (3.10) using a Lagrangian approach. After premultiplying the contact forces in Fig. 5.1 by the corresponding Jacobians, we get for each of the bodies one additional term which has to be added to eq. (3.10):

$$\sum_{i=1}^{n} \left\{ \begin{pmatrix} \boldsymbol{J}_S \\ \boldsymbol{J}_R \end{pmatrix}^T \begin{pmatrix} \dot{\boldsymbol{p}} \\ \dot{\boldsymbol{L}} \end{pmatrix} - \begin{pmatrix} \boldsymbol{J}_A \\ \boldsymbol{J}_R \end{pmatrix}^T \begin{pmatrix} \boldsymbol{F}_A \\ \boldsymbol{M}_A \end{pmatrix} \right\}_i \tag{5.1}$$
$$-\boldsymbol{J}_{C1}^T (\boldsymbol{F}_{N1} + \boldsymbol{F}_{T1}) - \boldsymbol{J}_{C2}^T (\boldsymbol{F}_{N2} + \boldsymbol{F}_{T2}) = \boldsymbol{0}\,.$$

Referring to Fig. 5.1, we see that the normal and tangential contact forces have the same direction as the corresponding unit vectors. Thus we can write

$$\boldsymbol{F}_{N1} = \boldsymbol{n}_1 \lambda_N ; \quad \boldsymbol{F}_{N2} = \boldsymbol{n}_2 \lambda_N ; \quad \boldsymbol{F}_{T1} = \boldsymbol{t}_1 \lambda_T ; \quad \boldsymbol{F}_{T2} = \boldsymbol{t}_2 \lambda_T , \tag{5.2}$$

where we should keep in mind that $|\boldsymbol{n}_i| = |\boldsymbol{t}_i| = 1$. With it, the terms λ_N and λ_T represent the scalar magnitudes of the normal and tangential contact forces, respectively. Putting (5.2) into (5.1) we obtain

$$\begin{aligned} &\sum_{i=1}^{n} \left\{ \begin{pmatrix} \boldsymbol{J}_S \\ \boldsymbol{J}_R \end{pmatrix}^T \begin{pmatrix} \dot{\boldsymbol{p}} \\ \dot{\boldsymbol{L}} \end{pmatrix} - \begin{pmatrix} \boldsymbol{J}_A \\ \boldsymbol{J}_R \end{pmatrix}^T \begin{pmatrix} \boldsymbol{F}_A \\ \boldsymbol{M}_A \end{pmatrix} \right\}_i \\ &\qquad - \left(\boldsymbol{J}_{C1}^T \boldsymbol{n}_1 + \boldsymbol{J}_{C2}^T \boldsymbol{n}_2 \right) \lambda_N - \left(\boldsymbol{J}_{C1}^T \boldsymbol{t}_1 + \boldsymbol{J}_{C2}^T \boldsymbol{t}_2 \right) \lambda_T = \boldsymbol{0} . \end{aligned} \tag{5.3}$$

The first part of eq. (5.3) corresponds to eq. (3.11), whereas the second part containing the contact terms can be rewritten by the results of the contact kinematic chapter, eq. (4.39),

$$\boldsymbol{M}\ddot{\boldsymbol{q}} - \boldsymbol{h} - \boldsymbol{w}_N \lambda_N - \boldsymbol{w}_T \lambda_T = \boldsymbol{0} . \tag{5.4}$$

Equation (5.4) describes our system under the influence of pairs of normal and tangential contact forces which are proportional to the multipliers λ_N and λ_T, respectively. These multipliers have to be determined by certain contact laws, which we can state in terms of the kinematical relations (4.23), (4.38), (4.45).

Now we consider a system with n_G contact points and introduce four index sets which describe the kinematical state of each of the contacts. Let

$$\begin{aligned} I_G &= \{1, 2, \ldots, n_G\} \\ I_S &= \{i \in I_G \mid g_{Ni} = 0\} \quad \text{with } n_S \text{ elements} \\ I_N &= \{i \in I_S \mid \dot{g}_{Ni} = 0\} \quad \text{with } n_N \text{ elements} \\ I_H &= \{i \in I_N \mid \dot{g}_{Ti} = 0\} \quad \text{with } n_H \text{ elements} . \end{aligned} \tag{5.5}$$

The set I_G consists of the n_G numbers of all contact points. I_S contains the n_S indices of the constraints with vanishing distance but arbitrary relative velocities, I_N describes the constraints which fulfill the necessary conditions for continuous contact (vanishing distance and zero relative velocity in the normal direction), and I_H those which are possibly sticking. The numbers of elements of I_S, I_N, I_H are not constant because the contact configuration of the dynamical system changes with time due to stick-slip transitions, impacts, and contact loss.

Equation (5.4) can then be rewritten as

$$\boldsymbol{M}\ddot{\boldsymbol{q}} - \boldsymbol{h} - \sum_{i \in I_N} (\boldsymbol{w}_N \lambda_N + \boldsymbol{w}_T \lambda_T)_i = \boldsymbol{0} , \tag{5.6}$$

where the remaining elements of I_S with $\dot{g}_{Ni} < 0$ are not taken into account because these constraints correspond to collisions and must be separately treated, which is done in Chapter 8. For solving (5.6), $2n_N$ additional equations must be available to determine the unknowns λ_{Ni} and λ_{Ti}. First, we express the tangential forces of the sliding contacts (elements of $I_N \backslash I_H$) by their normal forces, using Coulomb friction:

$$\lambda_{Ti} = -\mu_i \lambda_{Ni} \operatorname{sign}(\dot{g}_{Ti}) \quad \lambda_{Ni} \geq 0 \quad \forall i \in I_N \backslash I_H \,, \tag{5.7}$$

where μ_i is the coefficient of friction of the ith sliding contact, λ_{Ni} is the corresponding normal force which is assumed to act only as a compressive force ($\lambda_{Ni} \geq 0$), and the sign function regards the opposite direction of relative velocity and friction force. The remaining $n_N + n_H$ unknowns ($\lambda_{Ni} \in I_N, \lambda_{Ti} \in I_H$) are determined later by imposing some conditions on their relative accelerations (4.45):

$$\begin{aligned} \ddot{g}_{Ni} &= \boldsymbol{w}_{Ni}^T \ddot{\boldsymbol{q}} + \overline{w}_{Ni} & \qquad \forall i \in I_N \\ \ddot{g}_{Ti} &= \boldsymbol{w}_{Ti}^T \ddot{\boldsymbol{q}} + \overline{w}_{Ti} & \qquad \forall i \in I_H \,. \end{aligned} \tag{5.8}$$

Using the abbreviations

$$\begin{aligned} &\boldsymbol{\lambda}_N = \{\lambda_{Ni}\}\,, \;\; \boldsymbol{W}_N = \{\boldsymbol{w}_{Ni}\}\,, \;\; \overline{\boldsymbol{w}}_N = \{\overline{w}_{Ni}\}\,, \;\; \ddot{\boldsymbol{g}}_N = \{\ddot{g}_{Ni}\}\,, \;\; i \in I_N \\ &\boldsymbol{\lambda}_H = \{\lambda_{Ti}\}\,, \;\; \boldsymbol{W}_H = \{\boldsymbol{w}_{Ti}\}\,, \;\; \overline{\boldsymbol{w}}_H = \{\overline{w}_{Ti}\}\,, \;\; \ddot{\boldsymbol{g}}_H = \{\ddot{g}_{Ti}\}\,, \;\; i \in I_H \\ &\boldsymbol{\lambda}_G = \{\lambda_{Ti}\}\,, \;\; \boldsymbol{W}_G = \{\boldsymbol{w}_{Ti}\}\,, \;\; \overline{\overline{\boldsymbol{\mu}}}_G = \{-\mu_i \operatorname{sign}(\dot{g}_{Ti})\}\,, \quad i \in I_N \backslash I_H \end{aligned} \tag{5.9}$$

we write (5.6), (5.7), (5.8) in matrix notation

$$\begin{aligned} &\boldsymbol{M}\ddot{\boldsymbol{q}} - \boldsymbol{h} - \boldsymbol{W}_N \boldsymbol{\lambda}_N - \boldsymbol{W}_H \boldsymbol{\lambda}_H - \boldsymbol{W}_G \boldsymbol{\lambda}_G = \boldsymbol{0} \quad \in \mathbb{R}^f \\ &\boldsymbol{\lambda}_G = \overline{\overline{\boldsymbol{\mu}}}_G \boldsymbol{\lambda}_N \quad \in \mathbb{R}^{n_N - n_H} \\ &\ddot{\boldsymbol{g}}_N = \boldsymbol{W}_N^T \ddot{\boldsymbol{q}} + \overline{\boldsymbol{w}}_N \quad \in \mathbb{R}^{n_N} \\ &\ddot{\boldsymbol{g}}_H = \boldsymbol{W}_H^T \ddot{\boldsymbol{q}} + \overline{\boldsymbol{w}}_H \quad \in \mathbb{R}^{n_H} \end{aligned} \tag{5.10}$$

and eliminate the tangential forces of the sliding contacts $\boldsymbol{\lambda}_G$ from the first equation of (5.10) to get

$$\begin{aligned} &\boldsymbol{M}\ddot{\boldsymbol{q}} - \boldsymbol{h} - \left(\boldsymbol{W}_N + \boldsymbol{W}_G \overline{\overline{\boldsymbol{\mu}}}_G \mid \boldsymbol{W}_H\right) \begin{pmatrix} \boldsymbol{\lambda}_N \\ \boldsymbol{\lambda}_H \end{pmatrix} = \boldsymbol{0} \quad \in \mathbb{R}^f \\ &\begin{pmatrix} \ddot{\boldsymbol{g}}_N \\ \ddot{\boldsymbol{g}}_H \end{pmatrix} = \begin{pmatrix} \boldsymbol{W}_N^T \\ \boldsymbol{W}_H^T \end{pmatrix} \ddot{\boldsymbol{q}} + \begin{pmatrix} \overline{\boldsymbol{w}}_N \\ \overline{\boldsymbol{w}}_H \end{pmatrix} \quad \in \mathbb{R}^{n_N + n_H} . \end{aligned} \tag{5.11}$$

A more condensed representation of (5.11) is achieved by substituting the terms

$$\ddot{g} = \begin{pmatrix} \ddot{g}_N \\ \ddot{g}_H \end{pmatrix} \qquad \boldsymbol{\lambda} = \begin{pmatrix} \boldsymbol{\lambda}_N \\ \boldsymbol{\lambda}_H \end{pmatrix} \qquad \overline{\boldsymbol{w}} = \begin{pmatrix} \overline{\boldsymbol{w}}_N \\ \overline{\boldsymbol{w}}_H \end{pmatrix}$$
$$\boldsymbol{W} = (\boldsymbol{W}_N | \boldsymbol{W}_H) \qquad \boldsymbol{N}_G = \left(\boldsymbol{W}_G \overline{\overline{\boldsymbol{\mu}}}_G | \boldsymbol{0}\right) , \tag{5.12}$$

which leads to (see related works such as [1, 3, 12, 18, 20, 51, 52, 71])

$$\begin{aligned} &\boldsymbol{M}\ddot{\boldsymbol{q}} - \boldsymbol{h} - (\boldsymbol{W} + \boldsymbol{N}_G)\boldsymbol{\lambda} = \boldsymbol{0} \quad \in \mathbb{R}^f \\ &\ddot{\boldsymbol{g}} = \boldsymbol{W}^T \ddot{\boldsymbol{q}} + \overline{\boldsymbol{w}} \quad \in \mathbb{R}^{n_N + n_H} . \end{aligned} \tag{5.13}$$

Because the symmetric mass matrix $\boldsymbol{M}$ is almost always invertible, we can eliminate the accelerations $\ddot{\boldsymbol{q}}$ from the second equation in (5.13) and finally achieve a system of $n_N + n_H$ equations:

$$\ddot{\boldsymbol{g}} = \boldsymbol{W}^T \boldsymbol{M}^{-1} (\boldsymbol{W} + \boldsymbol{N}_G)\boldsymbol{\lambda} + (\boldsymbol{W}^T \boldsymbol{M}^{-1} \boldsymbol{h} + \overline{\boldsymbol{w}}) \quad \in \mathbb{R}^{n_N + n_H} , \tag{5.14}$$

which is linear in the $2(n_N + n_H)$ unknowns $(\ddot{\boldsymbol{g}}, \boldsymbol{\lambda})$. Assuming that each of the elements of I_N and I_H continue to keep contact and stick, the relative accelerations $\ddot{\boldsymbol{g}} = \left(\ddot{\boldsymbol{g}}_N^T \ddot{\boldsymbol{g}}_H^T\right)^T$ must be equal to zero. Thus, (5.14) becomes

$$\boldsymbol{W}^T \boldsymbol{M}^{-1} (\boldsymbol{W} + \boldsymbol{N}_G)\boldsymbol{\lambda} = -(\boldsymbol{W}^T \boldsymbol{M}^{-1} \boldsymbol{h} + \overline{\boldsymbol{w}}) \tag{5.15}$$

which is a linear system of equations of the form

$$\boldsymbol{A}\boldsymbol{\lambda} = -\boldsymbol{b} \tag{5.16}$$

and with which we can determine the contact forces $\boldsymbol{\lambda} = \left(\boldsymbol{\lambda}_N^T \boldsymbol{\lambda}_T^T\right)^T$. The assumption $\ddot{\boldsymbol{g}} = \boldsymbol{0}$ replaces the unilateral nature of the contact constraints by a bilateral formulation which excludes possible transitions from sticking to sliding and contact to separation, respectively. These transitions are dependent on the values of the contact forces. Contact, for example, is lost if the normal force becomes negative; transitions from sticking to sliding arise since the tangential force is bounded by Coulomb's frictional law. These additional conditions are used in Chapter 6 to determine simultaneousely both unknowns $(\ddot{\boldsymbol{g}}, \boldsymbol{\lambda})$ in (5.14). Then eq. (5.14) holds during continual contact and sticking and includes all possible transitions to sliding or separation.

By the contact model of Fig. 5.1 only forces have been taken into account in the equations of motion (5.1). This, however, may be too rough an approximation for some systems. The modeling of contact points where only forces are transmitted results from a neglected contact area due to the rigid body assumption. Generally, the bodies are deformable, and the contact point

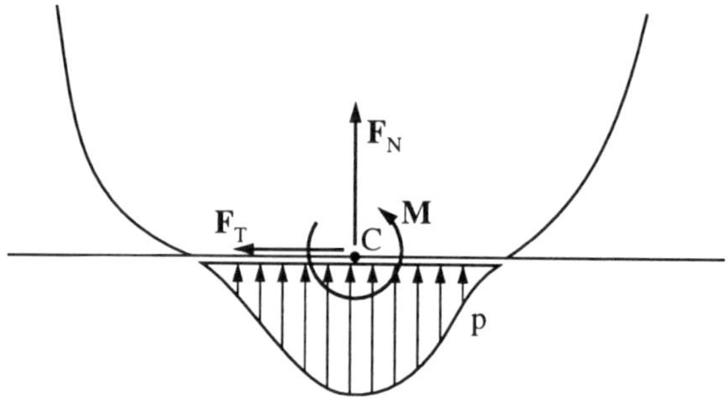

Figure 5.2: Asymmetric Stress Distribution

must be regarded as an approximation of a certain contact area. For example an asymmetric normal stress distribution in the contact area (see Fig. 5.2) leads to an additional moment $\boldsymbol{M}$ with respect to the contact point C. In the general three-dimensional case, six force magnitudes have to be considered at the contact point: one normal force, two tangential forces which take into account the possible directions at planar sliding, two moments which are induced by the normal stress distribution at the contact area, and one moment which considers the drilling friction.

5.2 Minimal Coordinates and Friction

For some applications, such as in guidances of mechanisms, the normal constraint is bilateral and sticking is not possible due to kinematics. These systems are not able to undergo state transitions, except for a change in the sliding direction, and allow a special evaluation. In this section we restrict ourselves to systems without state transitions and show that a description using minimal coordinates is possible even in the presence of sliding friction [21]. We start our discussion with eqs. (5.13) and (5.15):

$$\boldsymbol{M}\ddot{\boldsymbol{q}} - \boldsymbol{h} - (\boldsymbol{W} + \boldsymbol{N}_G)\boldsymbol{\lambda} = \boldsymbol{0} \quad \in \mathbb{R}^f \tag{5.17}$$

$$\boldsymbol{W}^T\boldsymbol{M}^{-1}(\boldsymbol{W} + \boldsymbol{N}_G)\boldsymbol{\lambda} = -(\boldsymbol{W}^T\boldsymbol{M}^{-1}\boldsymbol{h} + \overline{\boldsymbol{w}}) \quad \in \mathbb{R}^{n_N+n_H} \tag{5.18}$$

where we already have assumed only bilateral contact constraints

$$\ddot{\boldsymbol{g}} = \boldsymbol{W}^T\ddot{\boldsymbol{q}} + \overline{\boldsymbol{w}} = 0 \quad \in \mathbb{R}^{n_N+n_H}, \tag{5.19}$$

and compressive forces $\lambda_{Ni} \geq 0 \; \forall i \in I_N \backslash I_H$ for the sliding contacts (see eq. 5.7). Furthermore, in this section we only look at nonoverconstrained systems, which are characterized by a linear independency of the $n_N + n_H$ columns of the matrix $\boldsymbol{W}$. It is most important to notice that eq. (5.18) does *not* depend on any accelerations and can be solved independently. Thus, from now on the contact forces $\boldsymbol{\lambda}$ can be regarded as known, and the dynamical system is completely determined by the eqs. (5.17) and (5.19). Now we choose a set of minimal coordinates $\boldsymbol{p} \in \mathbb{R}^{f-(n_N+n_H)}$ which fulfills the constraints on the acceleration level (5.19). Because (5.19) defines an $(n_N + n_H)$-dimensional affine subset of $\mathbb{R}^f$, this can be done by the affine transformation

$$\ddot{\boldsymbol{q}} = \boldsymbol{Q}\ddot{\boldsymbol{p}} + \overline{\boldsymbol{\rho}}; \quad \boldsymbol{Q} \in \mathbb{R}^{f,f-(n_N+n_H)}. \tag{5.20}$$

Putting (5.20) into (5.19) we get

$$\ddot{\boldsymbol{g}} = \boldsymbol{W}^T\boldsymbol{Q}\ddot{\boldsymbol{p}} + (\boldsymbol{W}^T\overline{\boldsymbol{\rho}} + \overline{\boldsymbol{w}}) = 0 \quad \forall \ddot{\boldsymbol{p}} \in \mathbb{R}^{f-(n_N+n_H)} \tag{5.21}$$

which has to be fulfilled for *every* $\ddot{\boldsymbol{p}} \in \mathbb{R}^{f-(n_N+n_H)}$. This leads to the conditions

$$\boldsymbol{W}^T\overline{\boldsymbol{\rho}} + \overline{\boldsymbol{w}} = 0; \quad \boldsymbol{W}^T\boldsymbol{Q} = 0. \tag{5.22}$$

The second equation in (5.22) demands that the $f - (n_N + n_H)$ columns of $\boldsymbol{Q}$ have to be perpendicular to the $n_N + n_H$ linearly independent columns of $\boldsymbol{W}$. If we choose them to be linearly independent, then the overall matrix $(\boldsymbol{Q}\,\boldsymbol{W})$ forms a basis of $\mathbb{R}^f$ with $\boldsymbol{Q}$ representing the orthogonal space to $\boldsymbol{W}$. Now we premultiply the dynamics equation (5.17) by the transpose of this matrix and insert $\ddot{\boldsymbol{q}}$ from (5.20):

$$\begin{pmatrix} \boldsymbol{Q}^T \\ \boldsymbol{W}^T \end{pmatrix} [\boldsymbol{M}(\boldsymbol{Q}\ddot{\boldsymbol{p}} + \overline{\boldsymbol{\rho}}) - \boldsymbol{h} - (\boldsymbol{W} + \boldsymbol{N}_G)\boldsymbol{\lambda}] = 0. \tag{5.23}$$

In order to obtain the unknowns $\ddot{\boldsymbol{p}}$ only the first $f - (n_N + n_H)$ equations of (5.23) are needed. Regarding $\boldsymbol{Q}^T\boldsymbol{W} = 0$ by (5.22), we get

$$\boldsymbol{Q}^T\boldsymbol{M}\boldsymbol{Q}\ddot{\boldsymbol{p}} + \boldsymbol{Q}^T(\boldsymbol{M}\overline{\boldsymbol{\rho}} - \boldsymbol{h} - \boldsymbol{N}_G\boldsymbol{\lambda}) = 0 \tag{5.24}$$

which is a representation of a dynamical system with sliding friction using *minimal coordinates* $\boldsymbol{p}$. All the additional $n_N + n_H$ bilateral constraints (5.19) are automatically taken into account by the set of coordinates $\boldsymbol{p}$. Note that one part of the contact forces, $\boldsymbol{N}_G\boldsymbol{\lambda}$, remains in the equations of motion, whereas the other part $\boldsymbol{W}\boldsymbol{\lambda}$ has dropped from it due to the orthogonality $\boldsymbol{Q}^T\boldsymbol{W} = 0$. This is an important property of sliding friction where the overall sliding friction force, consisting of a normal and a tangential part, is not collinear to the constrained displacements. By the definition in Chapter 3 only the normal portion, which is contained in $\boldsymbol{W}\boldsymbol{\lambda}$, corresponds to a passive force. The tangential friction forces influence the dynamics of the system and therefore have to remain in the equations of motion. Additionally, they depend on the passive normal forces, which results in the expression $\boldsymbol{N}_G\boldsymbol{\lambda}$, where, by eq. (5.12), $\boldsymbol{N}_G$ contains the coefficients of friction. In the frictionless case, $\boldsymbol{N}_G = 0$; all contact forces vanish in (5.24). Otherwise, they are determined by the linear system (5.18). Even this set of equations should be commented on. In the frictionless case ($\boldsymbol{N}_G = 0$) the system matrix $\boldsymbol{W}^T\boldsymbol{M}^{-1}\boldsymbol{W}$ is invertible because $\boldsymbol{M}^{-1}$ is symmetric and $\boldsymbol{W} \in \mathbb{R}^{f, n_N+n_H}$ consists of $n_N + n_H \leq f$ linearly independent columns by the presupposition of non overconstrained systems. Sliding friction, however, may cause a decreasing of the determinant of $\boldsymbol{W}^T\boldsymbol{M}^{-1}(\boldsymbol{W} + \boldsymbol{N}_G)$ during the motion. Simultaneously the contact forces $\boldsymbol{\lambda}$ increase to infinite values when the determinant vanishes. Such a situation is given in the example of a sliding rod where the physical background of this phenomenon and the consequences on modeling are discussed.

Up to now we have assumed independent constraints. Sometimes, however, dependent constraints cannot be avoided, which means that the system is overconstrained. For example, one can imagine a complicated dynamical system with many possible contacts. Each contact has to be taken into account by one constraint. During the evolution in time, the contacts may be open or closed, which depends on the dynamics. Situations may occur where the closed contacts form kinematic chains within the dynamical system and the corresponding constraints become dependent. In order to discuss such systems we look once more at the sets of equations (5.17)–(5.19). The $n_N + n_H$ constraints on the acceleration level are given by eq. (5.19) and are said to be independent if rank $\boldsymbol{W} = n_N + n_H$. In the other case, rank $\boldsymbol{W} < n_N + n_H$, the system is overconstrained with respect to some coordinates and therefore the constraints become dependent. At that point we remember that eq. (5.19) is a kinematic condition on the system. If we have derived the kinematics with respect to Chapter 4, then this description is always consistent. There exists at least one $\ddot{\boldsymbol{q}} \in \mathbb{R}^f$ which fulfills (5.19), and thus the maximal rank of $\boldsymbol{W}$ is f. With it, the constraints are always dependent if $n_N + n_H > f$. Now we look at the equation of the contact forces (5.18) for the frictionless case, $\boldsymbol{N}_G = 0$. If the constraints are dependent, then rank $(\boldsymbol{W}^T\boldsymbol{M}^{-1}\boldsymbol{W}) < n_N + n_H$. The matrix $(\boldsymbol{W}^T\boldsymbol{M}^{-1}\boldsymbol{W})$ is not invertible, and the contact forces $\boldsymbol{\lambda}$ can no longer be determined uniquely. The values of $\ddot{\boldsymbol{q}} \in \mathbb{R}^f$, however, remain unique, which can be easily seen from optimization theory. Equations (5.17) and (5.19) are nothing else than the necessary and sufficient optimality conditions of the quadratic program ($\boldsymbol{N}_G = 0$)

$$\begin{aligned} \ddot{\boldsymbol{q}} &= \arg\min \left\{ f(\ddot{\boldsymbol{q}}^*) \mid \boldsymbol{W}^T\ddot{\boldsymbol{q}}^* + \overline{\boldsymbol{w}} = 0 \right\} \\ f(\ddot{\boldsymbol{q}}^*) &= \tfrac{1}{2}(\ddot{\boldsymbol{q}}^* - \boldsymbol{M}^{-1}\boldsymbol{h})^T\boldsymbol{M}(\ddot{\boldsymbol{q}}^* - \boldsymbol{M}^{-1}\boldsymbol{h}) \end{aligned} \tag{5.25}$$

which is called the "principle of least constraints." With $\boldsymbol{M}$ symmetric and positive definite, the quadratic cost function $f(\ddot{\boldsymbol{q}}^*)$ is strictly convex. Furthermore, the constraints are feasible by presupposition. Thus, we conclude that an optimal value $f(\ddot{\boldsymbol{q}})$ exists, and the optimal solution $\ddot{\boldsymbol{q}}$ is unique. Frictionless overconstrained systems always lead to unique solutions $\ddot{\boldsymbol{q}}$. This is not valid for the general case: The matrix $\boldsymbol{N}_G$ containing the coefficients of friction for the sliding contacts destroys the structure of the optimality conditions. No strictly convex optimization problem can be found which is equivalent to eq. (5.17) and (5.19). We will givw an example of a pantograph mechanism where the accelerations $\ddot{\boldsymbol{q}}$ are not unique under certain conditions.

5.3 Example: The Sliding Rod

In this section we present an example where the matrix $\boldsymbol{W}^T\boldsymbol{M}^{-1}(\boldsymbol{W}+\boldsymbol{N}_G)$ in eq. (5.15) becomes singular due to sliding friction [21]. The left part of Fig. 5.3 shows a rod (length $2s$, mass m, center of mass S, moment of inertia $J_S = \frac{1}{3}ms^2$) which has surface contact at point C and is sliding under Coulomb friction from left to right. In the no-contact case the system has three degrees of freedom, which we describe by the coordinates $\boldsymbol{q} = (x,\, y,\, \varphi)^T$.

In the right part of Fig. 5.3 we see the free-body diagram of our system. The rod is under the influence of gravity mg. At the contact point C the normal and tangential contact forces λ_N and λ_T act on the body. Now we state the balances in the x-, y-, and φ-directions:

$$\begin{aligned} m\ddot{x} &= \lambda_T \\ m\ddot{y} &= \lambda_N - mg \\ \tfrac{1}{3}ms^2\ddot{\varphi} &= -\lambda_T s\sin\varphi - \lambda_N s\cos\varphi\,. \end{aligned} \tag{5.26}$$

Next we must derive the distance g_N, the relative velocities at the contact point $(\dot{g}_N,\, \dot{g}_T)$, and their time derivatives $(\ddot{g}_N,\, \ddot{g}_T)$. From Fig. 5.3 the

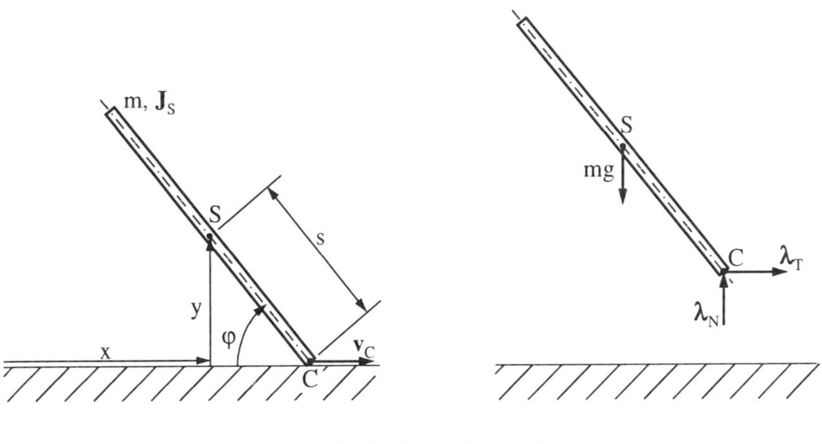

Figure 5.3: The Sliding Rod

distance is given by

$$g_N = y - s\sin\varphi\,, \tag{5.27}$$

and the relative velocities are

$$\dot{g}_N = \dot{y} - s\dot{\varphi}\cos\varphi\,; \quad \dot{g}_T = \dot{x} - s\dot{\varphi}\sin\varphi\,. \tag{5.28}$$

Differentiation of (5.28) leads to the constraints on the acceleration level:

$$\ddot{g}_N = \ddot{y} - s\ddot{\varphi}\cos\varphi + s\dot{\varphi}^2\sin\varphi\,; \quad \ddot{g}_T = \ddot{x} - s\ddot{\varphi}\sin\varphi - s\dot{\varphi}^2\cos\varphi\,. \tag{5.29}$$

Now we have to evaluate the assumed contact state by using eqs. (5.27)–(5.29). The rod has contact and is sliding to the right. Contact means that the distance (5.27) and the relative velocity in the normal direction (5.28) must vanish ($g_N = 0$, $\dot{g}_N = 0$), whereas sliding to the right is regarded by the condition (5.28), $\dot{g}_T > 0$. Thus, the tangential sliding friction force is given by (5.7):

$$\lambda_T = -\mu\lambda_N\ \mathrm{sign}(\dot{g}_T) = -\mu\lambda_N\,; \quad \dot{g}_T > 0\,. \tag{5.30}$$

Equations (5.26), (5.29) and (5.30) provide six equations for the unknown accelerations ($\ddot{x}$, $\ddot{y}$, $\ddot{\varphi}$, $\ddot{g}_N$, $\ddot{g}_T$) and forces (λ_N, λ_T). One condition is still missing and will be introduced later in order to solve the system.

Now we substitute the generalized accelerations ($\ddot{x}$, $\ddot{y}$, $\ddot{\varphi}$) from (5.26) into (5.29) and express with (5.30) all occurring tangential forces λ_T by their normal forces λ_N:

$$\ddot{g}_N = \frac{1}{m}\left[1 + 3\cos\varphi(\cos\varphi - \mu\sin\varphi)\right]\lambda_N + (s\dot{\varphi}^2\sin\varphi - g) \tag{5.31}$$

$$\ddot{g}_T = \frac{1}{m}\left[-\mu + 3\sin\varphi(\cos\varphi - \mu\sin\varphi)\right]\lambda_N - s\dot{\varphi}^2\cos\varphi. \tag{5.32}$$

For further discussion we only look at eq. (5.31). Equation (5.32) may be used to calculate the tangential contact acceleration $\ddot{g}_T$ after eq. (5.31) has been solved, but is not needed for the following evaluations. With (5.31) we have achieved the structure of eq. (5.14) in Section 5.1. It is a linear equation with two unknowns, $\ddot{g}_N$ and λ_N. In eq. (5.15) we have assumed that the system maintains contact, which is only possible if the normal force acts with a compressive magnitude, $\lambda_N \geq 0$, and the acceleration in the normal direction is equal to zero, $\ddot{g}_N = 0$. But this is only one of two permitted situations. The second one is given by the take-off transition where the rod loses contact. In that case the contact force must be equal to zero, $\lambda_N = 0$, and the separation process can only start with values $\ddot{g}_N \geq 0$. Both situations can be stated by two inequalities and one complementarity condition:

$$\ddot{g}_N \geq 0\,; \quad \lambda_N \geq 0\,; \quad \ddot{g}_N\lambda_N = 0\,, \tag{5.33}$$

where the term $\ddot{g}_N \lambda_N = 0$ switches between the two admissible states by demanding at least one of both factors to be zero. This complementarity idea is the main point of Chapter 6, but it is introduced here to advance the discussion of the example. With (5.33) we have found the missing conditions for determining the unknowns of eq. (5.31). Now we rewrite (5.31) and (5.33) by using the abbreviations

$$a := \frac{1}{m}\left[1 + 3\cos\varphi(\cos\varphi - \mu\sin\varphi)\right]\,; \quad b := g\left(\frac{s\dot{\varphi}^2}{g}\sin\varphi - 1\right) \tag{5.34}$$

and note that both terms may have positive or negative values depending on the parameters φ, μ and $\frac{s\dot{\varphi}^2}{g}$. The resulting condition

$$\ddot{g}_N = a\,\lambda_N + b\,; \quad \ddot{g}_N \geq 0\,; \quad \lambda_N \geq 0\,; \quad \ddot{g}_N \lambda_N = 0\,, \tag{5.35}$$

forms a one-dimensional Linear Complementarity Problem which we will discuss with respect to existence and uniqueness of the solutions. Hereby, we distinguish the cases $a > 0$ and $a < 0$.

Case I: $a > 0$

(i) Let $b > 0$: In this case the solution obviously is

$$\ddot{g}_N = b > 0\,; \quad \lambda_N = 0$$

because the second possibility ($\ddot{g}_N = 0$, $\lambda_N = -\frac{b}{a} < 0$) would contradict $\lambda_N \geq 0$. Here, the rod leaves the surface.

(ii) Let $b < 0$: This situation corresponds to continual sliding, which is given by the solution

$$\ddot{g}_N = 0\,; \quad \lambda_N = -\frac{b}{a} > 0\,.$$

Summarizing, we see that for $a > 0$ either separation or continual sliding takes place, depending on the sign of b.

Case II: $a < 0$

(i) Let $b > 0$: As in Case I(i),

$$\ddot{g}_N = b > 0\,; \quad \lambda_N = 0$$

is a solution of the LCP. But now a is negative and thus a *second* solution can be found:

$$\ddot{g}_N = 0\,; \quad \lambda_N = -\frac{b}{a} > 0\,.$$

Continual sliding *and* separation are possible solutions for the rod. So we no longer have unique solutions as in Case I.

(ii) Let $b < 0$: Clearly, ($\ddot{g}_N = b < 0,\ \lambda_N = 0$) cannot be a solution of the LCP. Even the possibility ($\ddot{g}_N = 0,\ \lambda_N = -\frac{b}{a} < 0$) fails because both a and b are negative. In that case, sliding as well as separation are contradictory to the equations of motion and the contact law; thus no solution exists.

For $a < 0$ either nonuniqueness or nonexistence occurs, depending on the values of b.

Obviously, the sign of a determines whether unique solutions are obtained. Figure 5.4 shows the areas of positive and negative values of a depending on the coefficient of friction μ and the angle of the rod φ in the range $0° \le \varphi \le 90°$. With respect to eq. (5.34) the boundary $a = 0$ is given by the function

$$\mu = \frac{1 + 3\cos^2\varphi}{3\sin\varphi\cos\varphi}\,. \tag{5.36}$$

From Fig. 5.4 we see that for $\mu < \frac{4}{3}$ the values of a are always positive. If the coefficient of friction is chosen small enough, then we always get a unique solution. Now we consider a friction coefficient greater than $\frac{4}{3}$, for example $\frac{5}{3}$. For small ($\varphi < 45°$) and large ($\varphi > 76°$) values of the inclination of the rod, the solution is unique; in the remaining area, however, ($45° < \varphi < 76°$) we either get nonuniqueness or nonexistence, which depends on b.

Now we return to our first question, the sliding rod, and discuss under which conditions the singularity $a = 0$ occurs and how it can be interpreted mechanically. We consider an initial condition where the rod is sliding from left to right with a small inclination angle φ and a coefficient of friction greater than $\frac{4}{3}$. Under that assumption, $a > 0$. Additionally, we assume that the rod is turning to the right by some values $\dot{\varphi}$ which are not too large in order to fulfill $b < 0$. Now we have ($a > 0,\ b < 0$), which is sliding by Case I(ii). Due

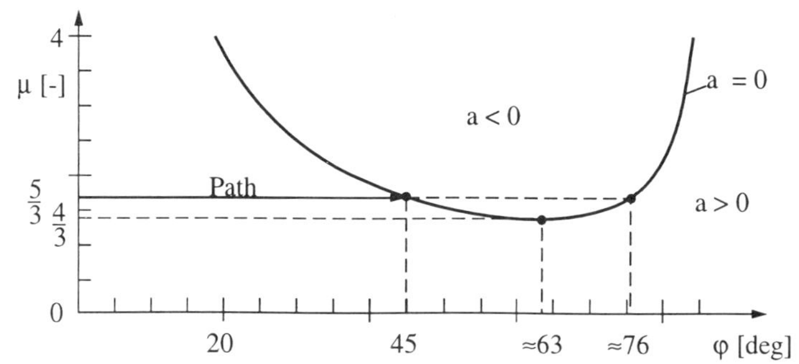

Figure 5.4: Critical Values of a

to $\dot{\varphi} > 0$ the inclination values φ of the rod are increasing during the sliding motion, which occurs by eq. (5.34) and Fig. 5.4 in decreasing values of the function a ("Path"!). This, however, yields a growing of the contact force λ_N by eq. (5.35). Now we let the process continue and observe a further increase of the inclination angle φ where the values of a tend more and more to zero. Simultaneously λ_N grows to infinite values. There we have $(a = 0,\ b < 0)$. In the next step, we would get $(a < 0,\ b < 0)$ (see Fig. 5.4), where no solution exists due to Case II(ii). In that situation, however, it is most important to notice that we *must not* try to pass the singularity point by using eq. (5.35). Infinite forces lead to infinite accelerations, due to the couplings in eq. (5.26), and hence to velocity jumps. These jumps are subject to impact theory and are *not* included in eqs. (5.26)–(5.35). Physically, infinite forces never occur. In this example they result from combining Coulomb's law with rigid body theory. Under the action of infinite (or very large) forces no body remains rigid. At that singularity point we have to drop the rigid body assumption, which means we have to change our model. This can be done either by introducing some compliance elements, for example, at the contact point or by applying a frictional impact which globally allows "deformation" by dropping the impenetrability condition on the acceleration level, $\ddot{g}_N \geq 0$ [64].

Summarizing the main results of this section, we see that sliding friction may cause singularities $(a = 0)$ where the contact forces together with the accelerations tend to infinity. The solution of the scalar Linear Complementarity Problem (5.35) determines whether contact is maintained or separation takes place. The zeros of the function $a(\varphi, \mu)$, eq. (5.34), separate the region of unique solutions from the area of nonexistence or nonuniqueness. In the opinion of the authors, Coulomb's law describes correctly the physical behavior of the rod up to the singularity point. The singularity does *not* mean that Coulomb's Law fails, but only expresses that the rigid body assumption is an insufficient modeling of the system [2].

5.4 Example: A Pantograph Mechanism

In this example we show that sliding of the pantograph mechanism in Fig. 5.5 leads to nonunique accelerations [21]. The mechanism consists of two masses (m_1, m_2) with neglected rotational inertias which are coupled in such a manner that a translational 2:1 transmission is achieved: When we move mass 1 in the direction of r by a certain value, say δr, then mass 2 is moved in the same direction by a value $2\delta r$.

The whole mechanism is linked by a rotational joint at point O and is under the influence of gravity g. To describe the system we use coordinates r and φ. The values of r give the radial displacements of mass 1 as well as those of mass 2 by the 2:1 transmission. The angle φ denotes the inclination

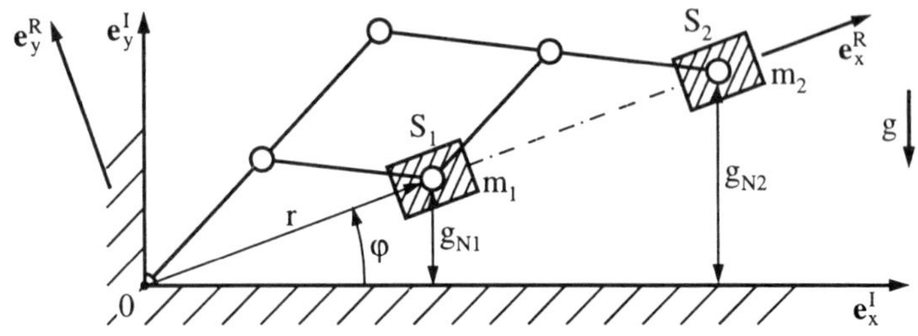

Figure 5.5: The Pantograph Mechanism

of the mechanism relative to a horizontal surface. For $\varphi = 0$ we assume that both bodies contact the environment and may slide on this surface under the influence of Coulomb friction.

In the first step we have to derive the equations of motion (3.11) of the system in the no-contact case. This has to be done in dependence on a set of generalized coordinates $\boldsymbol{q}$, which we choose as

$$\boldsymbol{q} := \begin{pmatrix} r \\ \varphi \end{pmatrix}; \quad \begin{array}{l} r > 0\,, \\ 0^\circ \leq \varphi \leq 90^\circ\,. \end{array} \tag{5.37}$$

Now we have to evaluate the kinematics of the system in terms of $\boldsymbol{q}$. We start with the displacements for each body i, $i = \{1, 2\}$, using the reference frame R. The displacements of the centers of masses S_i, $i = \{1, 2\}$, are given by

$${}_R\boldsymbol{r}_{OSi} = \begin{pmatrix} ir \\ 0 \\ 0 \end{pmatrix}. \tag{5.38}$$

With (2.18) we get the absolute velocities of the points S_i from (5.38):

$$\begin{aligned} {}_R\boldsymbol{v}_{Si} &= {}_R\dot{\boldsymbol{r}}_{OSi} + {}_R\boldsymbol{\omega}_{IR} \times {}_R\boldsymbol{r}_{OSi} \\ &= \begin{pmatrix} i\dot{r} \\ 0 \\ 0 \end{pmatrix} + \begin{pmatrix} 0 \\ 0 \\ \dot{\varphi} \end{pmatrix} \times \begin{pmatrix} ir \\ 0 \\ 0 \end{pmatrix} \\ &= \begin{pmatrix} i\dot{r} \\ ir\dot{\varphi} \\ 0 \end{pmatrix} \end{aligned} \tag{5.39}$$

where ${}_R\boldsymbol{\omega}_{IR}$ denotes the angular velocity of frame R relative to the environment. Now we use (2.24) in order to derive the absolute accelerations,

$$\begin{aligned} {}_R\boldsymbol{a}_{Si} &= {}_R\dot{\boldsymbol{v}}_{Si} + {}_R\boldsymbol{\omega}_{IR} \times {}_R\boldsymbol{v}_{Si} \\ &= \begin{pmatrix} i\ddot{r} \\ ir\ddot{\varphi} + i\dot{r}\dot{\varphi} \\ 0 \end{pmatrix} + \begin{pmatrix} 0 \\ 0 \\ \dot{\varphi} \end{pmatrix} \times \begin{pmatrix} i\dot{r} \\ ir\dot{\varphi} \\ 0 \end{pmatrix} \\ &= \begin{pmatrix} i\ddot{r} - ir\dot{\varphi}^2 \\ ir\ddot{\varphi} + 2i\dot{r}\dot{\varphi} \\ 0 \end{pmatrix} \end{aligned} \tag{5.40}$$

and state them in matrix notation,

$${}_R\boldsymbol{a}_{Si} = \begin{pmatrix} i & 0 \\ 0 & ir \\ 0 & 0 \end{pmatrix} \begin{pmatrix} \ddot{r} \\ \ddot{\varphi} \end{pmatrix} + \begin{pmatrix} -ir\dot{\varphi}^2 \\ 2i\dot{r}\dot{\varphi} \\ 0 \end{pmatrix} \tag{5.41}$$

to identify the Jacobian of translation by comparing (5.41) and (3.4):

$${}_R\boldsymbol{J}_{Si} = \begin{pmatrix} i & 0 \\ 0 & ir \\ 0 & 0 \end{pmatrix} . \tag{5.42}$$

The only active forces which have to be considered result from gravity and act on the centers of mass:

$${}_R\boldsymbol{F}_{Si} = \begin{pmatrix} -m_i g \sin\varphi \\ -m_i g \cos\varphi \\ 0 \end{pmatrix} . \tag{5.43}$$

Finally we state the equations of motion by using (3.10), where only the translational part is needed due to the neglected rotational inertias. By substituting the terms of (5.41), (5.42) and (5.43) we get

$$\begin{aligned} 0 &= \sum_{i=1}^{2} {}_R\boldsymbol{J}_{Si}^T \left(m_i \, {}_R\boldsymbol{a}_{Si} - {}_R\boldsymbol{F}_{Si}\right) \\ &= \sum_{i=1}^{2} \begin{pmatrix} i & 0 & 0 \\ 0 & ir & 0 \end{pmatrix} m_i \begin{pmatrix} i & 0 \\ 0 & ir \\ 0 & 0 \end{pmatrix} \begin{pmatrix} \ddot{r} \\ \ddot{\varphi} \end{pmatrix} \\ &\quad - \begin{pmatrix} i & 0 & 0 \\ 0 & ir & 0 \end{pmatrix} \begin{pmatrix} im_i r\dot{\varphi}^2 - m_i g \sin\varphi \\ -2im_i\dot{r}\dot{\varphi} - m_i g \cos\varphi \\ 0 \end{pmatrix} \\ &= \begin{pmatrix} (m_1 + 4m_2) & 0 \\ 0 & (m_1 + 4m_2) r^2 \end{pmatrix} \begin{pmatrix} \ddot{r} \\ \ddot{\varphi} \end{pmatrix} \\ &\quad - \begin{pmatrix} (m_1 + 4m_2) r\dot{\varphi}^2 - (m_1 + 2m_2) g \sin\varphi \\ -2(m_1 + 4m_2) r\dot{r}\dot{\varphi} - (m_1 + 2m_2) g r \cos\varphi \end{pmatrix} . \end{aligned} \tag{5.44}$$

With the abbreviations

$$M_1 := m_1 + 4m_2\,; \quad M_2 := m_1 + 2m_2 \tag{5.45}$$

the equations of motion for the no-contact case are

$$\underbrace{\begin{pmatrix} M_1 & 0 \\ 0 & M_1 r^2 \end{pmatrix}}_{\boldsymbol{M}} \underbrace{\begin{pmatrix} \ddot{r} \\ \ddot{\varphi} \end{pmatrix}}_{\ddot{\boldsymbol{q}}} - \underbrace{\begin{pmatrix} M_1 r\dot{\varphi}^2 - M_2 g \sin\varphi \\ -2M_1 r\dot{r}\dot{\varphi} - M_2 g r \cos\varphi \end{pmatrix}}_{\boldsymbol{h}} = 0 \tag{5.46}$$

which have the exact structure of eq. (3.11).

The second step consists of deriving the contact kinematics which can be done very easily for this example. The distances g_{Ni} between the two bodies and the horizontal surface are

$$g_{Ni} = ir \sin\varphi\,. \tag{5.47}$$

We need the relative velocities in the normal and tangential directions. In this example the first half of the contact points corresponds to the centers of mass S_i which have velocities $\boldsymbol{v}_{Si}$ from (5.39). The second half contains rigid body points of the surface with zero velocity. Thus, with respect to eq. (4.36) we only have to project the velocities $\boldsymbol{v}_{Si}$ onto the normal and tangential unit vectors of the surface to get $\dot{g}_N$ and $\dot{g}_T$. The unit vectors in the reference frame R are

$${}_R\boldsymbol{n}_i = \begin{pmatrix} \sin\varphi \\ \cos\varphi \\ 0 \end{pmatrix}\;; \quad {}_R\boldsymbol{t}_i = \begin{pmatrix} \cos\varphi \\ -\sin\varphi \\ 0 \end{pmatrix}, \tag{5.48}$$

and the projection leads to

$$\begin{aligned} \dot{g}_{Ni} &= {}_R\boldsymbol{n}_i^T\, {}_R\boldsymbol{v}_{Si} = i\dot{r}\sin\varphi + ir\dot{\varphi}\cos\varphi \\ \dot{g}_{Ti} &= {}_R\boldsymbol{t}_i^T\, {}_R\boldsymbol{v}_{Si} = i\dot{r}\cos\varphi - ir\dot{\varphi}\sin\varphi\,. \end{aligned} \tag{5.49}$$

Now we differentiate (5.49) with respect to time and get the changes of the relative velocities:

$$\begin{aligned} \ddot{g}_{Ni} &= i\ddot{r}\sin\varphi + ir\ddot{\varphi}\cos\varphi + 2i\dot{r}\dot{\varphi}\cos\varphi - ir\dot{\varphi}^2\sin\varphi \\ \ddot{g}_{Ti} &= i\ddot{r}\cos\varphi - ir\ddot{\varphi}\sin\varphi - 2i\dot{r}\dot{\varphi}\sin\varphi - ir\dot{\varphi}^2\cos\varphi\,. \end{aligned} \tag{5.50}$$

Rewriting eqs. (5.49) and (5.50) in matrix notation yields

$$\begin{aligned} \dot{g}_{Ni} &= \underbrace{(i\sin\varphi \quad ir\cos\varphi)}_{\boldsymbol{w}_{Ni}^T} \underbrace{\begin{pmatrix} \dot{r} \\ \dot{\varphi} \end{pmatrix}}_{\dot{\boldsymbol{q}}} \\ \dot{g}_{Ti} &= \underbrace{(i\cos\varphi \quad -ir\sin\varphi)}_{\boldsymbol{w}_{Ti}^T} \underbrace{\begin{pmatrix} \dot{r} \\ \dot{\varphi} \end{pmatrix}}_{\dot{\boldsymbol{q}}} \end{aligned} \tag{5.51}$$

$$
\begin{aligned}
\ddot{g}_{Ni} &= \boldsymbol{w}_{Ni}^T \ddot{\boldsymbol{q}} + \underbrace{(+2i\dot{r}\dot{\varphi}\cos\varphi - ir\dot{\varphi}^2 \sin\varphi)}_{\overline{w}_{Ni}} \\
\ddot{g}_{Ti} &= \boldsymbol{w}_{Ti}^T \ddot{\boldsymbol{q}} + \underbrace{(-2i\dot{r}\dot{\varphi}\sin\varphi - ir\dot{\varphi}^2 \cos\varphi)}_{\overline{w}_{Ti}}
\end{aligned}
\tag{5.52}
$$

which is the representation of the constraints given in eqs. (4.38) and (4.45). At this point note that the constraint vectors $\boldsymbol{w}_{N1}$ and $\boldsymbol{w}_{N2}$ are linearly dependent, which results from the model of our mechanism. Both contacts are always closed simultaneously. It is not possible to close only one of the contacts while keeping the other one open.

In the last step we have to define the contact situation and to derive the resulting contact forces and accelerations $\ddot{\boldsymbol{q}}$. We assume contact which simultaneously occurs at both bodies. By eq. (5.47) and the restrictions on r and φ in eq. (5.37) this results in the condition

$$
g_{Ni} = 0 \Leftrightarrow \varphi = 0\,. \tag{5.53}
$$

If contact is maintained, then the relative velocity in the normal direction (5.49) must vanish. Together with (5.53) we get

$$
g_{Ni} = 0\,; \quad \dot{g}_{Ni} = 0 \quad \Rightarrow \dot{\varphi} = 0\,. \tag{5.54}
$$

With $\varphi = \dot{\varphi} = 0$ the mass matrix and the vector $\boldsymbol{h}$ from (5.46) are simplified to

$$
\boldsymbol{M} = \begin{pmatrix} M_1 & 0 \\ 0 & M_1 r^2 \end{pmatrix}; \quad \boldsymbol{h} = \begin{pmatrix} 0 \\ -M_2 g r \end{pmatrix}, \tag{5.55}
$$

and the constraint magnitudes in eqs. (5.51), (5.52) change to

$$
\begin{aligned}
&\boldsymbol{w}_{N1} = \begin{pmatrix} 0 \\ r \end{pmatrix}; \quad \boldsymbol{w}_{N2} = \begin{pmatrix} 0 \\ 2r \end{pmatrix}; \quad \boldsymbol{w}_{T1} = \begin{pmatrix} 1 \\ 0 \end{pmatrix}; \quad \boldsymbol{w}_{T2} = \begin{pmatrix} 2 \\ 0 \end{pmatrix} \\
&\overline{w}_{N1} = \overline{w}_{N2} = \overline{w}_{T1} = \overline{w}_{T2} = 0\,.
\end{aligned}
\tag{5.56}
$$

Further, we assume that the masses are sliding from left to right, which correspond by eq. (5.49) to a tangential relative velocity greater than zero:

$$
\dot{g}_{Ti} > 0 \Rightarrow \dot{r} > 0\,. \tag{5.57}
$$

Under that assumption the equations of motion have to be completed by the two normal and tangential contact forces, according to eq. (5.6):

$$
\boldsymbol{M}\ddot{\boldsymbol{q}} - \boldsymbol{h} - \boldsymbol{w}_{N1}\lambda_{N1} - \boldsymbol{w}_{N2}\lambda_{N2} - \boldsymbol{w}_{T1}\lambda_{T1} - \boldsymbol{w}_{T2}\lambda_{T2} = 0\,. \tag{5.58}
$$

Due to sliding the tangential forces can be expressed by the corresponding normal forces (5.7):

$$\begin{aligned} \lambda_{T1} &= -\mu_1 \lambda_{N1}\,; \quad \lambda_{N1} \geq 0\,; \quad \dot{g}_{T1} > 0 \\ \lambda_{T2} &= -\mu_2 \lambda_{N2}\,; \quad \lambda_{N2} \geq 0\,; \quad \dot{g}_{T2} > 0 \end{aligned} \tag{5.59}$$

and finally we demand continual contact by

$$\begin{aligned} \ddot{g}_{N1} &= \boldsymbol{w}_{N1}^T \ddot{\boldsymbol{q}} = 0 \\ \ddot{g}_{N2} &= \boldsymbol{w}_{N2}^T \ddot{\boldsymbol{q}} = 0\,. \end{aligned} \tag{5.60}$$

Now we put (5.55), (5.56) into (5.58), express all tangential forces by their normal forces, and rewrite (5.58), (5.59) and (5.60):

$$\begin{gathered} \begin{pmatrix} M_1 & 0 \\ 0 & M_2 r^2 \end{pmatrix} \begin{pmatrix} \ddot{r} \\ \ddot{\varphi} \end{pmatrix} - \begin{pmatrix} 0 \\ -M_2 g r \end{pmatrix} - \begin{pmatrix} -\mu_1 & -2\mu_2 \\ r & 2r \end{pmatrix} \begin{pmatrix} \lambda_{N1} \\ \lambda_{N2} \end{pmatrix} = 0\,; \\ \begin{pmatrix} \lambda_{N1} \\ \lambda_{N2} \end{pmatrix} \geq 0\,; \quad \begin{pmatrix} 0 & r \\ 0 & 2r \end{pmatrix} \begin{pmatrix} \ddot{r} \\ \ddot{\varphi} \end{pmatrix} = 0\,. \end{gathered} \tag{5.61}$$

Equation (5.61) is a complete set of relations for our contact state. Altogether we have four equations for the unknowns $(\ddot{r}, \ddot{\varphi}, \lambda_{N1}, \lambda_{N2})$ and two inequalities which restrict the problem to the continual-contact case. However, the constraints are dependent, and thus no unique solution for the *contact forces* can be expected. By the results of Section 5.2, eq. (5.25), we know that, at least for the frictional case, the generalized accelerations $(\ddot{r}, \ddot{\varphi})$ remain unique.

We continue our discussion using eq. (5.61) and try to show under which conditions unique solutions are obtained. From the last set of equations in (5.61),

$$r\ddot{\varphi} = 0\,; \quad 2r\ddot{\varphi} = 0 \quad \Rightarrow \quad \ddot{\varphi} = 0 \tag{5.62}$$

we immediately see that $\ddot{\varphi}$ must be equal to zero ($r > 0!$). Putting this value into the first set of equations, we get the two conditions

$$M_1 \ddot{r} = -\mu_1 \lambda_{N1} - 2\mu_2 \lambda_{N2}\,; \quad \lambda_{N1} + 2\lambda_{N2} = M_2 g\,. \tag{5.63}$$

Now we eliminate $2\lambda_{N2}$ from the first equation in (5.63),

$$M_1 \ddot{r} = -(\mu_1 - \mu_2)\lambda_{N1} - \mu_2 M_2 g \tag{5.64}$$

and express the inequalities by using (5.63):

$$\begin{aligned} &\lambda_{N1} \geq 0\,; \quad \lambda_{N2} \geq 0\,; \quad \lambda_{N1} = M_2 g - 2\lambda_{N2} \\ &\Rightarrow \quad 0 \leq \lambda_{N1} \leq M_2 g\,. \end{aligned} \tag{5.65}$$

In the frictionless case, $\mu_1 = \mu_2 = 0$, eq. (5.64) yields $M_1 \ddot{r} = 0$. A unique solution $\ddot{r} = 0$ is obtained where both masses move to the right with constant velocities. Even in the case $\mu := \mu_1 = \mu_2$, eq. (5.64) results in a unique solution where the masses are decelerated from $\dot{r} > 0$ (eq. 5.57) by values of $\ddot{r} = -\mu \frac{M_2}{M_1} g < 0$. The general case, however, remains nonunique: By combining (5.64) and (5.65) we see that every value

$$\ddot{r} = -\mu_2 \frac{M_2}{M_1} g \left[1 + \alpha \left(\frac{\mu_1}{\mu_2} - 1 \right) \right] < 0; \quad \forall\, 0 \leq \alpha \leq 1 \tag{5.66}$$

fulfills the system (5.61). The direction of this acceleration, $\ddot{r} < 0$, is physically correct, but no unique value of $\ddot{r}$ can be given. This behavior results from the dependent normal constraints. The right equation in (5.63) expresses the balance of moments which are induced by the gravity and the normal contact forces, and is the only available equation for their determination. Either contact 1 carries the whole weight, or contact 2, or they share it in any arbitrary manner. Therefore the normal forces remain underdetermined but generate frictional forces due to Coulomb's law, which are consequently not unique either. These friction forces influence the dynamics of our system, in this case the acceleration $\ddot{r}$, and yield the uniqueness problem in (5.66). As a consequence, when modeling systems with sliding friction we have to look carefully at the choice of the constraints in order to avoid such problems. In this example vertical guidance instead of the revolute joint at point O would make the constraints independent.

6

DETACHMENT AND STICK-SLIP TRANSITIONS

In this section the conditions for transitions from contact to separation and sticking to sliding are formulated for coupled multibody systems [22, 23, 76, 77, 89]. The main difficulty results from instantaneous changes in the contact forces at transitions from sliding to sticking or reversed sliding. The jumps in the contact forces are easy to understand through the following example: We consider an oblique plane and an object sliding upward on the plane with some initial condition and acted upon only by gravity. Due to sliding, the tangential constraint is an element of $I_N \backslash I_H$ (eq. 5.5) with a given tangential friction force $\lambda_T = \mu\lambda_N$. The gravitation and frictional forces cause the object to slow down until the tangential relative velocity is zero. At that moment the tangential constraint has to be changed from the set of sliding contacts $I_N \backslash I_H$ to the set of possible sticking contacts I_H. Further evolution of the system now depends only on the angle of inclination of the plane. For very steep planes the object will not continue sticking, but will immediately start to slide downward. Thus, the constraint has to be quickly changed from I_H to $I_N \backslash I_H$, and the tangential friction force shows an instantaneous jump from $+\mu\lambda_N$ to $-\mu\lambda_N$. In the case of moderate angles of inclination the object will continue sticking and thus remain an element of I_H. Even this case leads to a jump in the tangential contact force, depending on the steepness of the plane.

Generalizing these considerations, we note that a vanishing tangential relative velocity does not necessarily lead to sticking and that each new element of I_H generally produces an unsteady, changing contact force. We now look at systems with more than one contact point. If the contacts are coupled kinematically, the contact forces influence each other. Thus, each new element in I_H generally affects all of the other elements in I_N and produces jumps in their contact forces. Due to these jumps, induced transitions in their contact state may occur which are transitions either to sliding or to take off. The problem is now that we do not know how many and which constraints change their contact state, influenced by the new elements in I_H.

In this section the conditions of transition are stated and then formulated as a Linear Complementarity Problem, which allows the evaluation of the transition problem, avoiding the combinatorial problem of testing all possible contact state combinations for the solution without contradicting the contact laws. We start our discussion with the equations of motion and the constraints on the acceleration level (eq. 5.11):

$$\begin{aligned} \boldsymbol{M}\ddot{\boldsymbol{q}} - \boldsymbol{h} - \left[\boldsymbol{W}_N + \boldsymbol{W}_G\overline{\overline{\boldsymbol{\mu}}}_G \mid \boldsymbol{W}_H\right] \begin{pmatrix} \boldsymbol{\lambda}_N \\ \boldsymbol{\lambda}_H \end{pmatrix} &= \boldsymbol{0} \qquad \in \mathbb{R}^f \\ \begin{pmatrix} \ddot{\boldsymbol{g}}_N \\ \ddot{\boldsymbol{g}}_H \end{pmatrix} = \begin{pmatrix} \boldsymbol{W}_N^T \\ \boldsymbol{W}_H^T \end{pmatrix} \ddot{\boldsymbol{q}} + \begin{pmatrix} \overline{\boldsymbol{w}}_N \\ \overline{\boldsymbol{w}}_H \end{pmatrix} & \qquad \in \mathbb{R}^{n_N+n_H} . \end{aligned} \tag{6.1}$$

The vectors $\boldsymbol{\lambda}_N \in \mathbb{R}^{n_N}$ and $\ddot{\boldsymbol{g}}_N \in \mathbb{R}^{n_N}$ contain the contact forces and accelerations of the closed contacts $\in I_N$ in the normal directions, where $\boldsymbol{\lambda}_H \in \mathbb{R}^{n_H}$ and $\ddot{\boldsymbol{g}}_H \in \mathbb{R}^{n_H}$ consist of the tangential forces and accelerations of the potentially sticking constraints $\in I_H$. The friction forces of the sliding constraints are already expressed by their corresponding normal forces, $\boldsymbol{\lambda}_G = \overline{\overline{\boldsymbol{\mu}}}_G\boldsymbol{\lambda}_N$ and occur in eq. (6.1) by the matrix expression $\boldsymbol{W}_G\overline{\overline{\boldsymbol{\mu}}}_G$. Altogether (6.1) consists of $f+n_N+n_H$ equations for the $f+2(n_N+n_H)$ unknowns $\ddot{\boldsymbol{q}}, \ddot{\boldsymbol{g}}_N, \ddot{\boldsymbol{g}}_H, \boldsymbol{\lambda}_N, \boldsymbol{\lambda}_H$.

If the contact situation remains unchanged (i.e., each element of I_N and I_H keeps in contact and is sticking), then the relative accelerations in (6.1) are equal to zero, $(\ddot{\boldsymbol{g}}_N^T \ddot{\boldsymbol{g}}_H^T) = 0$, and the system (6.1) is solvable (see also eq. 5.16). In the other case, transitions in the contact states prevent a predetermination of the relative accelerations. Thus, $n_N + n_H$ conditions are missing in order to solve eq. (6.1). These conditions are the contact laws of the system which are valid during continual contact and sticking as well as for transitions to sliding or separation.

6.1 Contact Law for Normal Constraints

Each closed contact constraint $i \in I_N$ is characterized by a vanishing distance $g_{Ni} = 0$ and normal relative velocity $\dot{g}_{Ni} = 0$. Under the assumption of impenetrability ($g_{Ni} \geq 0$), only two situations may occur: Contact is maintained or a transition to separation takes place. In the first case we know that the relative acceleration must be equal to zero and the normal contact forces must act with a compressive magnitude due to the unilateralness of the normal constraint:

$$\ddot{g}_{Ni} = 0 \quad \wedge \quad \lambda_{Ni} \geq 0 \, ; \quad i \in I_N \, . \tag{6.2}$$

The condition $\ddot{g}_{Ni} = 0$ ensures that contact is maintained. Since $\dot{g}_{Ni} = \ddot{g}_{Ni} = 0$ the values of $g_{Ni} = 0$ never change. The admissible values of the contact

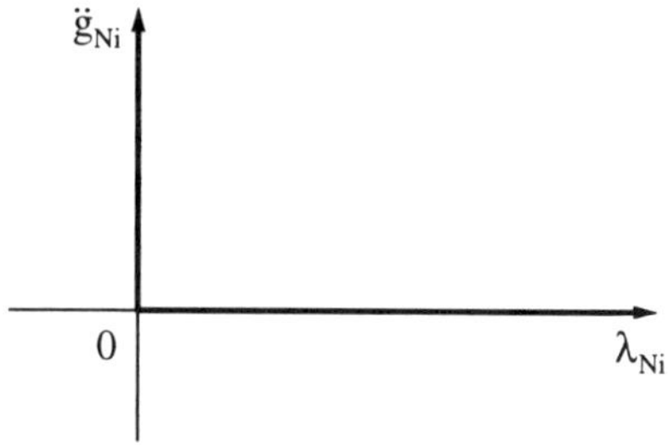

Figure 6.1: Complementarity of Normal Contacts

forces are given by $\lambda_{Ni} \geq 0$. A contact force $\lambda_{Ni} < 0$ would correspond to a pulling force and must be excluded.

The second case describes the separation of the bodies. Separation is only achieved by positive values of the normal relative acceleration and vanishing normal forces:

$$\ddot{g}_{Ni} \geq 0 \quad \wedge \quad \lambda_{Ni} = 0 ; \quad i \in I_N . \tag{6.3}$$

We remember that the state of separation is indicated by distances greater than zero, $g_{Ni} > 0$. Starting from $g_{Ni} = \dot{g}_{Ni} = 0$ such a transition only occurs for $\ddot{g}_{Ni} > 0$. Necessarily, compressive contact forces must not be transmitted at separation, because otherwise we would have continual contact. This leads to the second condition, $\lambda_{Ni} = 0$.

From (6.2) and (6.3) we see that the normal contact law shows a complementary behavior: The product of λ_{Ni} and $\ddot{g}_{Ni}$ is always zero:

$$\ddot{g}_{Ni} \lambda_{Ni} = 0 ; \quad i \in I_N . \tag{6.4}$$

With eqs. (6.2) and (6.3) the contact law in the normal direction is completely defined for a single contact, where the impenetrability condition and the occurrence of only compressive forces (no adhesive forces) are taken into account.

The relationship between the constraint force and the relative acceleration of the normal contact i is shown in Fig. 6.1. The characteristic consists of two branches which form a rectangular corner. The horizontal part allows values $\lambda_{Ni} \geq 0$ by a vanishing relative acceleration $\ddot{g}_{Ni} = 0$ and corresponds to eq. (6.2) where contact is maintained. The other branch describes the take-off transition of eq. (6.3) where the bodies start separation with values $\ddot{g}_{Ni} \geq 0$ and zero-contact forces $\lambda_{Ni} = 0$. We see that the corner characteristic of Fig. 6.1 is an equivalent representation of the contact laws (6.2) and (6.3).

In our system we have n_N contact constraints with each of them determined by the contact laws of eqs. (6.2) and (6.3). In vector notation,

$$\ddot{\boldsymbol{g}}_N \geq \mathbf{0} ; \quad \boldsymbol{\lambda}_N \geq \mathbf{0} ; \quad \ddot{\boldsymbol{g}}_N^T \boldsymbol{\lambda}_N = 0 , \tag{6.5}$$

we obtain n_N inequalities for $\ddot{\boldsymbol{g}}_N$ and $\boldsymbol{\lambda}_N$ and one complementarity condition. The notation for an arbitrary vector $\boldsymbol{x} \geq \mathbf{0}$ denotes that all components of $\boldsymbol{x}$ are nonnegative ($x_i \geq 0$). Equation (6.5) together with (6.1) completely describes the continual contact state as well as the detachment transition of our dynamical system.

At this point only n_H conditions are missing for determining the tangential stick-slip transitions, which are the subject of the following sections. Before starting this discussion, we shall investigate the cases covered by the formulation of eqs. (6.5) and (6.1). For this purpose we assume that all contacts $\in I_N$ are sliding. Then the set of the possible sticking contacts is empty, $n_H = 0$, and eqs. (6.1) and (6.5) transform to

$$\begin{aligned} &\boldsymbol{M}\ddot{\boldsymbol{q}} - \boldsymbol{h} - (\boldsymbol{W}_N + \boldsymbol{W}_G\overline{\overline{\boldsymbol{\mu}}}_G)\boldsymbol{\lambda}_N = \mathbf{0} \\ &\ddot{\boldsymbol{g}}_N = \boldsymbol{W}_N^T\ddot{\boldsymbol{q}} + \overline{\boldsymbol{w}}_N \\ &\ddot{\boldsymbol{g}}_N \geq \mathbf{0}\,; \quad \boldsymbol{\lambda}_N \geq \mathbf{0}\,; \quad \ddot{\boldsymbol{g}}_N^T\boldsymbol{\lambda}_N = 0\,. \end{aligned} \tag{6.6}$$

Now we eliminate $\ddot{\boldsymbol{q}}$ from the second equation by using the first relation,

$$\begin{aligned} &\ddot{\boldsymbol{g}}_N = \boldsymbol{W}_N^T\boldsymbol{M}^{-1}(\boldsymbol{W}_N + \boldsymbol{W}_G\overline{\overline{\boldsymbol{\mu}}}_G)\boldsymbol{\lambda}_N + (\boldsymbol{W}_N^T\boldsymbol{M}^{-1}\boldsymbol{h} + \overline{\boldsymbol{w}}_N) \\ &\ddot{\boldsymbol{g}}_N \geq \mathbf{0}\,; \quad \boldsymbol{\lambda}_N \geq \mathbf{0}\,; \quad \ddot{\boldsymbol{g}}_N^T\boldsymbol{\lambda}_N = 0\,, \end{aligned} \tag{6.7}$$

and note that eq. (6.7) has the structure

$$\boldsymbol{y} = \boldsymbol{A}\boldsymbol{x} + \boldsymbol{b}\,; \quad \boldsymbol{y} \geq \mathbf{0}\,; \quad \boldsymbol{x} \geq \mathbf{0}\,; \quad \boldsymbol{y}^T\boldsymbol{x} = 0 \tag{6.8}$$

which is known as a Linear Complementarity Problem in its standard form. In the case without sliding friction ($\overline{\overline{\boldsymbol{\mu}}}_G = \mathbf{0}$) eqs. (6.6) can be identified as the Kuhn-Tucker conditions of the Quadratic Program

$$\begin{aligned} &\ddot{\boldsymbol{q}} = \arg\min\left\{f(\ddot{\boldsymbol{q}}^*) \mid \boldsymbol{W}_N^T\ddot{\boldsymbol{q}}^* + \overline{\boldsymbol{w}}_N \geq 0\right\} \\ &f(\ddot{\boldsymbol{q}}^*) = \tfrac{1}{2}(\ddot{\boldsymbol{q}}^* - \boldsymbol{M}^{-1}\boldsymbol{h})^T\boldsymbol{M}(\ddot{\boldsymbol{q}}^* - \boldsymbol{M}^{-1}\boldsymbol{h})\,. \end{aligned} \tag{6.9}$$

Compared to eq. (5.25), it is an expansion of the principle of least constraints to the unilateral case where both continual contact *and* the take-off transition are included.

Summarizing the results of this section, we see that characteristics as in Fig. 6.1 combined with the dynamics equations (6.1) result in a Linear Complementarity Problem (LCP); see eqs. (6.7) and (6.8). Moreover, a Quadratic Program related to the LCP can be stated in the frictionless case which is an expansion of the principle of least constraints.

6.2 Coulomb's Friction Law

The set of equations (6.1) and (6.5) consists of $f + 2n_N + n_H$ conditions for the $f + 2n_N + 2n_H$ unknowns $\ddot{\boldsymbol{q}}, \ddot{\boldsymbol{g}}_N, \ddot{\boldsymbol{g}}_H, \boldsymbol{\lambda}_N, \boldsymbol{\lambda}_H$. It describes the dynamics of our system and includes continual contact and the detachment transition. However, n_H conditions are still missing, which determine the tangential contact state of the possible sticking constraints. This evaluation will be done with respect to the frictional law of Coulomb, which, however, does not mean a restriction of generality. With regard to friction the following three cases are possible:

$$\begin{array}{llll} \dot{g}_{Ti} = 0 & \Rightarrow & |\lambda_{Ti}| \leq \mu_{0i}\lambda_{Ni} & \text{(sticking)} \\ \dot{g}_{Ti} < 0 & \Rightarrow & \lambda_{Ti} = +\mu_i\lambda_{Ni} & \text{(negative sliding)} \\ \dot{g}_{Ti} > 0 & \Rightarrow & \lambda_{Ti} = -\mu_i\lambda_{Ni} & \text{(positive sliding)} \\ i \in I_N & & & \end{array} \tag{6.10}$$

This law states that during sliding the friction force is proportional to the normal force of a contact. For sliding friction,

$$\lambda_{Ti} = -\mu_i\lambda_{Ni} \operatorname{sign}(\dot{g}_{Ti})$$

(cp. eq. 5.7), where the coefficient of friction may depend additionally on the tangential relative velocity, $\mu_i = \mu_i(\dot{g}_{Ti})$. During sliding the tangential friction force is directed oppositely to the relative velocity $\dot{g}_T$. For static friction ($\dot{g}_{Ti} = 0$) the values of the tangential force are bounded by the normal force, $|\lambda_{Ti}| \leq \mu_{0i}\lambda_{Ni}$.

As stated in Chapter 1 most material pairings in technical practice possess a friction characteristic similar to those in Fig. 1.1. From this we may realistically assume that in transition points from stiction to sliding, and vice versa, the friction coefficients are the same, $\mu_{0i} = \mu_i(\dot{g}_{Ti} = 0)$. We shall therefore use $\mu(0) = \mu_0$ for all transition events, keeping in mind that for normal sliding any dependency $\mu_i(\dot{g}_{Ti})$ may be applied.

In eq. (6.10) Coulomb's law is stated in the original version which connects the contact forces and the tangential relative velocities. For the evaluation of our transition problem, however, we need a representation of (6.10) on the acceleration level in order to determine the tangential relative accelerations $\ddot{\boldsymbol{g}}_H$ in eq. (6.1). Thus, we split the index set I_N from eq. (6.10) into $I_N \backslash I_H$ and I_H, which corresponds to the sets of the sliding ($\dot{g}_{Ti} \neq 0$; $i \in I_N \backslash I_H$) and the potentially sticking ($\dot{g}_{Ti} = 0$; $i \in I_H$) contacts.

The friction forces of the sliding contacts are already taken into account in eq. (6.1) by $\boldsymbol{W}_G\overline{\overline{\boldsymbol{\mu}}}_G$ (see also eq. 5.7), so we only have to transform Coulomb's law to the acceleration level for the contacts in I_H. This is possible because the tangential relative velocities and accelerations have the same sign for

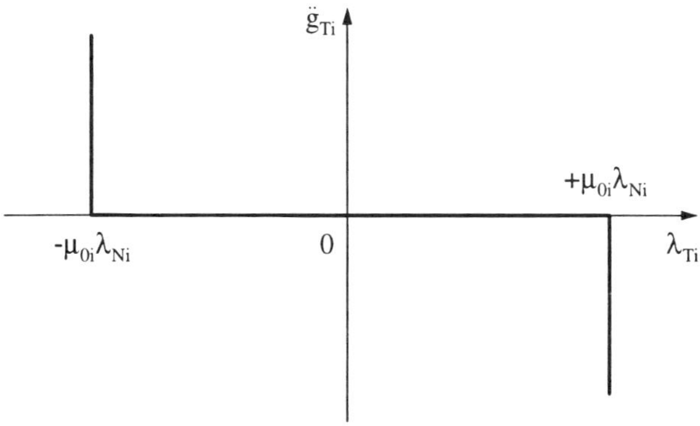

Figure 6.2: Conditions for Tangential Constraints according to Coulomb's Friction Law

transitions from $\dot{g}_{Ti} = 0$ to $\dot{g}_{Ti} \neq 0$. Thus, we get from eq. (6.10)

$$\begin{array}{l} \lambda_{Ni} \geq 0\,; \quad i \in I_H\,; \\ |\lambda_{Ti}| \leq \mu_{0i}\lambda_{Ni}\,; \end{array} \left\{ \begin{array}{lcl} |\lambda_{Ti}| < \mu_{0i}\lambda_{Ni} & \Rightarrow & \ddot{g}_{Ti} = 0 \\ \lambda_{Ti} = +\mu_{0i}\lambda_{Ni} & \Rightarrow & \ddot{g}_{Ti} \leq 0 \\ \lambda_{Ti} = -\mu_{0i}\lambda_{Ni} & \Rightarrow & \ddot{g}_{Ti} \geq 0\,. \end{array} \right. \tag{6.11}$$

The characteristic of eq. (6.11) is depicted in Fig. 6.2. The admissible values of the tangential force λ_{Ti} form a convex set C_{Ti} which is bounded by the values of the normal force:

$$C_{Ti} = \{\lambda^*_{Ti}\,|-\mu_{0i}\lambda_{Ni} \leq \lambda^*_{Ti} \leq +\mu_{0i}\lambda_{Ni}\}\ .$$

If the tangential force is in the interior of C_{Ti}, then we have continual sticking, $\ddot{g}_{Ti} = 0$. Otherwise λ_{Ti} lies at the boundary of C_{Ti} and allows a transition to sliding by arbitrary values of $\ddot{g}_{Ti}$ in the opposing direction. The characteristic in Fig. 6.2 is more complicated than that of the unilateral normal constraint in Fig. 6.1. It consists of two rectangular corners which provide a total of three different branches. From the results in Section 6.1 we know that the dynamics equations together with some characteristics such as in Fig. 6.1 can be stated as a LCP. This direct approach cannot be applied to the friction characteristic, but initially demands its decomposition. One possibility of such a decomposition is shown in Fig. 6.3, where the tangential relative acceleration is split into its positive and negative parts [42]:

$$\ddot{g}^+_{Ti} = \frac{1}{2}(|\ddot{g}_{Ti}| + \ddot{g}_{Ti})\,; \quad \ddot{g}^-_{Ti} = \frac{1}{2}(|\ddot{g}_{Ti}| - \ddot{g}_{Ti})\,; \quad \ddot{g}_{Ti} = \ddot{g}^+_{Ti} - \ddot{g}^-_{Ti}\,.$$

The resulting two characteristics are similar to Fig. 6.1 and enable a formulation of the stick-slip transition by using a LCP. The decomposition presented

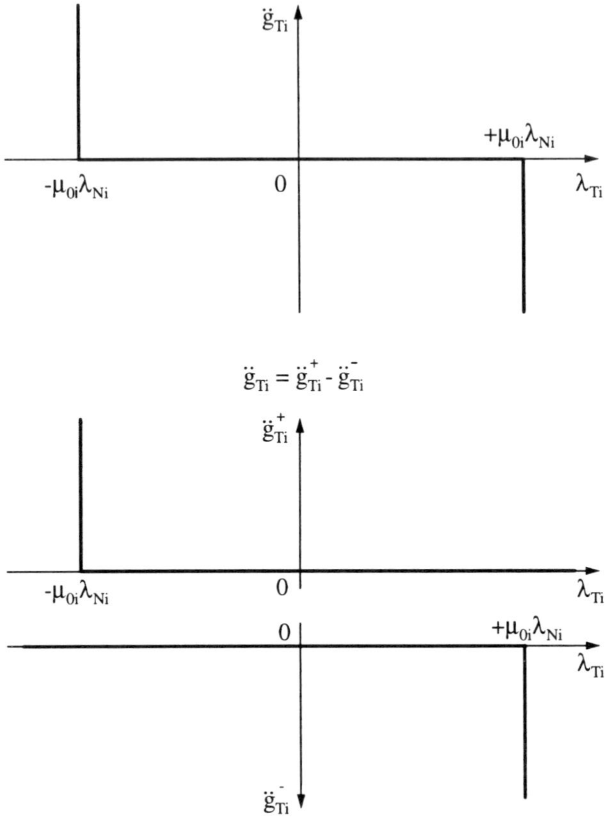

Figure 6.3: Possible Decomposition of the Friction Characteristic

here is the simplest one, but it is not further discussed because the resulting equations for building up the LCP matrix $\boldsymbol{A}$ in (6.8) demand an inversion of a matrix which becomes singular in the presence of dependent constraints. A more general decomposition will be introduced in Section 6.3 in which even overconstrained systems can be handled.

6.3 Decomposition of the Tangential Characteristic

The fundamental idea for the decomposition of the friction characteristic in Fig. 6.2 is to formulate each of the tangential constraints by two simultaneously appearing constraints. Both constraints transmit the tangential force λ_{Ti} by sharing it in portions $\lambda_{Ti}^{(+)}$ and $\lambda_{Ti}^{(-)}$ in positive and negative tangential directions, respectively.

Before continuing our discussion, some remarks on the notation used throughout this section must be given. The upper indices + and − denote the positive and negative parts of a scalar: $a^+ = \frac{1}{2}(|a| + a) \geq 0$, $a^- = \frac{1}{2}(|a| - a) \geq 0$, $a = a^+ - a^-$. Each value of a then can be expressed by either $(a^+ \geq 0\,, \quad a^- = 0)$ or $(a^+ = 0\,, \quad a^- \geq 0)$. The case that both values are greater than zero, $(a^+ > 0\,, \quad a^- > 0)$ is excluded by that definition. In contrast the superscript indices (+) and (−) are only chosen for distinctive reasons and do not have the above meaning. For example, if a scalar a is stated as the difference $a = a^{(+)} - a^{(-)}$, $a^{(+)} \geq 0$, $a^{(-)} \geq 0$, then both $a^{(+)}$ and $a^{(-)}$ may be greater than zero at the same time. A vector magnitude using one of these indices consists of elements which all have the corresponding properties defined above.

In the upper part of Fig. 6.4 once more the characteristic of the stick-slip transition is depicted. Its decomposition is done by mainly two steps which correspond to the lower two parts of Fig. 6.4. In order to derive the *first decomposition step* we look at the admissible values of the tangential force λ_{Ti}, $i \in I_H$. With respect to eq. (6.11), λ_{Ti} is an element of a convex set C_{Ti}:

$$\lambda_{Ti} \in C_{Ti} = \{\lambda^*_{Ti}\,|-\mu_{0i}\lambda_{Ni} \leq \lambda^*_{Ti} \leq +\mu_{0i}\lambda_{Ni}\}\ ; \quad i \in I_H\,. \tag{6.12}$$

Now we assume that this force is transmitted by two simultaneously appearing constraints with each of them transferring only a part of λ_{Ti}. We denote these parts by $\lambda^{(+)}_{Ti}$ and $\lambda^{(-)}_{Ti}$, and state λ_{Ti} as

$$\lambda_{Ti} = \lambda^{(+)}_{Ti} - \lambda^{(-)}_{Ti}\ ; \quad i \in I_H\,. \tag{6.13}$$

The values of $\lambda^{(+)}_{Ti}$ and $\lambda^{(-)}_{Ti}$ are not arbitrary but must be chosen in such a manner that the tangential force λ_{Ti} resulting from eq. (6.13) always lies in the convex set C_{Ti} defined in (6.12). This can be ensured by restricting the values of $\lambda^{(+)}_{Ti}$ and $\lambda^{(-)}_{Ti}$ to

$$\left.\begin{aligned} \lambda^{(+)}_{Ti} \in C^{(+)}_{Ti} &= \left\{\lambda^{(+)*}_{Ti}\,\middle|\,0 \leq \lambda^{(+)*}_{Ti} \leq \mu_{0i}\lambda_{Ni}\right\} \\ \lambda^{(-)}_{Ti} \in C^{(-)}_{Ti} &= \left\{\lambda^{(-)*}_{Ti}\,\middle|\,0 \leq \lambda^{(-)*}_{Ti} \leq \mu_{0i}\lambda_{Ni}\right\} \end{aligned}\right\} ; \quad i \in I_H\,. \tag{6.14}$$

Next for each of the new variables $\lambda^{(+)}_{Ti}$ and $\lambda^{(-)}_{Ti}$ we have to define a characteristic which connects them to the tangential relative acceleration $\ddot{g}_{Ti}$. This is done in the middle diagrams of Fig. 6.4. Finally, we have to prove that these characteristics are equivalent to the upper diagram of Fig. 6.4, which corresponds to the friction law in eq. (6.11). From the first equation in (6.11) we see that $|\lambda_{Ti}| < \mu_{0i}\lambda_{Ni}$ must result in an acceleration $\ddot{g}_{Ti} = 0$. The term λ_{Ti} obtained by (6.13) with admissible values of $\lambda^{(+)}_{Ti}$, $\lambda^{(-)}_{Ti}$ from (6.14)

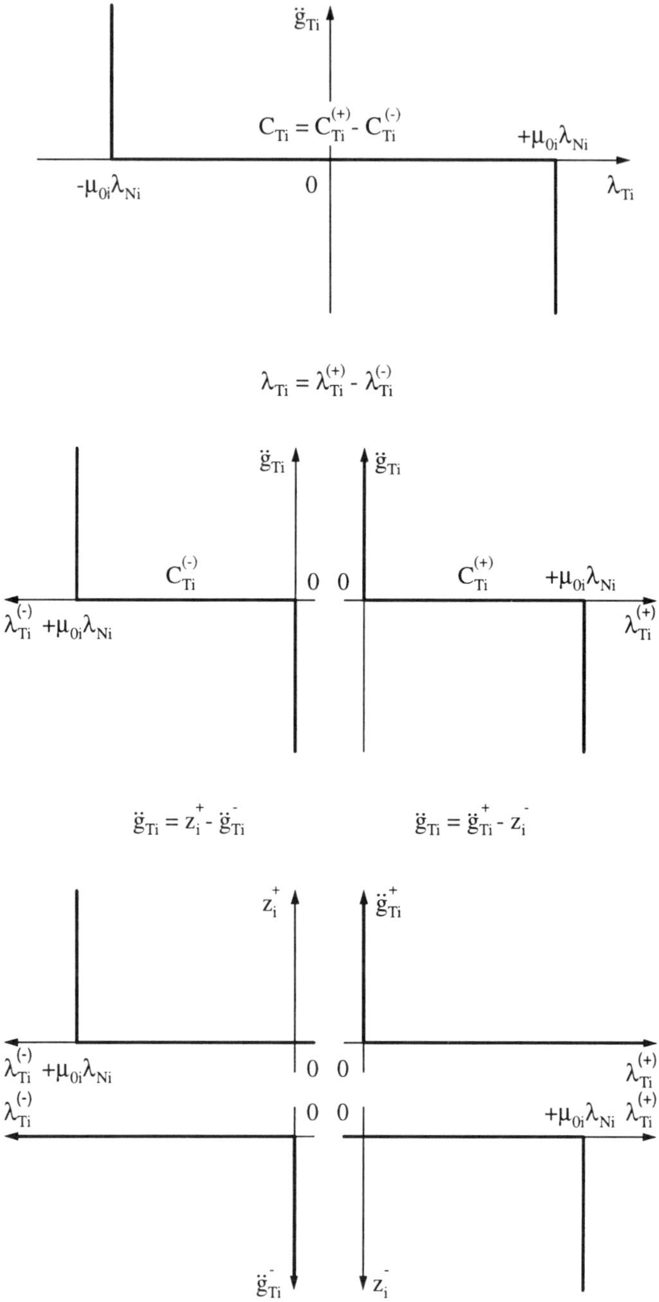

Figure 6.4: Decomposition of the Friction Characteristic

only fulfills this *strict* inequality if the two combinations, $(\lambda_{Ti}^{(+)} = +\mu_{0i}\lambda_{Ni}; \lambda_{Ti}^{(-)} = 0)$ and $(\lambda_{Ti}^{(+)} = 0; \lambda_{Ti}^{(-)} = +\mu_{0i}\lambda_{Ni})$ are excluded from (6.14). But these combinations are the only two which allow an acceleration $\ddot{g}_{Ti} \neq 0$: For $|\lambda_{Ti}| < \mu_{0i}\lambda_{Ni}$ at least one of the elements $(\lambda_{Ti}^{(+)}, \lambda_{Ti}^{(-)})$ must lie in the interior of the sets $(C_{Ti}^{(+)}, C_{Ti}^{(-)})$. Such an element, however, demands an acceleration $\ddot{g}_{Ti} = 0$ by the middle diagrams of Fig. 6.4. The other element may lie in the interior or at the boundary, and therefore may allow $\ddot{g}_{Ti} = 0$ or $\ddot{g}_{Ti} \geq 0$ or $\ddot{g}_{Ti} \leq 0$. Because both conditions must be fulfilled $(\ddot{g}_{Ti} = 0 \wedge (\ddot{g}_{Ti} = 0 \vee \ddot{g}_{Ti} \geq 0 \vee \ddot{g}_{Ti} \leq 0))$, the resulting value of the acceleration is indeed $\ddot{g}_{Ti} = 0$.

By the second equation in (6.11), negative values of the acceleration $\ddot{g}_{Ti}$ demand a tangential force λ_{Ti} equal to $+\mu_{0i}\lambda_{Ni}$. From the restrictions on $\lambda_{Ti}^{(+)}$ and $\lambda_{Ti}^{(-)}$ in eq. (6.14), this value of λ_{Ti} only results from (6.13) by using the tuple $(\lambda_{Ti}^{(+)} = +\mu_{0i}\lambda_{Ni}\,; \quad \lambda_{Ti}^{(-)} = 0)$. No other combination of elements $(\lambda_{Ti}^{(+)}, \lambda_{Ti}^{(-)})$ leads to a tangential force equal to $+\mu_{0i}\lambda_{Ni}$. In that case, both characteristics in the middle diagram of Fig. 6.4 enable negative accelerations, $\ddot{g}_{Ti} \leq 0$, which also corresponds to eq. (6.11). For proving the last equation in (6.11) we argue in the same way. The tangential force $\lambda_{Ti} = -\mu_{0i}\lambda_{Ni}$ results from the unique decomposition $(\lambda_{Ti}^{(+)} = 0\,; \quad \lambda_{Ti}^{(-)} = +\mu_{0i}\lambda_{Ni})$ which allows positive values of $\ddot{g}_{Ti}$. Altogether we see that the two characteristics in the middle of Fig. 6.4 are equivalent to the friction law in the upper part of Fig. 6.4. Arbitrarily chosen elements $(\lambda_{Ti}^{(+)}\,; \; \lambda_{Ti}^{(-)})$ from the sets in (6.14) provide a tangential force λ_{Ti} by (6.13), which then defines the acceleration state corresponding to eq. (6.11).

In the *second decomposition step*, each characteristic in the middle of Fig. 6.4 has to be split into two unilateral constraints; see Fig. 6.1 and Fig. 6.2. As mentioned this enables the formulation of a LCP by using the resulting inequalities and complementarity conditions together with the dynamical equations. We restrict our discussion to the decomposition of the right characteristic in the middle diagram of Fig. 6.4, but keep in mind that the same decomposition steps must be applied to the left one. First we split the tangential acceleration into its positive and negative parts,

$$\ddot{g}_{Ti} = \ddot{g}_{Ti}^{+} - z_i^{-}\,; \quad \ddot{g}_{Ti}^{+} \geq 0\,; \quad z_i^{-} \geq 0\,; \quad i \in I_H\,, \tag{6.15}$$

where we define $\ddot{g}_{Ti}^{+} = \frac{1}{2}(|\ddot{g}_{Ti}| + \ddot{g}_{Ti})$, $z_i^{-} = \frac{1}{2}(|\ddot{g}_{Ti}| - \ddot{g}_{Ti})$. Note that the case of both values simultaneously being greater than zero is excluded and must not occur in the following decomposition steps. From eq. (6.14) the values of $\lambda_{Ti}^{(+)}$ are restricted to

$$0 \leq \lambda_{Ti}^{(+)} \leq \mu_{0i}\lambda_{Ni}\,; \quad i \in I_H\,, \tag{6.16}$$

which can be expressed by the two inequality conditions

$$\lambda_{Ti}^{(+)} \geq 0\,; \quad \mu_{0i}\lambda_{Ni} - \lambda_{Ti}^{(+)} \geq 0\,; \quad i \in I_H\,, \tag{6.17}$$

where we let

$$\lambda_{T0i}^{(-)} = \mu_{0i}\lambda_{Ni} - \lambda_{Ti}^{(+)} ; \quad i \in I_H . \tag{6.18}$$

Finally we state the complementarity conditions between the variables in (6.15) and (6.17):

$$\ddot{g}_{Ti}^{+}\lambda_{Ti}^{(+)} = 0 ; \quad z_i^{-}(\mu_{0i}\lambda_{Ni} - \lambda_{Ti}^{(+)}) = 0 ; \quad i \in I_H , \tag{6.19}$$

and we get the decomposition which is depicted in the right half of the lower part of Fig. 6.4. These two unilateral constraints are equivalent to the right characteristic of the middle diagram, which we can see by the following arguments. Because both unilateral constraints have to be fulfilled simultaneously, the values of $\lambda_{Ti}^{(+)}$ are restricted with respect to (6.17), and hence according to (6.16). If a $\lambda_{Ti}^{(+)}$ is chosen in such a manner that both conditions in (6.17) hold with a strict inequality, then $\ddot{g}_{Ti}^{+} = z_i^{-} = 0$ result from the characteristics of the unilateral constraints and provide, by (6.15), an acceleration $\ddot{g}_{Ti} = 0$. For $\lambda_{Ti}^{(+)} = 0$ we get $\ddot{g}_{Ti}^{+} \geq 0$, $z_i^{-} = 0$, and, hence, $\ddot{g}_{Ti} = \ddot{g}_{Ti}^{+} \geq 0$.

In the remaining case ($\lambda_{Ti}^{(+)} = +\mu_{0i}\lambda_{Ni}$) $\ddot{g}_{Ti}^{+} = 0$ and $z_i^{-} \geq 0$, which result in an acceleration $\ddot{g}_{Ti} = -z_i^{-} \leq 0$. Now we have shown that the right characteristic in Fig. 6.4 can be split into two unilateral constraints which correspond to the right half of the lower part of Fig. 6.4.

In the same manner the left characteristic has to be decomposed. As a result, four unilateral constraints are needed to express the friction characteristic at the top of Fig. 6.4, where the following equations, inequalities and complementarity conditions have been used: By (6.13) we express the tangential force by two simultaneously acting forces:

$$\lambda_{Ti} = \lambda_{Ti}^{(+)} - \lambda_{Ti}^{(-)} ; \quad i \in I_H . \tag{6.20}$$

In (6.15) the tangential acceleration has been split into its positive and negative parts. Furthermore, we have introduced in (6.18) the new variable $\lambda_{T0i}^{(-)}$. Repeating this procedure for the left part of Fig. 6.4, we obtain

$$\left.\begin{aligned} \ddot{g}_{Ti} &= z_i^{+} - \ddot{g}_{Ti}^{-} \\ \ddot{g}_{Ti} &= \ddot{g}_{Ti}^{+} - z_i^{-} \\ \lambda_{T0i}^{(+)} &= \mu_{0i}\lambda_{Ni} - \lambda_{Ti}^{(-)} \\ \lambda_{T0i}^{(-)} &= \mu_{0i}\lambda_{Ni} - \lambda_{Ti}^{(+)} \end{aligned}\right\} ; \quad i \in I_H . \tag{6.21}$$

The physical meaning of the auxiliary variables z_i^{+} and z_i^{-} is obvious by this set of equations: They denote nothing more than the positive and negative parts of the accelerations and are introduced for distinctive reasons only. From (6.21) we see the correspondence $z_i^{+} = \ddot{g}_{Ti}^{+}$, $z_i^{-} = \ddot{g}_{Ti}^{-}$.

The upper left and lower right characteristics in the lower part of Fig. 6.4 do not yet have the standard form of a unilateral constraint as in Fig. 6.1. This form, however, can be achieved by shifting the corners to the origin and reflecting the characteristics with respect to the ordinate axis, which is done by the transformations in the second line of (6.21). The terms $\lambda_{T0i}^{(+)}$ and $\lambda_{T0i}^{(-)}$ are called friction saturations and state the differences of the maximal transferable and actual tangential forces. When they vanish, a transition to sliding is possible. Finally, we state all of the inequalities and complementarity conditions (6.15), (6.17), (6.19), which also can be seen from the lower part of Fig. 6.4:

$$\left.\begin{array}{lll} \ddot{g}_{Ti}^{-} \geq 0\,; & \lambda_{Ti}^{(-)} \geq 0\,; & \ddot{g}_{Ti}^{-}\lambda_{Ti}^{(-)} = 0 \\ \ddot{g}_{Ti}^{+} \geq 0\,; & \lambda_{Ti}^{(+)} \geq 0\,; & \ddot{g}_{Ti}^{+}\lambda_{Ti}^{(+)} = 0 \\ \lambda_{T0i}^{(+)} \geq 0\,; & z_i^{+} \geq 0\,; & \lambda_{T0i}^{(+)} z_i^{+} = 0 \\ \lambda_{T0i}^{(-)} \geq 0\,; & z_i^{-} \geq 0\,; & \lambda_{T0i}^{(-)} z_i^{-} = 0 \end{array}\right\} \; ; \quad i \in I_H\,. \tag{6.22}$$

This set of relations completely describes the decomposed friction characteristics which correspond to four unilateral constraints. Equations (6.20)–(6.22) have to be used for each of the potentially sticking contacts $i \in I_H$. In order to take into account all these contacts, we rewrite eqs. (6.20)–(6.22) in vector notation, where the following symbols according to eq. (5.9) are introduced:

$$\begin{array}{l} \ddot{\boldsymbol{g}}_H^{+} = \{\ddot{g}_{Ti}^{+}\}\,;\; \boldsymbol{z}^{+} = \{z_i^{+}\}\,;\; \boldsymbol{\lambda}_H^{(+)} = \{\lambda_{Ti}^{(+)}\}\,;\; \boldsymbol{\lambda}_{H0}^{(+)} = \{\lambda_{T0i}^{(+)}\}\,; \\ \ddot{\boldsymbol{g}}_H^{-} = \{\ddot{g}_{Ti}^{-}\}\,;\; \boldsymbol{z}^{-} = \{z_i^{-}\}\,;\; \boldsymbol{\lambda}_H^{(-)} = \{\lambda_{Ti}^{(-)}\}\,;\; \boldsymbol{\lambda}_{H0}^{(-)} = \{\lambda_{T0i}^{(-)}\}\,; \\ \overline{\overline{\boldsymbol{\mu}}}_H = \{\mu_{0i}\}\;;\; i \in I_H \end{array} \tag{6.23}$$

The resulting equations are

$$\boldsymbol{\lambda}_H = \boldsymbol{\lambda}_H^{(+)} - \boldsymbol{\lambda}_H^{(-)} \tag{6.24}$$

$$\begin{array}{llll} \ddot{\boldsymbol{g}}_H & = \boldsymbol{z}^{+} - \ddot{\boldsymbol{g}}_H^{-}\,; & \ddot{\boldsymbol{g}}_H & = \ddot{\boldsymbol{g}}_H^{+} - \boldsymbol{z}^{-} \\ \boldsymbol{\lambda}_{H0}^{(+)} & = \overline{\overline{\boldsymbol{\mu}}}_H \boldsymbol{\lambda}_N - \boldsymbol{\lambda}_H^{(-)}\,; & \boldsymbol{\lambda}_{H0}^{(-)} & = \overline{\overline{\boldsymbol{\mu}}}_H \boldsymbol{\lambda}_N - \boldsymbol{\lambda}_H^{(+)} \end{array} \tag{6.25}$$

$$\begin{array}{lll} \ddot{\boldsymbol{g}}_H^{-} \geq 0\,; & \boldsymbol{\lambda}_H^{(-)} \geq 0\,; & \ddot{\boldsymbol{g}}_H^{-T}\boldsymbol{\lambda}_H^{(-)} = 0 \\ \ddot{\boldsymbol{g}}_H^{+} \geq 0\,; & \boldsymbol{\lambda}_H^{(+)} \geq 0\,; & \ddot{\boldsymbol{g}}_H^{+T}\boldsymbol{\lambda}_H^{(+)} = 0 \\ \boldsymbol{\lambda}_{H0}^{(+)} \geq 0\,; & \boldsymbol{z}^{+} \geq 0\,; & \boldsymbol{\lambda}_{H0}^{(+)T}\boldsymbol{z}^{+} = 0 \\ \boldsymbol{\lambda}_{H0}^{(-)} \geq 0\,; & \boldsymbol{z}^{-} \geq 0\,; & \boldsymbol{\lambda}_{H0}^{(-)T}\boldsymbol{z}^{-} = 0 \end{array} \tag{6.26}$$

which take into account the states of continual sticking as well as the transitions to sliding. In the next section these sets of conditions will be used together with the dynamics equations and the unilateral impenetrability condition in the normal direction to obtain a LCP formulation of the fully coupled state transition problem.

6.4 The Linear Complementarity Problem

To derive the Linear Complementarity Problem of the detachment and stick-slip transitions we have to use eqs. (6.1), (6.5), (6.24)–(6.26). The upper part of eq. (6.1) consists of the dynamical equations of our multibody system which connect the unknown generalized accelerations $\ddot{\boldsymbol{q}}$, the normal forces $\boldsymbol{\lambda}_N$ of all closed contacts, and the tangential forces $\boldsymbol{\lambda}_H$ of all possible sticking contacts. The tangential forces of the contacts which are sliding with relative velocities $\dot{g}_{Ti} \neq 0$ are already taken into account and expressed by their corresponding normal forces, $\boldsymbol{W}_G\overline{\overline{\boldsymbol{\mu}}}_G\boldsymbol{\lambda}_N$. The second set of equations in (6.1) results from the relative kinematics relationships and shows the dependency of the normal and tangential relative accelerations $(\ddot{\boldsymbol{g}}_N, \ddot{\boldsymbol{g}}_H)$ on the generalized accelerations $\ddot{\boldsymbol{q}}$. Equation (6.5) connects the normal accelerations $\ddot{\boldsymbol{g}}_N$ and the normal contact forces $\boldsymbol{\lambda}_N$ by an impenetrability condition on the acceleration level. The corresponding characteristic is depicted in Fig. 6.1 and takes into account the continual contact state as well as the take-off transition. The remaining equations (6.24)–(6.26) are equivalent to the friction characteristic in Fig. 6.2 and complete our description by allowing the contacts either to continue sticking or to switch into the sliding state.

In the following steps we eliminate from these equations all variables which are not complementary, i.e., not contained in the complementarity conditions of eqs. (6.5) and (6.26). We first replace, with the help of (6.24), the tangential forces $\boldsymbol{\lambda}_H$ in (6.1) and get the dynamics equation in the form

$$\boldsymbol{M}\ddot{\boldsymbol{q}} - \boldsymbol{h} - \underbrace{\left[\boldsymbol{W}_N + \boldsymbol{W}_G\overline{\overline{\boldsymbol{\mu}}}_G \,|\boldsymbol{W}_H| - \boldsymbol{W}_H\right]}_{\boldsymbol{W}+\boldsymbol{N}_G} \underbrace{\begin{pmatrix} \boldsymbol{\lambda}_N \\ \boldsymbol{\lambda}_H^{(+)} \\ \boldsymbol{\lambda}_H^{(-)} \end{pmatrix}}_{\boldsymbol{\lambda}} = 0 \quad \in \mathbb{R}^f \,. \tag{6.27}$$

The tangential force $\boldsymbol{\lambda}_H$ has been expressed by the pair of simultaneously acting forces $\boldsymbol{\lambda}_H^{(+)}$ and $\boldsymbol{\lambda}_H^{(-)}$. Each is connected to the same relative acceleration $\ddot{\boldsymbol{g}}_H$, which we take into account by writing the second set of equations in (6.1) as

$$\underbrace{\begin{pmatrix} \ddot{\boldsymbol{g}}_N \\ \ddot{\boldsymbol{g}}_H \\ -\ddot{\boldsymbol{g}}_H \end{pmatrix}}_{\ddot{\boldsymbol{g}}_M} = \underbrace{\begin{pmatrix} \boldsymbol{W}_N^T \\ \boldsymbol{W}_H^T \\ -\boldsymbol{W}_H^T \end{pmatrix}}_{\boldsymbol{W}^T} \ddot{\boldsymbol{q}} + \underbrace{\begin{pmatrix} \overline{\boldsymbol{w}}_N \\ \overline{\boldsymbol{w}}_H \\ -\overline{\boldsymbol{w}}_H \end{pmatrix}}_{\overline{\boldsymbol{w}}} \quad \in \mathbb{R}^{n_N+2n_H} \,. \tag{6.28}$$

The remaining equations, eqs. (6.25), are also written in matrix notation, which yields

$$\underbrace{\begin{pmatrix} \ddot{g}_N \\ \ddot{g}_H \\ -\ddot{g}_H \end{pmatrix}}_{\ddot{\boldsymbol{g}}_M} = \underbrace{\begin{pmatrix} \ddot{g}_N \\ \ddot{g}_H^+ \\ \ddot{g}_H^- \end{pmatrix}}_{\ddot{\boldsymbol{g}}} - \underbrace{\begin{pmatrix} \boldsymbol{0} & \boldsymbol{0} \\ \boldsymbol{E} & \boldsymbol{0} \\ \boldsymbol{0} & \boldsymbol{E} \end{pmatrix}}_{\boldsymbol{I}^T} \underbrace{\begin{pmatrix} \boldsymbol{z}^- \\ \boldsymbol{z}^+ \end{pmatrix}}_{\boldsymbol{z}} \in \mathbb{R}^{n_N + 2n_H} \tag{6.29}$$

$$\underbrace{\begin{pmatrix} \boldsymbol{\lambda}_{H0}^{(-)} \\ \boldsymbol{\lambda}_{H0}^{(+)} \end{pmatrix}}_{\boldsymbol{\lambda}_{H0}} = \underbrace{\begin{pmatrix} \overline{\overline{\boldsymbol{\mu}}}_H & -\boldsymbol{E} & \boldsymbol{0} \\ \overline{\overline{\boldsymbol{\mu}}}_H & \boldsymbol{0} & -\boldsymbol{E} \end{pmatrix}}_{\boldsymbol{N}_H - \boldsymbol{I}} \underbrace{\begin{pmatrix} \boldsymbol{\lambda}_N \\ \boldsymbol{\lambda}_H^{(+)} \\ \boldsymbol{\lambda}_H^{(-)} \end{pmatrix}}_{\boldsymbol{\lambda}} \in \mathbb{R}^{2n_H} . \tag{6.30}$$

Finally we state the inequalities and complementarity conditions from (6.5) and (6.26):

$$\underbrace{\begin{pmatrix} \ddot{g}_N \\ \ddot{g}_H^+ \\ \ddot{g}_H^- \end{pmatrix}}_{\ddot{\boldsymbol{g}}} \geq \boldsymbol{0}; \quad \underbrace{\begin{pmatrix} \boldsymbol{\lambda}_N \\ \boldsymbol{\lambda}_H^{(+)} \\ \boldsymbol{\lambda}_H^{(-)} \end{pmatrix}}_{\boldsymbol{\lambda}} \geq \boldsymbol{0}; \quad \underbrace{\begin{pmatrix} \ddot{g}_N \\ \ddot{g}_H^+ \\ \ddot{g}_H^- \end{pmatrix}^T}_{\ddot{\boldsymbol{g}}^T} \underbrace{\begin{pmatrix} \boldsymbol{\lambda}_N \\ \boldsymbol{\lambda}_H^{(+)} \\ \boldsymbol{\lambda}_H^{(-)} \end{pmatrix}}_{\boldsymbol{\lambda}} = 0 \tag{6.31}$$

$$\underbrace{\begin{pmatrix} \boldsymbol{\lambda}_{H0}^{(-)} \\ \boldsymbol{\lambda}_{H0}^{(+)} \end{pmatrix}}_{\boldsymbol{\lambda}_{H0}} \geq \boldsymbol{0}; \quad \underbrace{\begin{pmatrix} \boldsymbol{z}^- \\ \boldsymbol{z}^+ \end{pmatrix}}_{\boldsymbol{z}} \geq \boldsymbol{0}; \quad \underbrace{\begin{pmatrix} \boldsymbol{\lambda}_{H0}^{(-)} \\ \boldsymbol{\lambda}_{H0}^{(+)} \end{pmatrix}^T}_{\boldsymbol{\lambda}_{H0}^T} \underbrace{\begin{pmatrix} \boldsymbol{z}^- \\ \boldsymbol{z}^+ \end{pmatrix}}_{\boldsymbol{z}} = 0 \tag{6.32}$$

and we obtain from (6.27)–(6.32) six sets of relations, which we rewrite by using the terms under the braces:

$$\begin{aligned}
&\boldsymbol{M}\ddot{\boldsymbol{q}} - \boldsymbol{h} - (\boldsymbol{W} + \boldsymbol{N}_G)\boldsymbol{\lambda} = \boldsymbol{0} \quad \in \mathbb{R}^f \\
&\ddot{\boldsymbol{g}}_M = \boldsymbol{W}^T\ddot{\boldsymbol{q}} + \overline{\boldsymbol{w}} \quad \in \mathbb{R}^{n_N + 2n_H} \\
&\ddot{\boldsymbol{g}}_M = \ddot{\boldsymbol{g}} - \boldsymbol{I}^T\boldsymbol{z} \quad \in \mathbb{R}^{n_N + 2n_H} \\
&\boldsymbol{\lambda}_{H0} = (\boldsymbol{N}_H - \boldsymbol{I})\boldsymbol{\lambda} \quad \in \mathbb{R}^{2n_H} \\
&\ddot{\boldsymbol{g}} \geq \boldsymbol{0}; \quad \boldsymbol{\lambda} \geq \boldsymbol{0}; \quad \ddot{\boldsymbol{g}}^T\boldsymbol{\lambda} = 0 \\
&\boldsymbol{\lambda}_{H0} \geq \boldsymbol{0}; \quad \boldsymbol{z} \geq \boldsymbol{0}; \quad \boldsymbol{\lambda}_{H0}^T\boldsymbol{z} = 0 .
\end{aligned} \tag{6.33}$$

We now solve the first equation with respect to $\ddot{\boldsymbol{q}}$, substitute it into the second one, which is equal to the third one, and get

$$\begin{aligned}
&\ddot{\boldsymbol{g}} - \boldsymbol{I}^T \boldsymbol{z} = \boldsymbol{W}^T \boldsymbol{M}^{-1}(\boldsymbol{W} + \boldsymbol{N}_G)\boldsymbol{\lambda} + \boldsymbol{W}^T \boldsymbol{M}^{-1}\boldsymbol{h} + \overline{\boldsymbol{w}} \quad \in \mathbb{R}^{n_N + 2n_H} \\
&\boldsymbol{\lambda}_{H0} = (\boldsymbol{N}_H - \boldsymbol{I})\boldsymbol{\lambda} \quad \in \mathbb{R}^{2n_H} \\
&\ddot{\boldsymbol{g}} \geq \boldsymbol{0}\,; \quad \boldsymbol{\lambda} \geq \boldsymbol{0}\,; \quad \ddot{\boldsymbol{g}}^T \boldsymbol{\lambda} = 0 \\
&\boldsymbol{\lambda}_{H0} \geq \boldsymbol{0}\,; \quad \boldsymbol{z} \geq \boldsymbol{0}\,; \quad \boldsymbol{\lambda}_{H0}^T \boldsymbol{z} = 0\,.
\end{aligned} \tag{6.34}$$

Stating these equations, inequalities, and complementarity conditions in matrix notation,

$$\begin{aligned}
&\underbrace{\begin{pmatrix} \ddot{\boldsymbol{g}} \\ \boldsymbol{\lambda}_{H0} \end{pmatrix}}_{\boldsymbol{y}} = \underbrace{\begin{pmatrix} \boldsymbol{W}^T \boldsymbol{M}^{-1}(\boldsymbol{W} + \boldsymbol{N}_G) & \boldsymbol{I}^T \\ \boldsymbol{N}_H - \boldsymbol{I} & \boldsymbol{0} \end{pmatrix}}_{\boldsymbol{A}} \underbrace{\begin{pmatrix} \boldsymbol{\lambda} \\ \boldsymbol{z} \end{pmatrix}}_{\boldsymbol{x}} + \underbrace{\begin{pmatrix} \boldsymbol{W}^T \boldsymbol{M}^{-1}\boldsymbol{h} + \overline{\boldsymbol{w}} \\ \boldsymbol{0} \end{pmatrix}}_{\boldsymbol{b}} \\
&\underbrace{\begin{pmatrix} \ddot{\boldsymbol{g}} \\ \boldsymbol{\lambda}_{H0} \end{pmatrix}}_{\boldsymbol{y}} \geq \boldsymbol{0}\,; \quad \underbrace{\begin{pmatrix} \boldsymbol{\lambda} \\ \boldsymbol{z} \end{pmatrix}}_{\boldsymbol{x}} \geq \boldsymbol{0}\,; \quad \underbrace{\begin{pmatrix} \ddot{\boldsymbol{g}} \\ \boldsymbol{\lambda}_{H0} \end{pmatrix}^T}_{\boldsymbol{y}^T} \underbrace{\begin{pmatrix} \boldsymbol{\lambda} \\ \boldsymbol{z} \end{pmatrix}}_{\boldsymbol{x}} = 0
\end{aligned} \tag{6.35}$$

we achieve a Linear Complementarity Problem in its standard form:

$$\begin{aligned}
&\boldsymbol{y} = \boldsymbol{A}\boldsymbol{x} + \boldsymbol{b}\,; \quad \boldsymbol{y} \geq \boldsymbol{0}\,; \quad \boldsymbol{x} \geq \boldsymbol{0}\,; \quad \boldsymbol{y}^T \boldsymbol{x} = 0 \\
&\boldsymbol{y},\ \boldsymbol{x} \in \mathbb{R}^{n_N + 4n_H}\,.
\end{aligned} \tag{6.36}$$

The unknowns in (6.36) are the vectors $\boldsymbol{y}$ and $\boldsymbol{x}$ which contain the relative accelerations, the contact forces, and the friction saturations from eqs. (6.35), (6.13) and (6.32). The events of continual contact and sticking, as well as the transitions to sliding and separation, are included in this formulation. After solving (6.36) we obtain the values of $\boldsymbol{x}$ and $\boldsymbol{y}$ where the complementarity condition $x_i y_i$ is taken into account by either $(x_i \geq 0; y_i = 0)$ or $(x_i = 0; y_i \geq 0)$. The LCP formulation of a system with unilateral normal constraints and friction is much more general than the representation of a frictionless bilaterally constrained system, but leads to additional complications with respect to existence and uniqueness of solutions.

For instance, the examples presented in Sections 5.3 and 5.4 are completely contained in the general description (6.35) and are only representative of what can happen. There exist no statements about the solvability of a general LCP, i.e., a LCP with arbitrary matrix $\boldsymbol{A}$. At the moment, the only known algorithms which are guaranteed to solve it are the *enumerative methods* [48, 66, 88]. They treat the problem by a combinatorial evaluation: The complementarity conditions $x_i y_i = 0$ of a LCP of dimension n provide 2^n different combinations of n variables which are allowed to be greater than zero at the same time. Collecting for each of these 2^n combinations the corresponding variables in a vector $\boldsymbol{z}_k$, $z_{ki} \in \{x_i, y_i\}$, $i = 1, \ldots, n$, enables us to rewrite (6.36) as 2^n different Linear Programming Problems of the form $\boldsymbol{C}_k \boldsymbol{z}_k = \boldsymbol{b}$, $\boldsymbol{z}_k \geq 0$, $k = 1, \ldots, 2^n$. Each solution of these 2^n LP problems is then a solution of our LCP where the following cases can occur: The 2^n sets $\{\boldsymbol{v}_k | \boldsymbol{v}_k = \boldsymbol{C}_k \boldsymbol{z}_k \; ; \; \boldsymbol{z}_k \geq 0\}$ are called the complementary cones of the LCP. If the vector $\boldsymbol{b}$ lies in the interior of only one of these cones, and the cone is not degenerated (det $\boldsymbol{C} \neq 0$), then the solution of the LCP is unique. If $\boldsymbol{b}$ lies in a region which is covered by some overlapping non-degenerated cones, then we get a finite number of different solutions. The number of solutions may be infinite if the cones are degenerate, det $\boldsymbol{C} = 0$. In some cases, certain areas of $\mathbb{R}^n$ may not be covered by any cone. If $\boldsymbol{b}$ lies in such an area, the LCP clearly has no solution. In Section 6.5 a mechanical example in $\mathbb{R}^2$ is presented where the four complementary cones are drawn. The enumerative methods can be used for small values of n. For large n, however, these methods become impractical since 2^n grows very rapidly.

Another way of solving the LCP is given by a *complementary pivot algorithm* [16], which is based on the simplex method and usually referred to as Lemke's algorithm. Here the n-dimensional LCP in eq. (6.36) is rewritten by moving the term $\boldsymbol{Ax}$ to the left-hand side and introducing an artifical variable $z \geq 0$ together with a vector $\boldsymbol{e}$ whose entries are all $+1$. This results in the new problem $(\boldsymbol{E} \; - \; \boldsymbol{A} \; - \; \boldsymbol{e})(\boldsymbol{y}^T \, \boldsymbol{x}^T \, z)^T = \boldsymbol{b}$, $\boldsymbol{y} \geq 0$, $\boldsymbol{x} \geq 0$, $\boldsymbol{y}^T \boldsymbol{x} = 0$, $z \geq 0$, where $\boldsymbol{E} \in \mathbb{R}^{n,n}$ is the identity matrix. Obviously, by choosing z and a certain set of $n-1$ elements of $\boldsymbol{y}$ as variables, which are allowed to be greater than zero by keeping all of the other variables equal to zero, leads to a solution of that problem. The set of the possible nonnegative variables is called the feasible basis. Together with the remaining variables, all the inequalities and complementarity conditions in eq. (6.36) are fulfilled. But this is clearly not the solution of (6.36) because the feasible basis still contains the artificial variable $z \geq 0$.

From that initialization step the algorithm defines a unique sequence of new feasible bases by dropping certain variables with respect to a complementary pivot rule and entering the complements of the previously dropped variables. After a finite number of steps the algorithm terminates by two possible ways: In the first case, the artificial variable z may drop out from the feasible basis or may become equal to zero. Then one of the solutions of the LCP is given by the (nonnegative) values of the variables contained in the feasible basis together with the zero-valued remaining variables. In the second case, the artificial variable z is strictly positive and still remaining in the feasible basis, but no dropping variable can be determined without violating the complementary pivot rule. The algorithm then is not able to solve the LCP. This means that either no solution exists, or existing solutions cannot be found. However, for certain classes of matrices, for example positive-semidefinite matrices, it can be guaranteed that the algorithm finds a solution if solutions exist or terminates without a solution only if the LCP is not solvable. In our case we see that the LCP matrix $\boldsymbol{A}$ in (6.35) is positive semidefinite for frictionless systems ($\boldsymbol{N}_G = \boldsymbol{N}_H = 0$). This means that the algorithm is working at least for systems without or with moderate friction. For further details of LCPs and available algorithms, see [66], for example.

The most general case where friction occurs is the three-dimensional contact problem. By it we understand a plane contact where motion can take place in two directions. The basic problem of such a configuration consists in the fact that additionally the friction directions of the possible sticking contacts are unknown at the first moment after the transition from stiction to sliding. This situation leads to a Nonlinear Complementarity Problem, which is not solvable with a simplex algorithm. Furthermore, the time function of the tangential relative velocity $|\dot{\boldsymbol{g}}_{Ti}|$ shows no change of sign at the transition from sliding to stiction, which makes the search for switch points of the constraints more difficult. These problems are being researched.

6.5 Example: The Detachment Transition

We consider the detachment process of a frictionless two-point-contact problem. The mechanical model is depicted in the left part of Fig. 6.5. It consists of two bodies with masses m in contact with each other at the beginning. They are under the influence of gravity g and applied forces F_1 and F_2. We shall discuss the conditions under which contact is lost between the masses and the surface.

From the mechanical model in the right part of Fig. 6.5 we derive the

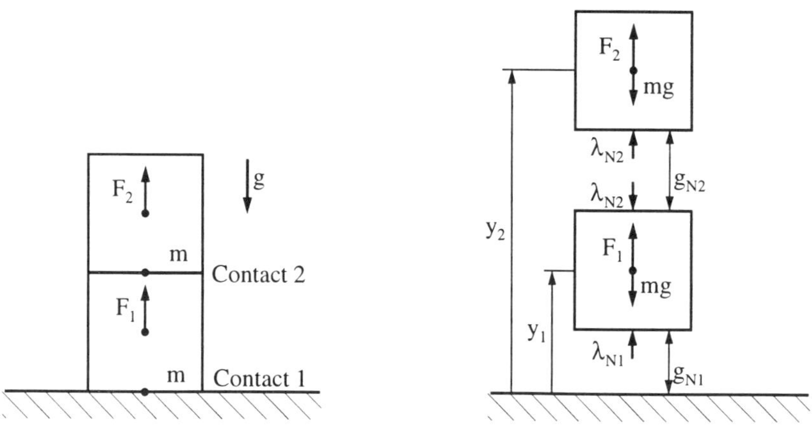

Figure 6.5: Mechanical Model and Free-Body Diagram

equations of motion in the following form:

$$
\begin{aligned}
m\ddot{y}_1 &= F_1 - mg + \lambda_{N1} - \lambda_{N2} \\
m\ddot{y}_2 &= F_2 - mg + \lambda_{N2}
\end{aligned}
\tag{6.37}
$$

where λ_{N1} and λ_{N2} are the normal forces in the contacts. As a next step we evaluate the contact state. The upper body rests on the lower body, which is in continual contact with the surface. Thus, both distances and normal relative velocities are equal to zero: $g_{Ni} = \dot{g}_{Ni} = 0$. $i = 1, 2$. Then the normal relative accelerations $\ddot{g}_{Ni}$ become

$$
\begin{aligned}
\ddot{g}_{N1} &= \ddot{y}_1 \\
\ddot{g}_{N2} &= \ddot{y}_2 - \ddot{y}_1 \,.
\end{aligned}
\tag{6.38}
$$

With respect to eq. (6.5) the contact situation is described by the following inequalities and complementarity conditions:

$$
\begin{aligned}
&\ddot{g}_{N1} \geq 0\,; \quad \lambda_{N1} \geq 0\,; \quad \ddot{g}_{N1}\lambda_{N1} = 0 \\
&\ddot{g}_{N2} \geq 0\,; \quad \lambda_{N2} \geq 0\,; \quad \ddot{g}_{N2}\lambda_{N2} = 0\,.
\end{aligned}
\tag{6.39}
$$

Substituting $\ddot{y}_1$ and $\ddot{y}_2$ from (6.37) into (6.38) we get

$$
\begin{aligned}
\ddot{g}_{N1} &= \frac{1}{m}(\lambda_{N1} - \lambda_{N2}) + \frac{F_1}{m} - g \\
\ddot{g}_{N2} &= \frac{1}{m}(2\lambda_{N2} - \lambda_{N1}) + \frac{F_2}{m} - \frac{F_1}{m}\,.
\end{aligned}
\tag{6.40}
$$

Putting $m = 1$ we rewrite (6.39) and (6.40) in matrix notation:

$$
\begin{aligned}
&\begin{pmatrix} \ddot{g}_{N1} \\ \ddot{g}_{N2} \end{pmatrix} = \begin{pmatrix} 1 & -1 \\ -1 & 2 \end{pmatrix} \begin{pmatrix} \lambda_{N1} \\ \lambda_{N2} \end{pmatrix} + \begin{pmatrix} F_1 - g \\ F_2 - F_1 \end{pmatrix} \\
&\begin{pmatrix} \ddot{g}_{N1} \\ \ddot{g}_{N2} \end{pmatrix} \geq 0\,; \quad \begin{pmatrix} \lambda_{N1} \\ \lambda_{N2} \end{pmatrix} \geq 0\,; \quad \begin{pmatrix} \ddot{g}_{N1} \\ \ddot{g}_{N2} \end{pmatrix}^T \begin{pmatrix} \lambda_{N1} \\ \lambda_{N2} \end{pmatrix} = 0
\end{aligned}
\tag{6.41}
$$

which is the LCP for a frictionless system with unilateral normal constraints according to eqs. (6.7) and (6.8). Generally the evaluation of (6.41) can be done numerically, but in the two-dimensional case a graphical approach might be used. For this purpose we rewrite the first equation in (6.41) as a linear combination of vector magnitudes by shifting all the variables to the left hand side,

$$
\underbrace{\begin{pmatrix} 1 \\ 0 \end{pmatrix}}_{\boldsymbol{e}_1} \ddot{g}_{N1} + \underbrace{\begin{pmatrix} 0 \\ 1 \end{pmatrix}}_{\boldsymbol{e}_2} \ddot{g}_{N2} - \underbrace{\begin{pmatrix} 1 \\ -1 \end{pmatrix}}_{\boldsymbol{a}_1} \lambda_{N1} - \underbrace{\begin{pmatrix} -1 \\ 2 \end{pmatrix}}_{\boldsymbol{a}_2} \lambda_{N2} = \underbrace{\begin{pmatrix} F_1 - g \\ F_2 - F_1 \end{pmatrix}}_{\boldsymbol{b}}
\tag{6.42}
$$

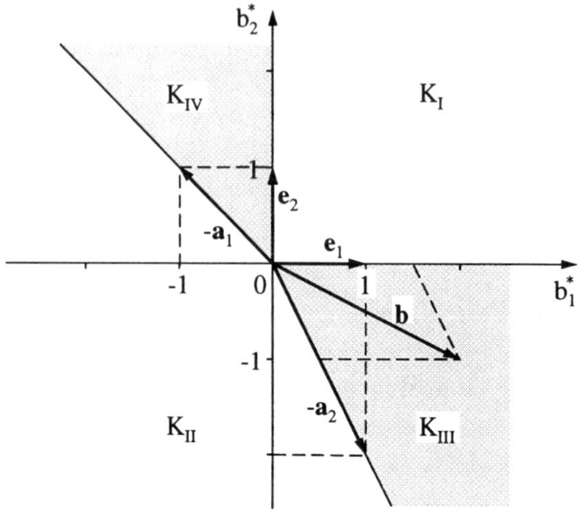

Figure 6.6: The Complementary Cones

We evaluate the combinatorial problem given by the inequalities and complementarity conditions of (6.41). A total of $2^2 = 4$ combinations results from those conditions, which are

$$
\begin{array}{ll}
\text{(I)} & \ddot{g}_{N1} \geq 0\,;\ \ddot{g}_{N2} \geq 0\,;\ \lambda_{N1} = 0\,;\ \lambda_{N2} = 0 \\
 & \text{(Take-off 1 ; Take-off 2)} \\
\text{(II)} & \ddot{g}_{N1} = 0\,;\ \ddot{g}_{N2} = 0\,;\ \lambda_{N1} \geq 0\,;\ \lambda_{N2} \geq 0 \\
 & \text{(Contact 1 ; Contact 2)} \\
\text{(III)} & \ddot{g}_{N1} \geq 0\,;\ \ddot{g}_{N2} = 0\,;\ \lambda_{N1} = 0\,;\ \lambda_{N2} \geq 0 \\
 & \text{(Take-off 1 ; Contact 2)} \\
\text{(IV)} & \ddot{g}_{N1} = 0\,;\ \ddot{g}_{N2} \geq 0\,;\ \lambda_{N1} \geq 0\,;\ \lambda_{N2} = 0 \\
 & \text{(Contact 1 ; Take-off 2)}
\end{array}
\tag{6.43}
$$

In each combination only two of the variables are allowed to be greater than zero, whereas the remaining pair must be zero. In context with the left-hand side of (6.42), each of the four cases describes a cone which is generated by all possible nonnegative linear combinations of two vectors. These cones are called the complementary cones [66] of the LCP and are depicted in Fig. 6.6. For example, the cone K_{III} results from the linear combinations $\boldsymbol{e}_1\ddot{g}_{N1} + (-\boldsymbol{a}_2)\lambda_{N2}$, $\ddot{g}_{N1} \geq 0$, $\lambda_{N2} \geq 0$, which can be seen from (6.42) and (6.43). The four complementary cones are given by

$$
\begin{aligned}
&K_i = \{\boldsymbol{b}^* \,|\, \boldsymbol{b}^* = \boldsymbol{C}_i \boldsymbol{z}\ ;\ \boldsymbol{z} \geq 0\}\ ;\ i = \text{I, II, III, IV} \\
&\quad \boldsymbol{C}_{\text{I}} \ = (\,\boldsymbol{e}_1\ \boldsymbol{e}_2\,)\,; \qquad \boldsymbol{C}_{\text{III}} = (\,\boldsymbol{e}_1\ -\boldsymbol{a}_2\,) \\
&\quad \boldsymbol{C}_{\text{II}} = (\,-\boldsymbol{a}_1\ -\boldsymbol{a}_2\,)\,; \quad \boldsymbol{C}_{\text{IV}} = (\,-\boldsymbol{a}_1\ \boldsymbol{e}_2\,)
\end{aligned}
\tag{6.44}
$$

We choose certain values for the parameters in the right-hand side of eq. (6.42) in order to specify the vector $\boldsymbol{b}$. For example, let

$$
g = 10\,; \quad F_1 = 12\,; \quad F_2 = 11\ \Rightarrow \boldsymbol{b} = \begin{pmatrix} 2 \\ -1 \end{pmatrix}. \tag{6.45}
$$

Drawing this vector in Fig. 6.6 shows that it lies in the interior of cone K_{III} and can therefore be expressed by the combination

$$
\begin{pmatrix} 1 \\ 0 \end{pmatrix} \ddot{g}_{N1} - \begin{pmatrix} -1 \\ 2 \end{pmatrix} \lambda_{N2} = \begin{pmatrix} 2 \\ -1 \end{pmatrix}, \tag{6.46}
$$

which enables us to state the solution of the LCP as

$$
\ddot{g}_{N1} = \frac{3}{2}\,; \quad \lambda_{N2} = \frac{1}{2}\,; \quad \ddot{g}_{N2} = \lambda_{N1} = 0\,. \tag{6.47}
$$

Resubstituting these values into eq. (6.37) yields together with (6.45) and $m = 1$ the generalized accelerations

$$
\ddot{y}_1 = \frac{3}{2}\,; \quad \ddot{y}_2 = \frac{3}{2}\,. \tag{6.48}
$$

In this case, the lower body leaves the surface while keeping contact with the upper body. The absolute vertical accelerations of both bodies are 3/2; the compressive normal force in the second contact is equal to 1/2.

At that point we should once more return to Fig. 6.6, where the four complementarity cones are splitting up the $\mathbb{R}^2$: Their structure contains no intersection possibilities, and they cover the whole $\mathbb{R}^2$. An arbitrarily chosen vector $\boldsymbol{b}$ can therefore only be arranged in two ways, either in the interior of exactly one cone or at the boundary between two neighboring cones. In both cases $\boldsymbol{b}$ can be expressed uniquely by a positive linear combination of at most two of the vectors forming the cones in (6.44). Thus we conclude that the LCP in this example always has a unique solution. This can also be seen very easily by another approach: A unilaterally constrained system without friction can be stated as a Quadratic Program with a strictly convex cost function; see eq. (6.9). We remember that eqs. (6.37)–(6.39) are nothing else than the Kuhn-Tucker conditions of that program. If the inequality constraints are feasible and independent as in this example, then a solution exists and is unique with respect to the Lagrange multipliers $(\lambda_{N1}, \lambda_{N2})$ and the optimal solutions $(\ddot{y}_1, \ddot{y}_2)$.

At the end of this example it would be interesting to evaluate the dependency of the contact states in (6.43) on the applied forces F_1 and F_2. Up to now we can conclude that a vector $\boldsymbol{b}$ lying in a cone K_i corresponds to the contact state i in (6.43). From the definition of $\boldsymbol{b}$ in (6.42) we easily write

$$\boldsymbol{b} = \begin{pmatrix} 1 & 0 \\ -1 & 1 \end{pmatrix} \begin{pmatrix} F_1 \\ F_2 \end{pmatrix} + \begin{pmatrix} -g \\ 0 \end{pmatrix} \tag{6.49}$$

which can be solved with respect to $(F_1\, F_2)^T$ by inversion:

$$\underbrace{\begin{pmatrix} F_1 \\ F_2 \end{pmatrix}}_{\boldsymbol{F}} = \underbrace{\begin{pmatrix} 1 & 0 \\ 1 & 1 \end{pmatrix}}_{\boldsymbol{I}} \boldsymbol{b} + \underbrace{\begin{pmatrix} g \\ g \end{pmatrix}}_{\boldsymbol{g}} . \tag{6.50}$$

Equations (6.49) and (6.50) describe a one-to-one correspondence between the vectors $\boldsymbol{b}$ and $\boldsymbol{F}$ given by the affine mapping

$$L: \ \boldsymbol{b} \to \boldsymbol{F} = \boldsymbol{I}\boldsymbol{b} + \boldsymbol{g} . \tag{6.51}$$

Now we apply this mapping onto the cones K_i in eq. (6.44) in order to get

$$\begin{aligned} V_i &= L(K_i) \\ &= \{\boldsymbol{F}^* \,|\, \boldsymbol{F}^* = L(\boldsymbol{b}^*)\} \\ &= \{\boldsymbol{F}^* \,|\, \boldsymbol{F}^* = \boldsymbol{I}\boldsymbol{C}_i\boldsymbol{z} + \boldsymbol{g}\,; \quad \boldsymbol{z} \geq 0\} \\ &= \{\boldsymbol{F}^* \,|\, (\boldsymbol{I}\boldsymbol{C}_i)^{-1}(\boldsymbol{F}^* - \boldsymbol{g}) \geq 0\} , \end{aligned} \tag{6.52}$$

and conclude with the equivalence

$$\begin{matrix} \boldsymbol{b} \in K_i & \Leftrightarrow & \boldsymbol{F} \in V_i \\ \boldsymbol{F} = L(\boldsymbol{b}) & & V_i = L(K_i) . \end{matrix} \tag{6.53}$$

The images of the cones K_i are depicted in Fig. 6.7. From the last equation in (6.52) we obtain the conditions on F_1 and F_2 under which the system states in (6.43) are maintained:

$$\begin{matrix} \text{(I)} & F_1 \geq g\,; & F_2 \geq F_1 & \text{(Take-off 1 ;} & \text{Take-off 2)} \\ \text{(II)} & F_2 \leq g\,; & F_1 \leq 2g - F_2 & \text{(Contact 1 ;} & \text{Contact 2)} \\ \text{(III)} & F_2 \leq F_1\,; & 2g - F_1 \leq F_2 & \text{(Take-off 1 ;} & \text{Contact 2)} \\ \text{(IV)} & F_1 \leq g\,; & F_2 \geq g & \text{(Contact 1 ;} & \text{Take-off 2)} \end{matrix} \tag{6.54}$$

Clearly, these conditions correspond to the areas V_i in Fig. 6.7 and could be stated directly without any calculations in this simple example.

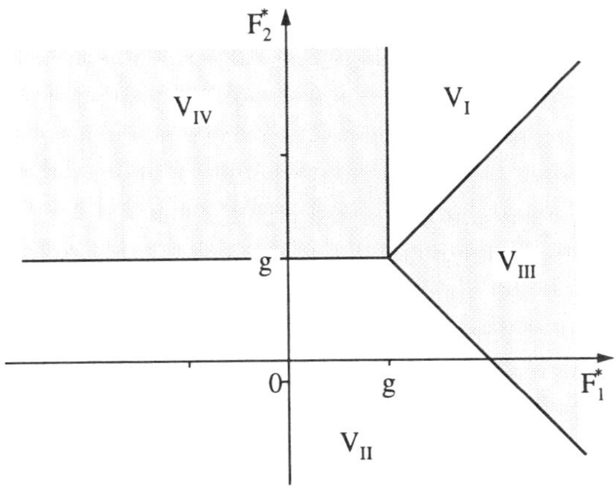

Figure 6.7: The Images of the Complementary Cones

6.6 Example: The Stick-Slip Transition

This example greatly resembles the one in Section 6.5. We consider the same mechanical model, applying forces F_1 and F_2 horizontally and assuming friction in both contacts. First we are interested in the conditions under which transitions from sticking to sliding take place. Second we investigate the dynamical behavior of the mechanical system which, in addition to other phenomena, is exerted on these stick-slip transitions.

The equations of motion in the horizontal direction are (see right model of Fig. 6.8)

$$\begin{aligned} m\ddot{x}_1 &= F_1 + \lambda_{T1} - \lambda_{T2} \\ m\ddot{x}_2 &= F_2 + \lambda_{T2} \,. \end{aligned} \tag{6.55}$$

They correspond to eq. (6.37) in the foregoing example. Additionally we state the balance of forces in the vertical direction, which leads to

$$\begin{aligned} \lambda_{N1} &= 2mg \\ \lambda_{N2} &= mg \,. \end{aligned} \tag{6.56}$$

As we can see from Fig. 6.8, λ_{Ni} and λ_{Ti} are the contact forces in the normal and tangential directions, respectively. The tangential relative accelerations are simply

$$\begin{aligned} \ddot{g}_{T1} &= \ddot{x}_1 \\ \ddot{g}_{T2} &= \ddot{x}_2 - \ddot{x}_1 \,. \end{aligned} \tag{6.57}$$

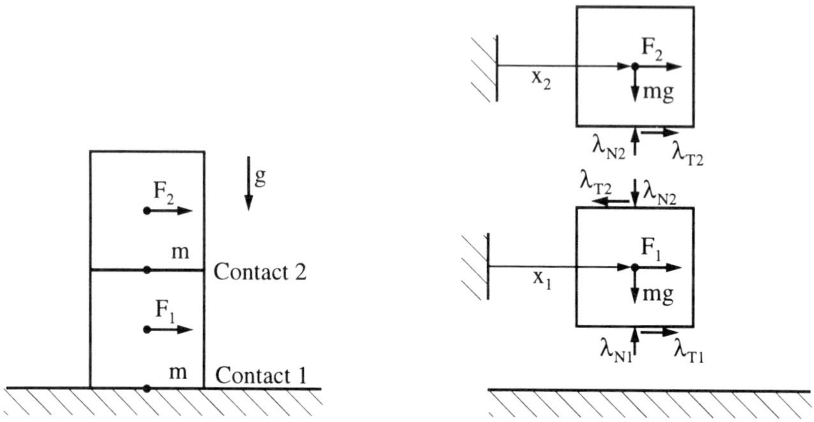

Figure 6.8: Mechanical Model and Free-Body Diagram

In order to get a complete description of the system we finally write the tangential contact laws by using the Coulomb friction model of eq. (6.11), where we assume the same coefficient of friction μ_0 in both contacts:

$$
\begin{aligned}
&|\lambda_{T1}| \leq \mu_0 \lambda_{N1}\,; \quad \begin{cases} |\lambda_{T1}| < \mu_0\lambda_{N1} & \Rightarrow \quad \ddot{g}_{T1} = 0 \\ \lambda_{T1} = +\mu_0\lambda_{N1} & \Rightarrow \quad \ddot{g}_{T1} \leq 0 \\ \lambda_{T1} = -\mu_0\lambda_{N1} & \Rightarrow \quad \ddot{g}_{T1} \geq 0 \end{cases} \\
&|\lambda_{T2}| \leq \mu_0 \lambda_{N2}\,; \quad \begin{cases} |\lambda_{T2}| < \mu_0\lambda_{N2} & \Rightarrow \quad \ddot{g}_{T2} = 0 \\ \lambda_{T2} = +\mu_0\lambda_{N2} & \Rightarrow \quad \ddot{g}_{T2} \leq 0 \\ \lambda_{T2} = -\mu_0\lambda_{N2} & \Rightarrow \quad \ddot{g}_{T2} \geq 0\,. \end{cases}
\end{aligned}
\tag{6.58}
$$

Now we proceed in the same manner as in Section 6.5. For an assumed value of $m = 1$ we rewrite (6.57) together with (6.58) in matrix notation after eliminating the generalized accelerations $(\ddot{x}_1, \ddot{x}_2)$ and the normal forces $(\lambda_{N1}, \lambda_{N2})$ by using (6.55) and (6.56). Corresponding to eq. (6.41) the resulting equations are

$$
\begin{aligned}
&\begin{pmatrix} \ddot{g}_{T1} \\ \ddot{g}_{T2} \end{pmatrix} = \begin{pmatrix} 1 & -1 \\ -1 & 2 \end{pmatrix} \begin{pmatrix} \lambda_{T1} \\ \lambda_{T2} \end{pmatrix} + \begin{pmatrix} F_1 \\ F_2 - F_1 \end{pmatrix} \\
&|\lambda_{T1}| \leq 2\mu_0 g\,; \quad \begin{cases} |\lambda_{T1}| < 2\mu_0 g & \Rightarrow \quad \ddot{g}_{T1} = 0 \\ \lambda_{T1} = +2\mu_0 g & \Rightarrow \quad \ddot{g}_{T1} \leq 0 \\ \lambda_{T1} = -2\mu_0 g & \Rightarrow \quad \ddot{g}_{T1} \geq 0 \end{cases} \\
&|\lambda_{T2}| \leq \mu_0 g\,; \quad \begin{cases} |\lambda_{T2}| < \mu_0 g & \Rightarrow \quad \ddot{g}_{T2} = 0 \\ \lambda_{T2} = +\mu_0 g & \Rightarrow \quad \ddot{g}_{T2} \leq 0 \\ \lambda_{T2} = -\mu_0 g & \Rightarrow \quad \ddot{g}_{T2} \geq 0 \end{cases}
\end{aligned}
\tag{6.59}
$$

which is not yet a LCP formulation. The friction characteristics first have to be decomposed, which can be done according to either Fig. 6.3 or Fig. 6.4. Even for the most compact decomposition with respect to Fig. 6.3, the equations in (6.59) would result in a LCP of order $n = 4$. A graphical solution by drawing the complementary cones as in Section 6.5 is no longer possible. Thus, we solve the problem numerically by a pointwise evaluation of the corresponding sets of eqs. (6.35). The resulting diagram for $\mu_0 = 1$ and $g = 10$ is shown in Fig. 6.9 and has a structure which is similar to the images of the complementarity cones in Fig. 6.7. A total of nine separate areas can be observed which correspond to the nine possible combinations of the states "sticking," "positive sliding," and "negative sliding" for each of the two contacts. These results can be obtained analytically, but numerous calculations are required.

In order to calculate the contact forces and accelerations we choose certain values for F_1 and F_2. For example, let

$$g = 10\,; \quad \mu_0 = 1\,; \quad F_1 = 20\,; \quad F_2 = 10\,. \tag{6.60}$$

With respect to Fig. 6.10 these values correspond to the state where the lower mass starts sliding to the right and the upper mass is sticking. From the contact laws in (6.59) we therefore get

$$\ddot{g}_{T1} \geq 0\,; \quad \lambda_{T1} = -20\,; \quad \ddot{g}_{T2} = 0\,; \quad |\lambda_{T2}| < 10\ . \tag{6.61}$$

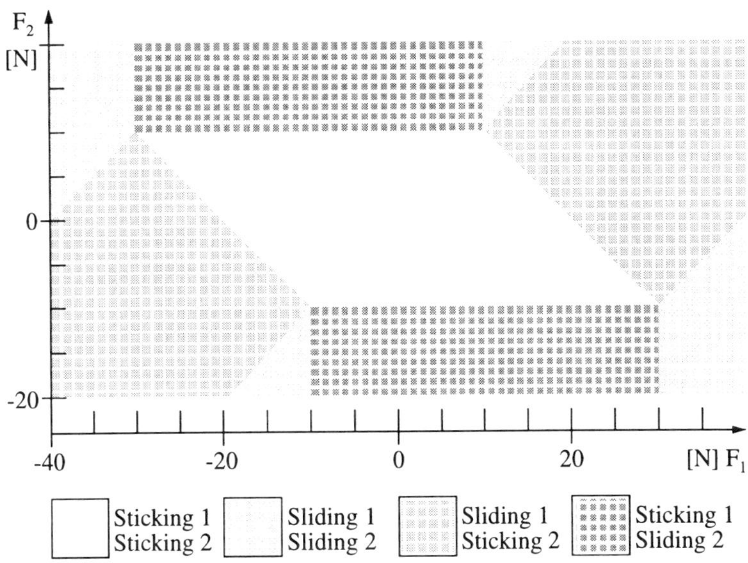

Figure 6.9: Possible State Combinations

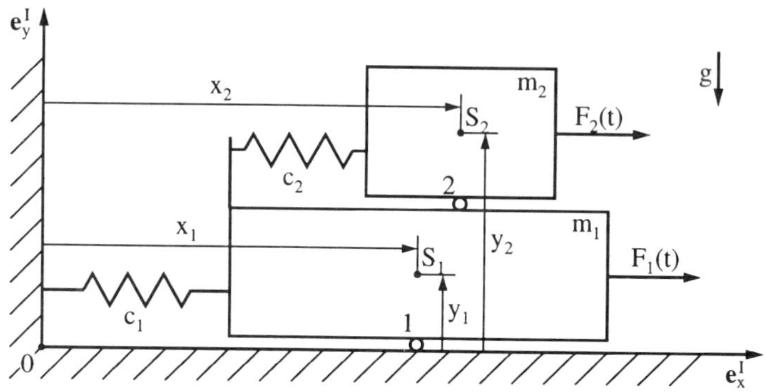

Figure 6.10: A Stick-Slip Oscillator

Inserting the values of λ_{T1} and $\ddot{g}_{T2}$ into the first equation of (6.59) yields, together with (6.60),

$$\begin{pmatrix} \ddot{g}_{T1} \\ 0 \end{pmatrix} = \begin{pmatrix} 1 & -1 \\ -1 & 2 \end{pmatrix} \begin{pmatrix} -20 \\ \lambda_{T2} \end{pmatrix} + \begin{pmatrix} 20 \\ -10 \end{pmatrix} . \tag{6.62}$$

The resulting values of $\ddot{g}_{T1}$ and λ_{T2} are then

$$\ddot{g}_{T1} = 5 \, ; \quad \lambda_{T2} = -5 \tag{6.63}$$

which obviously fulfill the restrictions in (6.61). Finally, from (6.55) we get the absolute accelerations

$$\ddot{x}_1 = 5 \, ; \quad \ddot{x}_2 = 5 \tag{6.64}$$

which must be of the same size in order to enable the sticking of mass 2 onto mass 1.

In Fig. 6.10 the bodies of Fig. 6.8 are connected together and to the environment by two springs with stiffnesses c_1 and c_2. The horizontal forces F_1 and F_2 are now assumed to be functions of time given by

$$F_1(t) = \hat{F}_1 \cos(\omega_1 t + \varphi_1) \, ; \quad F_1(t) = \hat{F}_2 \cos(\omega_2 t + \varphi_2) \, . \tag{6.65}$$

For the two-mass oscillator of Fig. 6.10 the contact situation as discussed above may occur if both masses stop moving at positions with the springs not under stresses. Then the combinatorial problem (6.59) must be solved because no information of the time history of the system can be used to determine the new system state: As mentioned in the introduction of Chapter 5, a vanishing tangential relative velocity generally allows the contact forces to change unsteadily. If we have, for instance, sticking in contact 1 and a

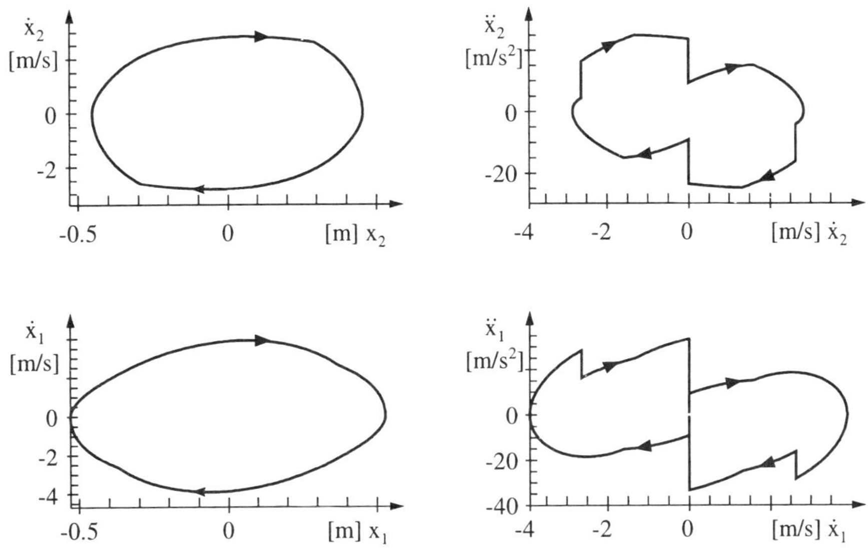

Figure 6.11: Phase Space Portraits

possible end of a sliding period indicated by $\dot{g}_{T2} = 0$ in contact 2, then the new contact state is *not* necessarily sticking in both contacts. Since the contact forces in our example influence each other (see eq. 6.59), a force jump in contact 2 may lead to another force jump in contact 1 and thus to an induced state transition there. The new contact situation depends *only* on the values of the topical tangential forces F_1 and F_2, and may be any of the nine possibilities depicted in Fig. 6.9.

The results of a numerical integration of the equations of motion and evaluation of the state transitions occurring are shown in Fig. 6.11, where the following parameters have been used: $m_1 = m_2 = 1$; $c_1 = 150$; $c_2 = 0$; $\hat{F}_1 = 60$; $\hat{F}_2 = 15$; $\omega_1 = \omega_2 = 2\pi$; $\varphi_1 = 0$; $\varphi_2 = \pi$; $g = 10$; $\mu_1 = \mu_2 = 1$. In the left part of Fig. 6.11 we see the phase space diagrams of the coordinates x_1 and x_2 (Fig. 6.10), in the right part their time derivatives. These curves contain two properties typical for stick-slip systems. Changes in the sliding directions and state transitions to sticking lead to jumps in the contact forces and hence in the accelerations. The transitions from sticking to sliding produce jump discontinuities in the derivatives of the accelerations. Thus, the velocities are continuous but not smooth. Both types of discontinuities, jumps and kinks, can be observed in the right diagrams of Fig. 6.11. They always occur simultaneously in both coordinates, which is due to the coupled contact forces.

7

FRICTIONLESS IMPACTS BY NEWTON'S LAW

In the previous chapters the methods of formulating the equations of constrained multibody systems were presented. The unilateralty of the contact constraints was taken into account on the acceleration level by stating a set of inequalities and complementarity conditions for each of the closed contacts. This formulation was shown to hold during contact as well as for the transition to separation. In a similar manner dry friction was handled. By decomposing the Coulomb friction characteristic into four unilateralty conditions, a formulation was found which contains the sticking state together with all possible transitions to sliding. Furthermore we showed that by using all of the normal and tangential unilateralty conditions together with the dynamics equations, a Linear Complementarity Problem can be formulated. Its solution provides all of the normal and tangential contact forces and relative accelerations. Summarizing these results, the states

separation, contact, sticking, sliding

together with the transitions

sliding $\rightarrow$ *sticking, sticking* $\rightarrow$ *sliding, contact* $\rightarrow$ *separation*

can now be handled in a general manner. Some transitions, however, are still missing: Bodies which are initially separated may come into contact. This process is called an impact and involves unsteady changes in the velocities. Normally, two bodies approach each other with a nonzero relative velocity in the normal direction until the distance between them has been vanished. Then there are several possibilities: The bodies may separate immediately after the collision with a finite positive normal relative velocity, or they may remain in contact. The first case is usually referred to a partly inelastic shock which might be fully elastic if the absolute values of the normal relative velocities before and after the impact are the same. In the second case, the normal relative velocity, which is negative before the impact, jumps to zero and thus enables the bodies to stay together. This type of an impact is called completely inelastic. In both cases, the velocity jumps are enforced by infinite values of the contact forces during the infinitely small time interval of the impact. These impulsive forces are, of course, not restricted to act only in the normal directions. Even in the tangential directions such impulses are transferred and cause velocity jumps there. Depending on the impact law, these phenomena can be taken into account, for example, by the theory of impacts with friction which is the subject of Chapter 8. In this chapter we start the discussion about collisions, but restrict ourself to the frictionless case [31, 43, 70, 72, 73, 74, 93, 94, 95].

7.1 Assumptions and Basic Equations

One of the most frequently used impact laws, Newton's law, relates the normal relative velocities before and after the impact through a kinematic condition. It is usually applied to frictionless systems with only one impact contact and sometimes used together with Coulomb friction in the tangential direction. In addition to the single-contact case there exist some expansions to multiple impacts where both bilateral and unilateral formulations are used. We apply Newton's impact law only in the frictionless case. Some reasons for avoiding the combination of Newton's law in the normal direction and, say, Coulomb's law in the tangential direction are given in Chapter 8. Furthermore, we treat multiple impacts but assume that only compressive impulses are transferred. An example where this assumption fails is given at the end of this chapter. In addition to these restrictions, which are special for this chapter, some general assumptions are made which also will be presupposed in Chapter 8:

- The duration of the impact is "very short."
- The impact can be divided into two phases: the compression phase (index C) and the expansion phase (index E).
- The compression phase starts at time t_A and ends at time t_C. The end of the compression equals the start of the expansion phase. Expansion is finished at time t_E, which is also the end of the impact.
- While the impact takes place all magnitudes of the multibody system for position and orientation as well as all nonimpulsive forces and torques remain constant.
- Wave effects are not taken into account.

In multiple-contact problems there might be one impact only in one of the contacts or several impacts in several contacts simultaneously. The theory presented will cover both possibilities. The locations of impacts are given by the n_G contact points of I_G. For each of them we can write the distance $g_{Ni}(\boldsymbol{q}, t)$ in the normal direction. If one or more of these indicators becomes zero at one time instant t_A and the corresponding relative velocities $\dot{g}_{Ni}$ are less than zero, an impact occurs. The impact contacts are then closed and the unilateral constraints are active. The set of constraints which participate in the impact is then given by

$$I_S^* = \{i \in I_G \mid g_{Ni} = 0\,; \quad \dot{g}_{Ni} \leq 0\} \quad \text{with } n_S^* \text{ elements.}$$

From eq. (5.6) we get the equations of motion for a constrained system without friction:

$$\boldsymbol{M}\ddot{\boldsymbol{q}} - \boldsymbol{h} - \sum_{i \in I_S^*} (\boldsymbol{w}_N \lambda_N)_i = 0 \quad \in \mathbb{R}^f \tag{7.1}$$

where the terms $(\boldsymbol{w}_N \lambda_N)_i$ result from a projection of the normal contact forces into the space of the generalized coordinates. As a second equation we use the relative velocity in the normal direction (eq. 4.38):

$$\dot{g}_{Ni} = \boldsymbol{w}_{Ni}^T \dot{\boldsymbol{q}} + \tilde{w}_{Ni}\,; \quad i \in I_S^*\,. \tag{7.2}$$

Equations (7.1) and (7.2) can be stated in matrix notation:

$$\boldsymbol{M}\ddot{\boldsymbol{q}} - \boldsymbol{h} - \boldsymbol{W}_N \boldsymbol{\lambda}_N = \boldsymbol{0}\,; \quad \dot{\boldsymbol{g}}_N = \boldsymbol{W}_N^T \dot{\boldsymbol{q}} + \tilde{\boldsymbol{w}}_N\,, \tag{7.3}$$

where we will assume independent constraints, rank $\boldsymbol{W}_N = n_S^* \leq f$. Furthermore $\boldsymbol{M} = \boldsymbol{M}(\boldsymbol{q}, t)$, $\boldsymbol{h} = \boldsymbol{h}(\dot{\boldsymbol{q}}, \boldsymbol{q}, t)$ and $\boldsymbol{W}_N = \boldsymbol{W}_N(\boldsymbol{q}, t)$, $\tilde{\boldsymbol{w}}_N = \tilde{\boldsymbol{w}}_N(\boldsymbol{q}, t)$.

Next, we integrate over the time interval of the impact to achieve a representation of the equations of motion on the impulse level. Let t_A and t_E denote the time instances at the beginning and end of the impact, respectively, and let

$$\dot{\boldsymbol{q}}_A = \dot{\boldsymbol{q}}(t_A)\,; \quad \dot{\boldsymbol{q}}_E = \dot{\boldsymbol{q}}(t_E)$$

be the generalized velocities at these instances. The relative velocities in the normal direction are then given by

$$\dot{\boldsymbol{g}}_{NA} = \dot{\boldsymbol{g}}_N(t_A)\,; \quad \dot{\boldsymbol{g}}_{NE} = \dot{\boldsymbol{g}}_N(t_E)\,,$$

and the integration of the dynamics equation (7.3) over the impact yields

$$\lim_{t_E \to t_A} \int_{t_A}^{t_E} (\boldsymbol{M}\ddot{\boldsymbol{q}} - \boldsymbol{h} - \boldsymbol{W}_N \boldsymbol{\lambda}_N)\, dt = \boldsymbol{0}\,. \tag{7.4}$$

During this integration, only terms that can rise to infinity have to be taken into account. The vector $\boldsymbol{h}$ consists of finite nonimpulsive terms and therefore vanishes. Under the assumption of constant displacements we get

$$\boldsymbol{M}\left(\dot{\boldsymbol{q}}_E - \dot{\boldsymbol{q}}_A\right) - \boldsymbol{W}_N \boldsymbol{\Lambda}_N = \boldsymbol{0} \quad \text{with} \quad \boldsymbol{\Lambda}_N = \lim_{t_E \to t_A} \int_{t_A}^{t_E} \boldsymbol{\lambda}_N \, dt \tag{7.5}$$

with $\boldsymbol{\Lambda}_N$ being the impulses transferred by the contacts during the impact. Finally, we state the relative velocities in eqs. (7.3) at the instances t_A and t_E to be

$$\dot{\boldsymbol{g}}_{NA} = \boldsymbol{W}_N^T \dot{\boldsymbol{q}}_A + \tilde{\boldsymbol{w}}_N \, ; \quad \dot{\boldsymbol{g}}_{NE} = \boldsymbol{W}_N^T \dot{\boldsymbol{q}}_E + \tilde{\boldsymbol{w}}_N \, , \tag{7.6}$$

and express (7.6) for convenience as a sum and a difference:

$$\begin{aligned} \dot{\boldsymbol{g}}_{NE} + \dot{\boldsymbol{g}}_{NA} &= \boldsymbol{W}_N^T \left(\dot{\boldsymbol{q}}_E + \dot{\boldsymbol{q}}_A\right) + 2\tilde{\boldsymbol{w}}_N \\ \dot{\boldsymbol{g}}_{NE} - \dot{\boldsymbol{g}}_{NA} &= \boldsymbol{W}_N^T \left(\dot{\boldsymbol{q}}_E - \dot{\boldsymbol{q}}_A\right) . \end{aligned} \tag{7.7}$$

7.2 Newton's Impact Law

After the elimination of the f-vector $(\dot{\boldsymbol{q}}_E - \dot{\boldsymbol{q}}_A)$ with the help of eq. (7.5), the second equation of (7.7) consists of n_S^* relations for the $2n_S^*$ unknowns $(\dot{\boldsymbol{g}}_{NE}, \boldsymbol{\Lambda}_N)$:

$$\dot{\boldsymbol{g}}_{NE} - \dot{\boldsymbol{g}}_{NA} = \boldsymbol{G}_N \boldsymbol{\Lambda}_N \, ; \quad \boldsymbol{G}_N = \boldsymbol{W}_N^T \boldsymbol{M}^{-1} \boldsymbol{W}_N \, , \tag{7.8}$$

thus n_S^* conditions are missing to determine the transferred impulses $\boldsymbol{\Lambda}_N$ and the relative velocities $\dot{\boldsymbol{g}}_{NE}$ at the end of the impact. These missing conditions are called the impact laws of the problem. Here we will use Newton's law, which connects the relative velocities before and after the impact by the relation

$$\dot{\boldsymbol{g}}_{NE} = -\bar{\bar{\boldsymbol{\varepsilon}}}_N \dot{\boldsymbol{g}}_{NA} \tag{7.9}$$

where $\bar{\bar{\boldsymbol{\varepsilon}}}_N$ is a diagonal matrix, $\bar{\bar{\boldsymbol{\varepsilon}}}_N = \operatorname{diag}\{\varepsilon_{Ni}\}$, which contains the n_S^* coefficients of restitution $0 \le \varepsilon_{Ni} \le 1$. The value $\varepsilon_{Ni} = 0$ means a completely inelastic shock where both collision partners remain in contact, and $\varepsilon_{Ni} = 1$ describes fully reversible behavior. Inserting (7.9) into (7.8) yields

$$-\left(\boldsymbol{E} + \bar{\bar{\boldsymbol{\varepsilon}}}_N\right) \dot{\boldsymbol{g}}_{NA} = \boldsymbol{G}_N \boldsymbol{\Lambda}_N \tag{7.10}$$

which determines the transferred impulses $\boldsymbol{\Lambda}_N$:

$$\boldsymbol{\Lambda}_N = -\boldsymbol{G}_N^{-1} \left(\boldsymbol{E} + \bar{\bar{\boldsymbol{\varepsilon}}}_N\right) \dot{\boldsymbol{g}}_{NA} \, . \tag{7.11}$$

If only one contact participates in the impact, then the matrix $\boldsymbol{G}_N^{-1}$ reduces to the scalar $\boldsymbol{G}_N^{-1} = 1/(\boldsymbol{w}_N^T \boldsymbol{M}^{-1} \boldsymbol{w}_N)$. Under these circumstances eq. (7.11) determines the impulses of an equivalent system where the condensed mass $\boldsymbol{G}_N^{-1}$ bounces against a rigid wall. The term $\boldsymbol{G}_N^{-1}$ corresponds to that reduced mass of our multibody system which is effective in the impact direction [99].

After resubstituting the impulses (7.11) into eq. (7.5) we get the generalized velocities $\dot{\boldsymbol{q}}_E$ at the end of the impact:

$$\dot{\boldsymbol{q}}_E = \dot{\boldsymbol{q}}_A - \boldsymbol{M}^{-1} \boldsymbol{W}_N \boldsymbol{G}_N^{-1} \left(\boldsymbol{E} + \bar{\bar{\boldsymbol{\epsilon}}}_N\right) \dot{\boldsymbol{g}}_{NA} , \tag{7.12}$$

where $\dot{\boldsymbol{g}}_{NA}$ is given with eq. (7.6).

Note that the impulses $\boldsymbol{\Lambda}_N$ must act with a compressive magnitude ($\boldsymbol{\Lambda}_N \geq 0$) in the physical sense. But nowhere in this section has this restriction been taken into consideration. Thus, eq. (7.11) must be checked after evaluation in order to overcome this problem. If any of the transferred impulses violates this condition, the corresponding constraint has to be removed from I_S^* and the evaluation of the impact laws has to be repeated by the new and smaller set I_S^*. In the worst case this procedure may result in a combinatorial problem where every imaginable impact configuration has to be evaluated. A more effective and elegant method of handling such problems was introduced in Chapter 6 through the concept of linear complementarity and will be applied on impacts in Chapter 8.

The use of Newton's impact law ensures that no penetration of the bodies after the impact can occur. This is easy to see, since by the construction of I_S^* only contacts with $\dot{g}_{NAi} \leq 0$ are considered. With eq. (7.9) and the property of the coefficients of restitution, $0 \leq \varepsilon_{Ni} \leq 1$, it is obvious that the bodies separate after the impact, $\dot{g}_{NEi} \geq 0$. Contacts, however, which have been removed from I_S^* due to the reasons described above, are excluded from the evaluation of Newton's law and therefore are not expected to have final relative velocities which avoid penetration. Thus, impulses of contacts in I_S^* and relative velocities of contacts which have been removed from I_S^* must be checked with respect to physical correctness. All these difficulties can be handled by a unilateral formulation of the impact laws which will be presented in Chapter 8, applied to the Poisson's impact hypothesis.

7.3 Energy Considerations

The impact law of Newton is always dissipative or energy preserving, depending on the values of the coefficients of restitution. The dissipated energy, lost while the impact happens, can be written as the difference between the kinetic energy T_A at the beginning of the impact and T_E at the end, because changes in the potential energy do not occur due to the premise of constant displacements.

$$
\begin{aligned}
T_E - T_A &= \tfrac{1}{2}\dot{\boldsymbol{q}}_E^T \boldsymbol{M} \dot{\boldsymbol{q}}_E - \tfrac{1}{2}\dot{\boldsymbol{q}}_A^T \boldsymbol{M} \dot{\boldsymbol{q}}_A \\
&= \tfrac{1}{2}\left(\dot{\boldsymbol{q}}_E + \dot{\boldsymbol{q}}_A\right)^T \boldsymbol{M} \left(\dot{\boldsymbol{q}}_E - \dot{\boldsymbol{q}}_A\right) \\
&\overset{(7.5)}{=} \tfrac{1}{2}\left(\dot{\boldsymbol{q}}_E + \dot{\boldsymbol{q}}_A\right)^T \boldsymbol{W}_N \boldsymbol{\Lambda}_N = \tfrac{1}{2}\boldsymbol{\Lambda}_N^T \boldsymbol{W}_N^T \left(\dot{\boldsymbol{q}}_E + \dot{\boldsymbol{q}}_A\right) \\
&\overset{(7.7)}{=} \tfrac{1}{2}\boldsymbol{\Lambda}_N^T \left(\dot{\boldsymbol{g}}_{NA} + \dot{\boldsymbol{g}}_{NE}\right) - \boldsymbol{\Lambda}_N^T \tilde{\boldsymbol{w}}_N \,.
\end{aligned}
\tag{7.13}
$$

The discussion of energy dissipation only makes sense if no external kinematic excitation is considered. This excitation is represented by the vector $\tilde{\boldsymbol{w}}_N$ in the kinematic equation (7.6). Thus we set $\tilde{\boldsymbol{w}}_N = 0$ in eq. (7.13) and get

$$
\begin{aligned}
T_E - T_A &= \tfrac{1}{2}\boldsymbol{\Lambda}_N^T \left(\dot{\boldsymbol{g}}_{NA} + \dot{\boldsymbol{g}}_{NE}\right) \\
&\overset{(7.9)}{=} \tfrac{1}{2}\boldsymbol{\Lambda}_N^T \left(\boldsymbol{E} - \overline{\overline{\boldsymbol{\epsilon}}}_N\right) \dot{\boldsymbol{g}}_{NA} \\
&\overset{(7.11)}{=} -\tfrac{1}{2}\dot{\boldsymbol{g}}_{NA}^T \left(\boldsymbol{E} + \overline{\overline{\boldsymbol{\epsilon}}}_N\right) \boldsymbol{G}_N^{-1} \left(\boldsymbol{E} - \overline{\overline{\boldsymbol{\epsilon}}}_N\right) \dot{\boldsymbol{g}}_{NA} \\
&= -\tfrac{1}{2}\dot{\boldsymbol{g}}_{NA}^T \boldsymbol{G}_N^{-1} \dot{\boldsymbol{g}}_{NA} + \tfrac{1}{2}\dot{\boldsymbol{g}}_{NA}^T \overline{\overline{\boldsymbol{\epsilon}}}_N \boldsymbol{G}_N^{-1} \overline{\overline{\boldsymbol{\epsilon}}}_N \dot{\boldsymbol{g}}_{NA}
\end{aligned}
\tag{7.14}
$$

Since $\Lambda_{Ni} \geq 0$ and $\dot{g}_{NAi} \leq 0$ by assumption, and $\overline{\overline{\boldsymbol{\epsilon}}}_N$ is a diagonal matrix consisting of elements $0 \leq \varepsilon_{Ni} \leq 1$, it follows from the second line in (7.14) that

$$
T_E - T_A \leq 0 \,. \tag{7.15}
$$

From (7.14) it follows readily that maximum dissipation is achieved at completely inelastic impacts, $\overline{\overline{\boldsymbol{\epsilon}}}_N = 0 \Rightarrow T_E - T_A = -\frac{1}{2}\dot{\boldsymbol{g}}_{NA}^T \boldsymbol{G}_N^{-1} \dot{\boldsymbol{g}}_{NA}$, and energy conservation is only possible if $\overline{\overline{\boldsymbol{\epsilon}}}_N = \boldsymbol{E} \Rightarrow T_E - T_A = 0$. This corresponds to completely elastic impacts in all of the contacts.

7.4 Example: Impact between Two Point Masses

In most standart textbooks this example is presented in connection with impacts [54]. Two point masses m_1 and m_2 are moving translationally with absolute velocities $\dot{x}_1$ and $\dot{x}_2$ until they hit each other (see Fig. 7.1). We calculate the absolute velocities after the collision by using Newton's law, which depend on the coefficient of restitution.
With respect to Fig. 7.1 the relative velocity between the masses is given by

$$
\dot{g}_N = \dot{x}_2 - \dot{x}_1 = \begin{pmatrix} -1 & 1 \end{pmatrix} \begin{pmatrix} \dot{x}_1 \\ \dot{x}_2 \end{pmatrix} \tag{7.16}
$$

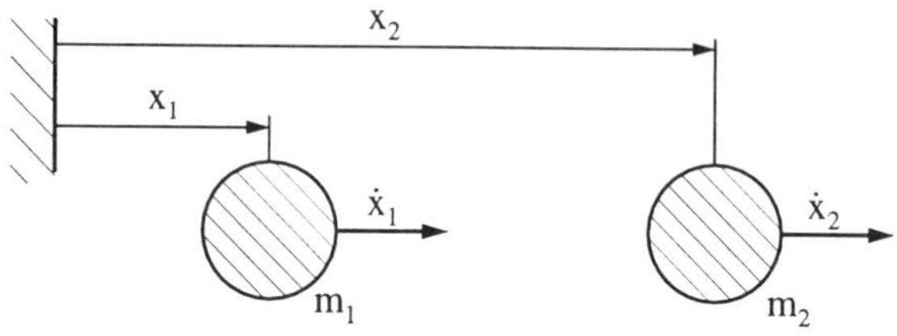

Figure 7.1: Impact between Two Point Masses

from which we identify the constraint vector $\boldsymbol{w}_N$ and the generalized coordinates $\boldsymbol{q}$. Comparing eq. (7.16) and eq. (4.38), we get

$$\dot{\boldsymbol{q}} = \begin{pmatrix} \dot{x}_1 \\ \dot{x}_2 \end{pmatrix} ; \quad \boldsymbol{w}_N = \begin{pmatrix} -1 \\ 1 \end{pmatrix} . \tag{7.17}$$

With respect to eq. (7.5) the balance of the impulses is formulated as

$$\begin{pmatrix} m_1 & 0 \\ 0 & m_1 \end{pmatrix} \begin{pmatrix} \dot{x}_{E1} - \dot{x}_{A1} \\ \dot{x}_{E2} - \dot{x}_{A2} \end{pmatrix} - \begin{pmatrix} -1 \\ 1 \end{pmatrix} \Lambda_N = 0 , \tag{7.18}$$

where the subscripts E and A denote the end and the beginning of the impact, and Λ_N describes the normal impulse which is transferred between the masses during the collision. Now we express the relative velocities before and after the impact as a difference by using the second equation in (7.7):

$$\dot{g}_{NE} - \dot{g}_{NA} = \begin{pmatrix} -1 & 1 \end{pmatrix} \begin{pmatrix} \dot{x}_{E1} - \dot{x}_{A1} \\ \dot{x}_{E2} - \dot{x}_{A2} \end{pmatrix} ; \tag{7.19}$$

we eliminate the terms $\dot{x}_{Ei} - \dot{x}_{Ai}$ with the help of eq. (7.18). The resulting relation is then

$$\begin{aligned} \dot{g}_{NE} - \dot{g}_{NA} &= \begin{pmatrix} -1 & 1 \end{pmatrix} \begin{pmatrix} \frac{1}{m_1} & 0 \\ 0 & \frac{1}{m_2} \end{pmatrix} \begin{pmatrix} -1 \\ 1 \end{pmatrix} \Lambda_N \\ &= \frac{m_1 + m_2}{m_1 m_2} \Lambda_N . \end{aligned} \tag{7.20}$$

By using Newton's impact law from (7.9), we express the relative velocity at the end of the impact in terms of the relative velocity before the impact:

$$\dot{g}_{NE} = -\varepsilon \dot{g}_{NA} \tag{7.21}$$

and we put it into eq. (7.20), which yields the value of the transferred impulse:

$$\Lambda_N = -(1+\varepsilon) \frac{m_1 m_2}{m_1 + m_2} \dot{g}_{NA} . \tag{7.22}$$

Since the relative velocity before the impact expresses an approaching process, $\dot{g}_{NA} < 0$, the transferred impulse acts with a compressive magnitude, $\Lambda_N > 0$. Substituting this impulse into eq. (7.18), and expressing $\dot{g}_{NA}$ in terms of $(\dot{x}_{A1}, \dot{x}_{A2})$, we solve eq. (7.18) with respect to $\dot{x}_{Ei}$:

$$\begin{aligned} \dot{x}_{E1} &= \dot{x}_{A1} + \frac{m_2}{m_1 + m_2}(1+\varepsilon)(\dot{x}_{A2} - \dot{x}_{A1}) \\ \dot{x}_{E2} &= \dot{x}_{A2} - \frac{m_1}{m_1 + m_2}(1+\varepsilon)(\dot{x}_{A2} - \dot{x}_{A1}) \end{aligned} \tag{7.23}$$

which are the absolute velocities after the impact. If $m_1 = m_2$, eq. (7.23) reduces for a completely elastic impact ($\varepsilon = 1$) to

$$\dot{x}_{E1} = \dot{x}_{A2}\,; \quad \dot{x}_{E2} = \dot{x}_{A1}\,. \tag{7.24}$$

In this case, the velocities of both bodies are exchanged by the collision. A completely inelastic impact ($\varepsilon = 0$) results in

$$\dot{x}_{E1} = \dot{x}_{E2} = \frac{1}{2}\left(\dot{x}_{A1} + \dot{x}_{A2}\right), \tag{7.25}$$

where both masses move together after the collision with the same absolute velocity.

7.5 Example: Double Impact on a Rod

In this example we examine a freely moving rod (length $2l$, mass m, moment of inertia $J_S = \frac{1}{3}ml^2$) which hits two obstacles simultaneously [21]. Figure 7.2 shows the general arrangement of the rod and the obstacles from which we derive the equations of motion and the distance functions:

$$m\ddot{x} = 0\,; \quad m\ddot{y} = -mg\,; \quad \frac{1}{3}ml^2\ddot{\varphi} = 0 \tag{7.26}$$

$$\begin{aligned} g_{N1} &= y\cos\varphi + (a - x)\sin\varphi \\ g_{N2} &= y\cos\varphi + (b - x)\sin\varphi\,. \end{aligned} \tag{7.27}$$

Defining a set of generalized coordinates $\boldsymbol{q}$, we write the mass matrix $\boldsymbol{M}$ and the force vector $\boldsymbol{h}$ by (7.26)

$$\boldsymbol{q} = \begin{pmatrix} x \\ y \\ \varphi \end{pmatrix}; \quad \boldsymbol{M} = \begin{pmatrix} m & 0 & 0 \\ 0 & m & 0 \\ 0 & 0 & \frac{1}{3}ml^2 \end{pmatrix}; \quad \boldsymbol{h} = \begin{pmatrix} 0 \\ -mg \\ 0 \end{pmatrix} \tag{7.28}$$

and the normal relative velocities are obtained from differentiating eq. (7.27):

$$\underbrace{\begin{pmatrix} \dot{g}_{N1} \\ \dot{g}_{N2} \end{pmatrix}}_{\dot{\boldsymbol{g}}_N} = \underbrace{\begin{pmatrix} -\sin\varphi & \cos\varphi & (a-x)\cos\varphi - y\sin\varphi \\ -\sin\varphi & \cos\varphi & (b-x)\cos\varphi - y\sin\varphi \end{pmatrix}}_{\boldsymbol{W}_N^T} \underbrace{\begin{pmatrix} \dot{x} \\ \dot{y} \\ \dot{\varphi} \end{pmatrix}}_{\dot{\boldsymbol{q}}}. \tag{7.29}$$

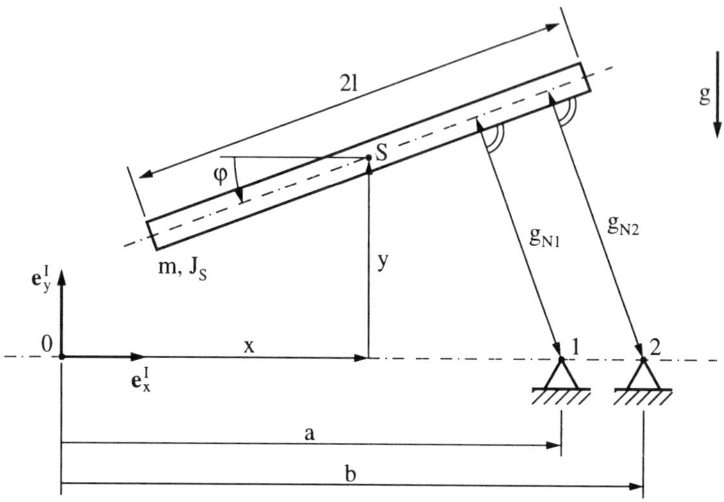

Figure 7.2: Impact of a Rod against Two Obstructions

In order to enable a simultaneous impact between the rod and the obstacles we choose the displacements of the rod as $x = y = \varphi = 0$ and demand $-l \leq a, b \leq l$. The resulting geometric configuration is depicted in Fig. 7.3 and corresponds to a horizontal rod touching both obstacles. The relative velocities before the impact are still undetermined but will be specified later in the example.

For the chosen impact configuration we state the balance of the impulses according to eq. (7.5) by using the matrices $\boldsymbol{M}$ and $\boldsymbol{W}_N$ from (7.28) and (7.29), which leads to

$$\begin{pmatrix} m & 0 & 0 \\ 0 & m & 0 \\ 0 & 0 & \frac{1}{3}ml^2 \end{pmatrix} \begin{pmatrix} \dot{x}_E - \dot{x}_A \\ \dot{y}_E - \dot{y}_A \\ \dot{\varphi}_E - \dot{\varphi}_A \end{pmatrix} - \begin{pmatrix} 0 & 0 \\ 1 & 1 \\ a & b \end{pmatrix} \begin{pmatrix} \Lambda_{N1} \\ \Lambda_{N2} \end{pmatrix} = 0 . \tag{7.30}$$

Since $x = y = \varphi = 0$, the relative velocities (7.29) at the beginning A and the end E of the impact can now be written as

$$\begin{aligned} \begin{pmatrix} \dot{g}_{NA1} \\ \dot{g}_{NA2} \end{pmatrix} &= \begin{pmatrix} 0 & 1 & a \\ 0 & 1 & b \end{pmatrix} \begin{pmatrix} \dot{x}_A \\ \dot{y}_A \\ \dot{\varphi}_A \end{pmatrix} \\ \begin{pmatrix} \dot{g}_{NE1} \\ \dot{g}_{NE2} \end{pmatrix} &= \begin{pmatrix} 0 & 1 & a \\ 0 & 1 & b \end{pmatrix} \begin{pmatrix} \dot{x}_E \\ \dot{y}_E \\ \dot{\varphi}_E \end{pmatrix} . \end{aligned} \tag{7.31}$$

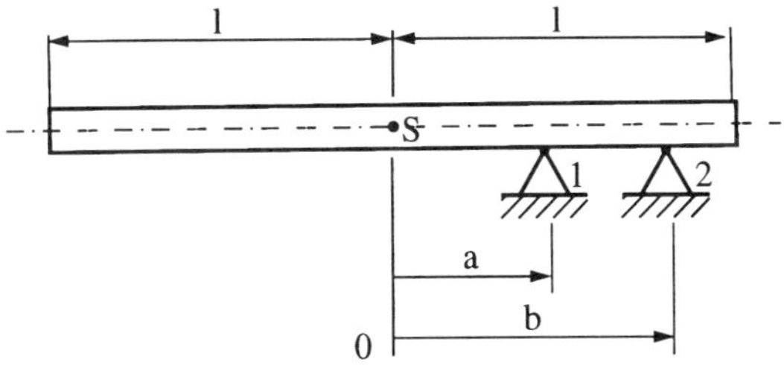

Figure 7.3: Special Impact Configuration

Finally we have to state the impact law with respect to eq. (7.9). In this example we assume a completely inelastic impact at both contacts, $\varepsilon_{N1} = \varepsilon_{N2} = 0$, which results in vanishing relative velocities after the collision:

$$\begin{pmatrix} \dot{g}_{NE1} \\ \dot{g}_{NE2} \end{pmatrix} = 0\,. \tag{7.32}$$

The relations (7.30)–(7.32) provide a complete set of equations for the impact process. After the choice of the initial velocities $\dot{x}_A, \dot{y}_A, \dot{\varphi}_A$, the general elimination process described in eqs. (7.5)–(7.12) could be applied. In this example, however, the results are obtained much faster by a direct approach. First we choose the velocities before the impact as

$$\dot{x}_A = \dot{\varphi}_A = 0\,; \quad \dot{y}_A < 0 \tag{7.33}$$

which describes a rod which moves vertically against the obstacles. With these values, the relative velocities before the impacts result from the left equation in (7.31):

$$\dot{g}_{NA1} = \dot{g}_{NA2} = \dot{y}_A\,. \tag{7.34}$$

The relative velocities after the impact are prescribed by Newton's law and are already known (see eq. 7.32):

$$\dot{g}_{NE1} = \dot{g}_{NE2} = 0\,. \tag{7.35}$$

The first equation in (7.30) is not affected by any impulse. With $\dot{x}_A = 0$ from (7.32), the horizontal velocity after the impact clearly is $\dot{x}_E = 0$. With $\dot{g}_{Ni} = 0$ from (7.35) the remaining values of the generalized velocities are obtained by the right equation in (7.31). For $a \neq b$ we get

$$\dot{x}_E = \dot{y}_E = \dot{\varphi}_E = 0\,. \tag{7.36}$$

All velocities have now been determined. From eq. (7.36) we see that the rod is not moving after the completely inelastic shocks against the obstacles. It keeps contact with both of them. Finally we calculate the transferred impulses by using the last two equations in (7.30). With the help of (7.33) and (7.36) we can write them as

$$-m\dot{y}_A - \Lambda_{N1} - \Lambda_{N2} = 0\,; \quad -a\Lambda_{N1} - b\Lambda_{N2} = 0 \tag{7.37}$$

which leads to the values

$$\Lambda_{N1} = -\frac{1}{1-\frac{a}{b}}\, m\dot{y}_A\,; \quad \Lambda_{N2} = -\frac{1}{1-\frac{b}{a}}\, m\dot{y}_A\,. \tag{7.38}$$

For $b = l$, obstacle 2 is resting at the right end of the rod. The position of obstacle 1 depends on the values of a. In the following we distinguish the cases $-l \leq a \leq 0$ where obstacle 1 lies at the left side with respect to the rod's center of mass, and $0 < a < l$ where both obstacles are located on the right half (see Fig. 7.4).

With $\dot{y}_A$ less than zero (see eq. 7.33), both transferred impulses are greater than zero in Case I, which corresponds to a compressive behavior. In Case II however, the impulse acting on obstacle 2 shows negative values, which is physically wrong: An impact configuration as in the right part of Fig. 7.4 cannot lead to the final velocities in eq. (7.36). In that case obstacle 2 is superfluous and must not be included in the evaluation of the impact equations. The whole (compressive) impulse is transferred at contact 1. In this example the decision of which of the contacts has to be regarded in the evaluation of the impact laws can be made very easily through physical understanding of the mechanical system. More complex models, however, demand an extended formulation of the impact equations which automatically check whether impulses are transferred, because the occurring impact configurations may become extremely complicated due to large numbers of simultaneously arising shocks and kinematical couplings between them. If Newton's law is used in the bilateral manner described above, one has to carefully verify the assumption of only compressive impulses. In Chapter 8 a unilateral formulation of

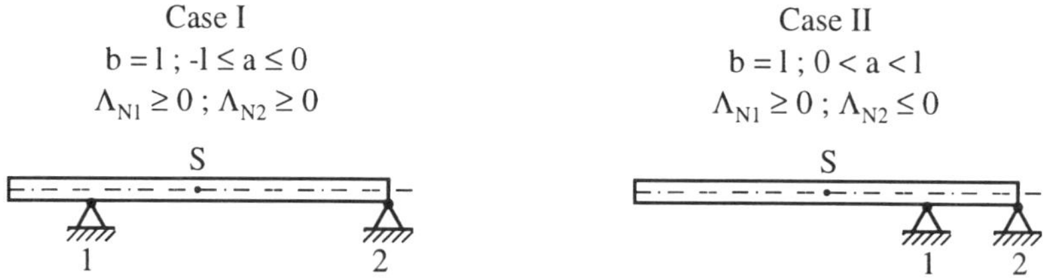

Figure 7.4: Transferred Impulses

the impact equations is presented where contacts which would transfer negative impulses are automatically removed. We will continue the discussion of this example in Section 8.11 with a numerical evaluation, and will present the correct results even for Case II in Fig. 7.4.

8

IMPACTS WITH FRICTION BY POISSON'S LAW*

Usually an impact in the normal direction is modeled by the hypotheses of Newton or Poisson. Both approaches make use of a coefficient of restitution ε, which is defined in the first case as the ratio of the relative velocities after and before the impact, $\varepsilon = -\dot{g}_{NE}/\dot{g}_{NA}$ (cp. eq. 7.9), and in the second case as the ratio of the normal impulses during expansion and compression, $\varepsilon = \Lambda_{NE}/\Lambda_{NC}$. A common method of regarding friction is given by the law of Coulomb (see Chapter 6). The tangential forces are bounded by the normal forces, $|\lambda_T| \leq \mu\lambda_N$, $\lambda_N \geq 0$, and allow the following states: If $|\lambda_T| < \mu\lambda_N$, then the system is sticking, $\dot{g}_T = 0$. If $|\lambda_T| = \mu\lambda_N$, then the system may slide with a relative velocity in the opposite direction of λ_T, $\dot{g}_T = -\alpha\lambda_T$, $\alpha \geq 0$. Impacts with friction demand a combination of the laws of Newton, Poisson and Coulomb, which result immediately in a mismatch in the dimensions of the variables in the governing equations. Velocities, forces and impulses have to be treated simultaneously and are, in addition, partially restricted.

Generally, two ways of solving the problem are possible. One could formulate all the equations on the force level, which may be achieved, for example, by a time-scaling method, and may assume constant displacements during the time interval of the collision. Coulomb's law could then be evaluated by the methods of Chapter 6, which would imply stick-slip transitions and therefore unsteady changes in the normal and tangential forces. Unfortunately, these changes could lead to premature separations of some of the contacts under the common assumption of a solely compressive character of the normal forces ($\lambda_N \geq 0$). Furthermore, no general statements about the admissible values of the accelerations in the normal directions could be made due to the physical fact that every rigid constraint becomes elastic in the presence of a shock. Therefore, this method seems to be applicable only for single shocks and would lead, in the case of multiple collisions, to a time-consuming evaluation which needs additional assumptions about the end of the impacts.

The second approach consists of an integration over the time interval of the impact and leads to a description on the impulse level. Problems arise from the fact that no representation of the friction impulses can be stated, which is equivalent to Coulomb's law. Possible stick–slip transitions during the collision with reversed sliding prevent an analytical general integration of the tangential forces over the impact interval.

In the literature both approaches are used by many authors and are evaluated by different methods. In the following a short survey about the wide class of publications in this area is given in order to highlight the main differences of the presented impact laws. Brach [4, 5] considers only single collisions and formulates the impact equations on the impulse level by using Newton's law. Conditions for sticking or reversed sliding after the impact are derived from the kinetic energy by requiring an always dissipative impact law. This method is restricted for single impacts because for multiple collisions such conditions cannot be derived from one scalar equation without additional assumptions. In [56, 96, 97] Wang and Mason apply a time-scaling method and solve the impact equations on the force level by using Poisson's and Coulomb's laws for a single contact. A classification of the possible impact configurations is given, which also includes the shocks without collisions. This approach coincides with the Newton-Euler equations during the impact under the assumption of constant displacements. The treatment of multiple impacts, however, regards additional assumptions and may be time consuming. In Chapter 7 we treated multiple frictionless impacts on the impulse level using Newton's law and assumed that impulses are transferred at each of the contacts. This replaced the unilateral character of the constraints in the normal direction by a bilateral formulation and led to a set of linear equations for the relative velocities after the impact. Jean and Moreau [33, 34, 62, 63, 64] reformulate Newton's law in a unilateral manner for multiple impacts with friction and introduce a nearly Coulomb friction element on the impulse level. This is an impressive approach since a very compact formulation has been developed, which includes all of the physical main effects with respect to friction and in the case of completely inelastic impacts it is fully compatible to unilaterally constrained motion with friction. This method seems to be one of the most promising concepts in dealing with multiple impacts and is used as a basis in the physical sense for what follows. Although we adopt their model of completely inelastic impacts, we treat partly elastic impacts using a modification of Poisson's law, due to some missing physical effects in Newton's hypothesis.

Even in the frictionless case both concepts, Newton and Poisson, are different for multiple impacts. Generally, Newton's law prescribes the relative velocity in the normal direction at the end of the impact by using the fixed proportion $\varepsilon = -\dot{g}_{NE}/\dot{g}_{NA}$; thus, tangential impulses cannot influence this direction, however strong they may be. In comparison, Poisson's hypothesis allows an energy transfer between the normal and tangential directions, which is more realistic in the mechanical sense. Thus, an impact model based on Poisson's hypothesis will be presented, where the absolute values of the tangential impulses are bounded by the frictional law of Coulomb. It will be proven that, under certain conditions, this model is dissipative or energy preserving and coincides with the results given by applying a time-scaling method and using the Newton-Euler equations during the impacts. The model also contains the special effect of the impacts without collisions and is applicable even if dependent constraints are present. It also includes the case when no impulses are transferred at nonzero approaching velocities and handles induced separations of existing contacts. Combined with the theory of unilaterally constrained motion of Chapter 6, locking effects in the static and dynamic sense can be treated. The evaluation of the problem is done by solving one set of complementarity conditions during compression and a nearly identical set of equations during expansion.

8.1 Assumptions and Basic Equations

For the formulation of the impacts the same assumptions are made as in Chapter 7, but we now allow friction at the contact points and take into account all active unilateral constraints, which means constraints that are elements of I_S in eq. (5.5):

$$I_S = \{i \in I_G \mid g_{Ni} = 0\} \quad \text{with } n_S \text{ elements.} \tag{8.1}$$

It is noteworthy that eq. (8.1) contains all the sliding and sticking continuous-contact constraints ($\dot{g}_{Ni} = 0$) as well as the impact contacts ($\dot{g}_{Ni} < 0$). This enables us to examine whether a contact separates under the influence of an impact at a different location in the multibody system. We start with the derivation of the impact equations as in Chapter 7 by the dynamical and kinematical equations (5.6) and (4.38), respectively:

$$\begin{aligned} &\boldsymbol{M}\ddot{\boldsymbol{q}} - \boldsymbol{h} - \sum_{i \in I_S} (\boldsymbol{w}_N \lambda_N + \boldsymbol{w}_T \lambda_T)_i = \mathbf{0} \\ &\dot{g}_{Ni} = \boldsymbol{w}_{Ni}^T \dot{\boldsymbol{q}} + \tilde{w}_{Ni}\,; \quad i \in I_S \\ &\dot{g}_{Ti} = \boldsymbol{w}_{Ti}^T \dot{\boldsymbol{q}} + \tilde{w}_{Ti}\,; \quad i \in I_S \end{aligned} \tag{8.2}$$

We write them in matrix notation:

$$\boldsymbol{M}\ddot{\boldsymbol{q}} - \boldsymbol{h} - (\boldsymbol{W}_N \boldsymbol{W}_T) \begin{pmatrix} \boldsymbol{\lambda}_N \\ \boldsymbol{\lambda}_T \end{pmatrix} = \mathbf{0} \tag{8.3}$$

$$\begin{pmatrix} \dot{\boldsymbol{g}}_N \\ \dot{\boldsymbol{g}}_T \end{pmatrix} = \begin{pmatrix} \boldsymbol{W}_N^T \\ \boldsymbol{W}_T^T \end{pmatrix} \dot{\boldsymbol{q}} + \begin{pmatrix} \tilde{\boldsymbol{w}}_N \\ \tilde{\boldsymbol{w}}_T \end{pmatrix} \tag{8.4}$$

and we integrate, as in eqs. (7.4, 7.5), over the impact phases of compression and expansion:

$$\begin{aligned} \boldsymbol{M}\left(\dot{\boldsymbol{q}}_C-\dot{\boldsymbol{q}}_A\right)-\left(\boldsymbol{W}_N \boldsymbol{W}_T\right)\begin{pmatrix}\boldsymbol{\Lambda}_{NC}\\ \boldsymbol{\Lambda}_{TC}\end{pmatrix} &= \mathbf{0} \\ \boldsymbol{M}\left(\dot{\boldsymbol{q}}_E-\dot{\boldsymbol{q}}_C\right)-\left(\boldsymbol{W}_N \boldsymbol{W}_T\right)\begin{pmatrix}\boldsymbol{\Lambda}_{NE}\\ \boldsymbol{\Lambda}_{TE}\end{pmatrix} &= \mathbf{0}\,. \end{aligned} \tag{8.5}$$

Here $\boldsymbol{\Lambda}_{NC}, \boldsymbol{\Lambda}_{TC}$ are the impulses in the normal and tangential directions which are transferred during compression, and $\boldsymbol{\Lambda}_{NE}, \boldsymbol{\Lambda}_{TE}$ are those of expansion. Remembering that $\dot{\boldsymbol{q}}_A = \dot{\boldsymbol{q}}(t_A)$, $\dot{\boldsymbol{q}}_C = \dot{\boldsymbol{q}}(t_C)$, $\dot{\boldsymbol{q}}_E = \dot{\boldsymbol{q}}(t_E)$, we express the relative velocities (eq. 8.4) at these instances as

$$\begin{aligned} \begin{pmatrix}\dot{\boldsymbol{g}}_{NA}\\ \dot{\boldsymbol{g}}_{TA}\end{pmatrix} &= \begin{pmatrix}\boldsymbol{W}_N^T\\ \boldsymbol{W}_T^T\end{pmatrix}\dot{\boldsymbol{q}}_A + \begin{pmatrix}\tilde{\boldsymbol{w}}_N\\ \tilde{\boldsymbol{w}}_T\end{pmatrix} \\ \begin{pmatrix}\dot{\boldsymbol{g}}_{NE}\\ \dot{\boldsymbol{g}}_{TE}\end{pmatrix} &= \begin{pmatrix}\boldsymbol{W}_N^T\\ \boldsymbol{W}_T^T\end{pmatrix}\dot{\boldsymbol{q}}_E + \begin{pmatrix}\tilde{\boldsymbol{w}}_N\\ \tilde{\boldsymbol{w}}_T\end{pmatrix} \\ \begin{pmatrix}\dot{\boldsymbol{g}}_{NC}\\ \dot{\boldsymbol{g}}_{TC}\end{pmatrix} &= \begin{pmatrix}\boldsymbol{W}_N^T\\ \boldsymbol{W}_T^T\end{pmatrix}\dot{\boldsymbol{q}}_C + \begin{pmatrix}\tilde{\boldsymbol{w}}_N\\ \tilde{\boldsymbol{w}}_T\end{pmatrix}. \end{aligned} \tag{8.6}$$

Equations (8.5) and (8.6) describe the dynamic and kinematic connections during the phases of compression and expansion and must be completed by the frictional impact laws.

8.2 Phase of Compression

For establishing the relations of the impact laws during compression we consider the difference of the relative velocities in (8.6) together with the corresponding dynamics equation (8.5):

$$\begin{aligned} \boldsymbol{M}\left(\dot{\boldsymbol{q}}_C-\dot{\boldsymbol{q}}_A\right) &= \left(\boldsymbol{W}_N \boldsymbol{W}_T\right)\begin{pmatrix}\boldsymbol{\Lambda}_{NC}\\ \boldsymbol{\Lambda}_{TC}\end{pmatrix}, \\ \begin{pmatrix}\dot{\boldsymbol{g}}_{NC}\\ \dot{\boldsymbol{g}}_{TC}\end{pmatrix} &= \begin{pmatrix}\boldsymbol{W}_N^T\\ \boldsymbol{W}_T^T\end{pmatrix}\left(\dot{\boldsymbol{q}}_C-\dot{\boldsymbol{q}}_A\right) + \begin{pmatrix}\dot{\boldsymbol{g}}_{NA}\\ \dot{\boldsymbol{g}}_{TA}\end{pmatrix}. \end{aligned} \tag{8.7}$$

After elimination of the f-vector $(\dot{\boldsymbol{q}}_C-\dot{\boldsymbol{q}}_A)$ eqs. (8.7) consist of $2n_S$ equations for the $4n_S$ unknowns $(\dot{\boldsymbol{g}}_{NC}, \dot{\boldsymbol{g}}_{TC}, \boldsymbol{\Lambda}_{NC}, \boldsymbol{\Lambda}_{TC})$. Thus, $2n_S$ conditions must yet be formulated in order to determine the transferred impulses and the relative velocities at the end of compression.

The *normal impulse* of compression results from an integration of the normal force over the phase of compression:

$$\Lambda_{NCi} = \lim_{t_C \to t_A} \int_{t_A}^{t_C} \lambda_{Ni}\, dt\,, \tag{8.8}$$

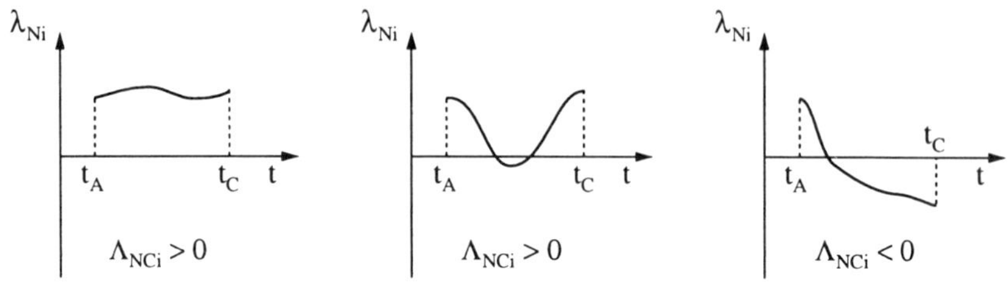

Figure 8.1: Normal Forces and Impulses

where due to the unilateral character of the contact constraint only compressive forces are possible:

$$\lambda_{Ni}(t) \geq 0 \quad \forall\ t \in [t_A, t_C]\ . \tag{8.9}$$

Thus, integrating (eq. 8.8) the normal forces with the property (8.9) results unambiguously in nonnegative values of the normal impulses:

$$\Lambda_{NCi} \geq 0\,, \tag{8.10}$$

but not vice versa: From a given impulse Λ_{NCi} being greater than zero we *cannot* conclude that the normal force λ_{Ni} acts in a compressive manner for all times within the interval $[t_A, t_C]$. There *might* be regions of tension (see Fig. 8.1). On the other hand, a negative impulse Λ_{NCi} *always* requires normal forces less than zero. A restriction of the admissible values of the normal impulse to $\Lambda_{NCi} \geq 0$ seems to be reasonable, because at least the case of only positive contact forces is contained, whereas the obviously wrong situation, resulting from values $\Lambda_{NCi} < 0$, can be excluded.

During the impact, the accelerations in the normal directions $\ddot{\boldsymbol{g}}_N$ are *not* restricted to any sign. This is one of the main differences between unilaterally constrained motion (Chapter 6) and impact behavior. It results from the fact that due to the infinitely strong impact forces each of the contact zones becomes elastic. The assumption of rigid constraints must be dropped. Anyway, we are able to demand that at the end of compression the approaching process of the bodies has to be completed. Thus, negative values of the relative velocities are forbidden:

$$\dot{g}_{NCi} \geq 0\,. \tag{8.11}$$

One additional relation which connects (8.10) and (8.11) can be stated from the physical behavior of the impact. If an impulse is transferred ($\Lambda_{NCi} > 0$), then the corresponding contact participates in the impact and the end of compression is given by $\dot{g}_{NCi} = 0$. If no impulse is transferred ($\Lambda_{NCi} = 0$),

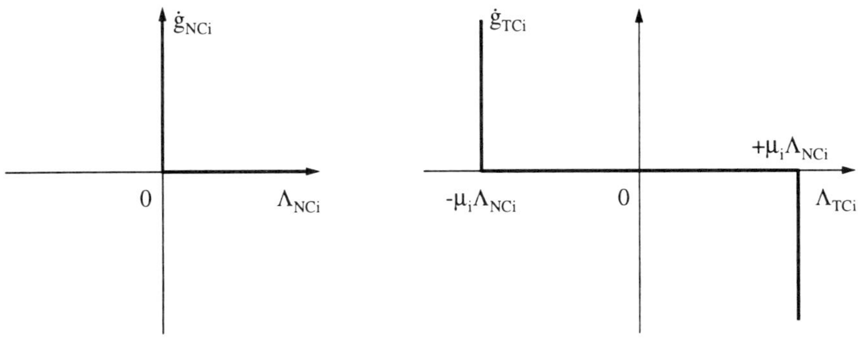

Figure 8.2: Impact Law for Compression

then the corresponding constraint is superfluous. Thus, we allow velocities $\dot{g}_{NCi} \geq 0$. This behaviour can be completely expressed by the single complementarity condition

$$\Lambda_{NCi}\, \dot{g}_{NCi} = 0\,, \tag{8.12}$$

and the impact law in the normal direction for compression is therefore given by the n_S conditions (8.10, 8.11, 8.12)

$$\Lambda_{NCi} \geq 0\,; \quad \dot{g}_{NCi} \geq 0\,; \quad \Lambda_{NCi}\, \dot{g}_{NCi} = 0\,; \quad i \in I_S \tag{8.13}$$

whose characteristics are shown in the left diagram of Fig. 8.2.

The bounds of the *tangential impulse* can be derived similarly to eq. (8.8) by integration:

$$\Lambda_{TCi} = \lim_{t_C \to t_A} \int_{t_A}^{t_C} \lambda_{Ti}\, dt\,. \tag{8.14}$$

Applying Coulomb's laws of friction we get restrictions similar to eq. (6.10):

$$|\lambda_{Ti}(t)| \leq \mu_i \lambda_{Ni}(t)\,; \quad \lambda_{Ni}(t) \geq 0\,; \quad \forall\ t \in [t_A, t_C]\,. \tag{8.15}$$

This immediately leads, together with (8.14), to the inequality

$$|\Lambda_{TCi}| \leq \mu_i \Lambda_{NCi}\,. \tag{8.16}$$

As in eq. (8.10) and Fig. 8.1 we cannot argue in the opposite direction: Though (8.16) is fulfilled there might be time instances within $[t_A, t_C]$ where the tangential forces $|\lambda_{Ti}|$ exceed the values of $\mu_i \lambda_{Ni}$. Nevertheless, the case (8.15) is included in the condition (8.16), and thus we will use this restriction in the following.

Coulomb's law originally stated is (cp. eq. 6.10)

$$|\lambda_{Ti}| \leq \mu_i \lambda_{Ni}\,; \quad \begin{cases} |\lambda_{Ti}| < \mu_i \lambda_{Ni} & \Rightarrow \quad \dot{g}_{Ti} = 0 \\ \lambda_{Ti} = +\mu_i \lambda_{Ni} & \Rightarrow \quad \dot{g}_{Ti} \leq 0 \\ \lambda_{Ti} = -\mu_i \lambda_{Ni} & \Rightarrow \quad \dot{g}_{Ti} \geq 0\,. \end{cases} \tag{8.17}$$

Possible stick-slip transitions during the collision with reversed sliding prevent an analytical general integration of (8.17) over the impact interval, so no representation of the friction impulses can be given to determine the relative velocities at the end of compression according to Coulomb's law. However, for special cases the integration can be performed:

1. Sticking at the end of compression:

   From Coulomb's law (8.17) we get

$$|\lambda_{Ti}(t_C)| < \mu_i \lambda_{Ni}(t_C) \quad \Rightarrow \quad \dot{g}_{TCi} = 0\,. \tag{8.18}$$

   On the other hand, we see from (8.14) that

$$|\lambda_{Ti}(t_C)| < \mu_i \lambda_{Ni}(t_C) \quad \Rightarrow \quad |\Lambda_{TCi}| < \mu_i \Lambda_{NCi} \tag{8.19}$$

   if there is an arbitrary short episode of sticking at the end of compression. This behavior is depicted in the upper two diagrams of Fig. 8.3. If continual sticking or a transition to sticking at the end of compression occurs, then the transferred tangential impulse is always less than the value of $\mu_i \Lambda_{NCi}$.

2. Sliding at the end of compression:

   From (8.17) we get

$$\begin{aligned} \lambda_{Ti}(t_C) &= +\mu_i \lambda_{Ni}(t_C) \quad \Rightarrow \quad \dot{g}_{TCi} \leq 0 \\ \lambda_{Ti}(t_C) &= -\mu_i \lambda_{Ni}(t_C) \quad \Rightarrow \quad \dot{g}_{TCi} \geq 0\,. \end{aligned} \tag{8.20}$$

   No general statement about the values of the tangential impulses can be given. Only in the case of continuous sliding in one direction during the compression phase,

$$\begin{aligned} \lambda_{Ti}(t) = +\mu_i \lambda_{Ni}(t) \quad &\Rightarrow \quad \dot{g}_{Ti}(t) \leq 0\,; \quad \forall\ t \in [t_A, t_C] \\ \lambda_{Ti}(t) = -\mu_i \lambda_{Ni}(t) \quad &\Rightarrow \quad \dot{g}_{Ti}(t) \geq 0\,; \quad \forall\ t \in [t_A, t_C]\,, \end{aligned} \tag{8.21}$$

   we can evaluate the integral (8.14) to give

$$\begin{aligned} \Lambda_{TCi} &= +\mu_i \Lambda_{NCi} \\ \Lambda_{TCi} &= -\mu_i \Lambda_{NCi}\,. \end{aligned} \tag{8.22}$$

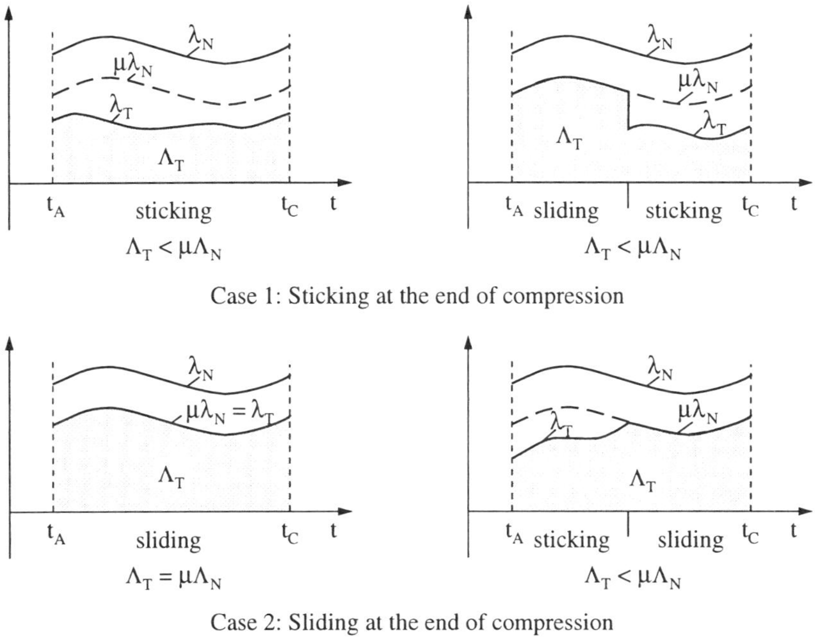

Figure 8.3: Normal and Tangential Impulses during Compression

In the lower two diagrams of Fig. 8.3 the connection between the tangential forces and impulses are shown when the compression phase ends with a sliding episode. For continual sliding we have $\Lambda_{TCi} = |\mu_i \Lambda_{NCi}|$. If there is any other additional state like sticking or sliding in the opposing direction (not depicted) within the compression phase, then the value of the transferred tangential impulse is clearly less than $\mu_i \Lambda_{NCi}$ when Coulomb friction is assumed during the impact interval. As an example, a transition from sticking to sliding is depicted in the lower right part of Fig. 8.3.

Now we state an impact law in the tangential direction which fulfills as much as possible the conditions described above:

$$
\begin{aligned}
&|\Lambda_{TCi}| \le \mu_i \Lambda_{NCi}\,; \\
&\left\{\begin{array}{lcl}
|\Lambda_{TCi}| < \mu_i \Lambda_{NCi} & \Rightarrow & \dot{g}_{TCi} = 0 \\
\Lambda_{TCi} = +\mu_i \Lambda_{NCi} & \Rightarrow & \dot{g}_{TCi} \le 0\,; \quad i \in I_S\,. \\
\Lambda_{TCi} = -\mu_i \Lambda_{NCi} & \Rightarrow & \dot{g}_{TCi} \ge 0
\end{array}\right.
\end{aligned}
\tag{8.23}
$$

Its characteristic is shown in the right diagram of Fig. 8.2. From the remarks above we see that the results produced by the impact law (8.23) coincide with

Coulomb friction (8.17) in the cases of continuous sliding during compression and of arbitrary transitions to sticking at the end of compression. Only the events of reversed sliding or transitions from sticking to sliding with a sliding phase at the end of compression are different from Coulomb's law. They can also be handled by (8.23), but the resulting impulses are greater than those coming from an integration with respect to Fig. 8.3. Thus eq. (8.23) should be regarded as an independent tangential impact law which coincides often to Coulomb's hypothesis and contains all the physical main effects of dry friction. Finally, note the always dissipative character of eq. (8.23) by the work-expression

$$\Lambda_{TCi}\,\dot{g}_{TCi} \leq 0\,; \quad i \in I_S\,. \tag{8.24}$$

With (8.13) and (8.23) the missing $2n_S$ impact conditions are found for completing the description of the phase of compression (eqs. 8.7).

The characteristics of the impact laws for compression are of the same type as those of the contact constraints and Coulomb friction in Chapter 6. Thus the evaluation of eqs. (8.7), (8.13) and (8.23) can be performed similarly to Chapter 6 by formulating the corresponding Linear Complementarity Problem, which will be done in Section 8.8.

8.3 Phase of Expansion

Similarly to eqs. (8.7) the kinetics and kinematics equations for expansion are given by [21]

$$\begin{aligned} &\boldsymbol{M}\,(\dot{\boldsymbol{q}}_E - \dot{\boldsymbol{q}}_C) = (\boldsymbol{W}_N \boldsymbol{W}_T) \begin{pmatrix} \boldsymbol{\Lambda}_{NE} \\ \boldsymbol{\Lambda}_{TE} \end{pmatrix} \\ &\begin{pmatrix} \dot{\boldsymbol{g}}_{NE} \\ \dot{\boldsymbol{g}}_{TE} \end{pmatrix} = \begin{pmatrix} \boldsymbol{W}_N^T \\ \boldsymbol{W}_T^T \end{pmatrix} (\dot{\boldsymbol{q}}_E - \dot{\boldsymbol{q}}_C) + \begin{pmatrix} \dot{\boldsymbol{g}}_{NC} \\ \dot{\boldsymbol{g}}_{TC} \end{pmatrix}, \end{aligned} \tag{8.25}$$

which have to be completed by $2n_S$ relations consisting of the impact laws for expansion. For the impact constraint in the normal direction we use Poisson's hypothesis which as originally stated is for a single impact $\Lambda_{NEi} = \varepsilon_{Ni}\Lambda_{NCi}$. The impulse Λ_{NCi} acting during compression is reduced by a dissipation coefficient, called the coefficient of restitution ε_{Ni} $(0 \leq \varepsilon_{Ni} \leq 1)$, and is then applied to the contact as an expansion impulse Λ_{NEi}. For multiple impacts, however, even this relation has to be modified to ensure the impenetrability condition at the end of the impact. After the process of expansion, each of the relative velocities must show positive values,

$$\dot{g}_{NEi} \geq 0\,, \tag{8.26}$$

because negative relative velocities at the end of the impact would lead to a further approach of the bodies and thus to penetration. The magnitudes of the relative velocities at the end of expansion depend on the strength of the expansion impulses. For coupled impact problems situations may occur where the original Poisson impulse $\varepsilon_{Ni}\Lambda_{NCi}$ in one of the contacts is *not* strong enough to prevent penetration because the contact partners are simultaneously under the influence of other impulsive forces. Thus, we generally have to allow impulses greater than the original Poisson impulse:

$$\Lambda_{NEi} \geq \varepsilon_{Ni}\Lambda_{NCi} \,. \tag{8.27}$$

If the Poisson impulse $\Lambda_{NEi} = \varepsilon_{Ni}\Lambda_{NCi}$ is strong enough to admit a separation, then any positive values of relative velocity are allowed, $\dot{g}_{NEi} \geq 0$. In the other case, the expansion impulse must be increased enough ($\Lambda_{NEi} > \varepsilon_{Ni}\Lambda_{NCi}$) that at least penetration can be avoided. Thus $\dot{g}_{NEi} = 0$. Both cases can be expressed by the complementarity condition

$$(\Lambda_{NEi} - \varepsilon_{Ni}\Lambda_{NCi})\,\dot{g}_{NEi} = 0 \,. \tag{8.28}$$

Using the new variable

$$\Lambda_{NPi} = \Lambda_{NEi} - \varepsilon_{Ni}\Lambda_{NCi} \tag{8.29}$$

the complete impact law during expansion as presented with eqs. (8.26), (8.27), (8.28) can be stated by the complementarities

$$\Lambda_{NPi} \geq 0\,; \quad \dot{g}_{NEi} \geq 0\,; \quad \Lambda_{NPi}\,\dot{g}_{NEi} = 0\,; \quad i \in I_S \,. \tag{8.30}$$

The corresponding characteristic is shown in the left diagram of Fig. 8.4, where it should be noted that one part of it crosses the positive orthant. This is equivalent to a gain of energy, expressed by the relation $\Lambda_{NEi}\dot{g}_{NEi} \geq 0$.

The tangential behavior during expansion may be formulated in the same way as during compression only if impulses related to dry friction act on the bodies. The resulting characteristic and impact law would be the same as in Fig. 8.3 and eq. (8.23) by simply replacing the indices C (compression) by E (expansion). An advanced modeling enables us to also take into account reversible portions of the tangential impulse which may occur, for example, in the highly elastic materials used in making superballs. These effects are considered when shifting the tangential characteristic by an amount $2\Lambda_{TSi}$ (see Fig. 8.4). The values of $2\Lambda_{TSi}$ are specified in eq. (8.37) and require the introduction of additional tangential impact parameters.

In order to state the tangential impact law we first reconsider the normal behavior for a single impact: The approaching process before the impact is characterized by *negative* relative velocities. During the compression phase, a *positive* impulse acts on the bodies and "decelerates" them until the relative velocity vanishes. This is the end of compression.

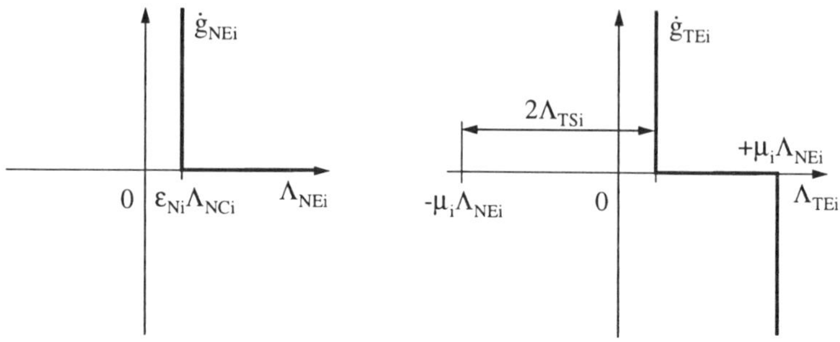

Figure 8.4: Impact Law for Expansion

The Poisson impulse applied during expansion acts in the same direction as the compression impulse and "accelerates" the bodies until the final *positive* relative velocity is achieved. During the impact, the relative velocity has changed its sign under the influence of always positive (compressive) impulses. This behavior has been taken into account by shifting the compression characteristic (Fig. 8.3) to the right in order to get the characteristic for expansion (Fig. 8.4). Now we transfer these results to the tangential direction and assume a *positive* tangential impulse during compression, which usually corresponds to a *negative* tangential approaching velocity. After compression, a certain amount of the transferred impulse may be "stored" in the bodies as elastic deformation, and may lead to a change in the sliding direction during expansion when acting in the *same* direction. Thus, we have to shift the left branch of the tangential impact characteristic ($\dot{g}_{TEi} \geq 0$) to the right until it crosses the positive orthant (see Fig. 8.4). On the other hand, if no reversed sliding occurs during the whole impact interval, the transferred tangential impulse should be bounded by the normal impulse just as in the compression phase. This leads to the right branch of the characteristic. According to Fig. 8.4 we now state the impact law for expansion in the case of positive tangential impulses during compression:

Case 1: $\Lambda_{TCi} \geq 0$; $\Lambda_{TSi} \geq 0$

$$-\mu_i\Lambda_{NEi} + 2\Lambda_{TSi} \leq \Lambda_{TEi} \leq +\mu_i\Lambda_{NEi}\,;$$

$$\begin{cases} -\mu_i\Lambda_{NEi} + 2\Lambda_{TSi} < \Lambda_{TEi} < +\mu_i\Lambda_{NEi} & \Rightarrow \quad \dot{g}_{TEi} = 0 \\ \Lambda_{TEi} = +\mu_i\Lambda_{NEi} & \Rightarrow \quad \dot{g}_{TEi} \leq 0 \\ \Lambda_{TEi} = -\mu_i\Lambda_{NEi} + 2\Lambda_{TSi} & \Rightarrow \quad \dot{g}_{TEi} \geq 0\,. \end{cases}$$

In the same way we proceed for negative tangential impulses during compression. A negative value of Λ_{TCi} analogously leads to a shifting of the righthand side of the characteristic to the negative direction,

Case 2: $\Lambda_{TCi} \le 0;\ \Lambda_{TSi} \le 0$

$$-\mu_i \Lambda_{NEi} \le \Lambda_{TEi} \le +\mu_i \Lambda_{NEi} + 2\Lambda_{TSi}\,;$$

$$\begin{cases} -\mu_i \Lambda_{NEi} < \Lambda_{TEi} < +\mu_i \Lambda_{NEi} + 2\Lambda_{TSi} & \Rightarrow \quad \dot{g}_{TEi} = 0 \\ \Lambda_{TEi} = +\mu_i \Lambda_{NEi} + 2\Lambda_{TSi} & \Rightarrow \quad \dot{g}_{TEi} \le 0 \\ \Lambda_{TEi} = -\mu_i \Lambda_{NEi} & \Rightarrow \quad \dot{g}_{TEi} \ge 0\,. \end{cases}$$

Both cases together can be stated as

$$\begin{aligned} & \operatorname{sign}(\Lambda_{TCi}) = \operatorname{sign}(\Lambda_{TSi}) =: \sigma_i \\ & -(\mu_i \Lambda_{NEi} - |\Lambda_{TSi}|) \le \Lambda_{TEi} - \Lambda_{TSi} \le +(\mu_i \Lambda_{NEi} - |\Lambda_{TSi}|)\,; \\ & \begin{cases} |\Lambda_{TEi} - \Lambda_{TSi}| < (\mu_i \Lambda_{NEi} - |\Lambda_{TSi}|) & \Rightarrow \quad \dot{g}_{TEi} = 0 \\ \Lambda_{TEi} - \Lambda_{TSi} = +(\mu_i \Lambda_{NEi} - |\Lambda_{TSi}|) & \Rightarrow \quad \dot{g}_{TEi} \le 0 \\ \Lambda_{TEi} - \Lambda_{TSi} = -(\mu_i \Lambda_{NEi} - |\Lambda_{TSi}|) & \Rightarrow \quad \dot{g}_{TEi} \ge 0 \end{cases} \end{aligned} \tag{8.31}$$

which is the impact law for expansion. The first equation in (8.31) takes into account the direction of the tangential compression impulse. The inequality in the second line defines the admissible values of the term $\Lambda_{TEi} - \Lambda_{TSi}$, and the statements to the right of the brace connect the impulses to certain velocities at the end of the impact.

In order to ensure a nonempty set of admissible values of $\Lambda_{TEi} - \Lambda_{TSi}$, the inequality

$$2|\Lambda_{TSi}| \le 2\mu_i \Lambda_{NEi} \tag{8.32}$$

resulting from the second condition in (8.31) has to be fulfilled. Furthermore, we choose

$$2|\Lambda_{TSi}| \le \mu_i \Lambda_{NEi} + \varepsilon_{Ni} |\Lambda_{TCi}| \tag{8.33}$$

as a physical condition on the values of Λ_{TSi}, which ensures that the amount of the impulse Λ_{TEi} applied during expansion does not exceed the amount stored during compression. Equation (8.32) holds if (8.33) is fulfilled. This can be seen easily by the estimation

$$\varepsilon_{Ni} |\Lambda_{TCi}| \le \varepsilon_{Ni} \mu_i \Lambda_{NCi} = \mu_i \varepsilon_{Ni} \Lambda_{NCi} \le \mu_i \Lambda_{NEi}, \tag{8.34}$$

where eqs. (8.23) and (8.27) have been used. Finally, note that for $\mu_i \Lambda_{NEi} < 2|\Lambda_{TSi}| \le \mu_i \Lambda_{NEi} + \varepsilon_{Ni} |\Lambda_{TCi}|$ energy might be gained during expansion because one branch of the characteristic crosses either the positive or the negative orthant.

Using the transformation

$$\Lambda_{TPi} := \Lambda_{TEi} - \varepsilon_{Ni}\Lambda_{TCi} \tag{8.35}$$

the impact law (8.31) can be stated in terms of Λ_{TPi}, which is more convenient with respect to its evaluation in Section 8.9 and enables us to derive the inequality

$$\Lambda_{TPi}\,\dot{g}_{TEi} \leq 0 \tag{8.36}$$

which will be used for proving the energy loss during the collision: For $\dot{g}_{TEi} > 0$ the value of Λ_{TPi} resulting from the impact law (8.31) and the transformation rule (8.35) is given by

$$\Lambda_{TPi} = -\mu_i\Lambda_{NEi} + |\Lambda_{TSi}| + \Lambda_{TSi} - \varepsilon_{Ni}\Lambda_{TCi}.$$

In order to verify $\Lambda_{TPi}\,\dot{g}_{TEi} \leq 0$ we only have to show that $\Lambda_{TPi} \leq 0$, which can be done by rewriting the terms Λ_{TSi} and Λ_{TCi} with respect to the first equation in (8.31),

$$\Lambda_{TCi} = \sigma_i|\Lambda_{TCi}|\,; \quad \Lambda_{TSi} = \sigma_i|\Lambda_{TSi}|$$

and putting them into the expression for Λ_{TPi}:

$$\Lambda_{TPi} = -\mu_i\Lambda_{NEi} + (1+\sigma_i)|\Lambda_{TSi}| - \varepsilon_{Ni}\sigma_i|\Lambda_{TCi}|\,.$$

Now we evaluate the two cases ($\sigma_i = \pm 1$) of the sign function:

$$\Lambda_{TPi} = \begin{cases} -\mu_i\Lambda_{NEi} + 2|\Lambda_{TSi}| - \varepsilon_{Ni}|\Lambda_{TCi}| \leq 0\,; & \sigma_i = +1 \\ -\mu_i\Lambda_{NEi} + \varepsilon_{Ni}|\Lambda_{TCi}| \leq 0\,; & \sigma_i = -1\,. \end{cases}$$

We now compare the results with eqs. (8.33) and (8.34), and see that in both cases the values of Λ_{TPi} are not positive. Thus, the inequality (8.36) holds under the assumption $\dot{g}_{TEi} > 0$. For $\dot{g}_{TEi} < 0$ we proceed in the same way, which yields $\Lambda_{TPi} \geq 0$ and completes the proof of eq. (8.36). At that point we note that due to (8.36) $\Lambda_{TPi} = 0$ is always included in the admissible values of Λ_{TPi}. This property will be used in Section 8.7 for the decomposition of an unsymmetric tangential impact law.

Finally, the value of Λ_{TSi} in eq. (8.31) has to be determined as a function of the terms during compression in such a manner that the conditions in the first equations of (8.31) and (8.33) are not violated. We have done this by using the very simple relation

$$2\Lambda_{TSi} = \mu_i\nu_i\Lambda_{NEi}\,\mathrm{sgn}(\Lambda_{TCi}) + \varepsilon_{Ni}\varepsilon_{Ti}\Lambda_{TCi}\,; \quad 0 \leq \nu_i, \varepsilon_{Ti} \leq 1, \tag{8.37}$$

which could be used as a basis for the development of special tangential impact laws in order to describe the behavior of certain loading and unloading

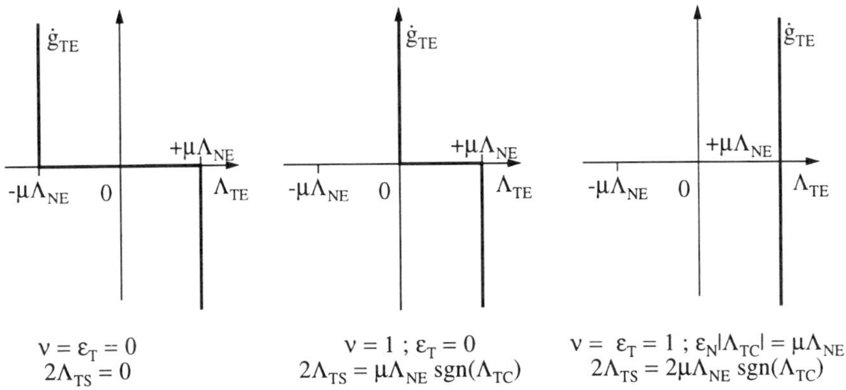

Figure 8.5: Tangential Impact Characteristics

processes. The magnitudes ν_i and ε_{Ti} are additional tangential impact parameters which specify the shifting amount Λ_{TSi}.

Figure 8.5 shows the tangential characteristic for certain values of ν_i and ε_{Ti}. The characteristic depicted on the left-hand side corresponds to Coulomb friction on the impulse level, since the figure on the right-hand side describes the behavior of a maximal and completely reversible tangential shock.

With respect to the evaluation of the expansion phase in Section 8.9 the impact law (8.31) has to be stated as a function of Λ_{TPi}. After the elimination of the terms Λ_{NEi}, Λ_{TSi}, Λ_{TEi} with the help of eqs. (8.29), (8.35), (8.37) and some computationally intensive calculations the impact law becomes

$$\begin{aligned} &-\mu_i^{(-)}(\Lambda_{NPi} + \varepsilon_{Ni}\Lambda_{NCi}) - \varepsilon_{Ti}^{(-)}\varepsilon_{Ni}\Lambda_{TCi} \leq \Lambda_{TPi} \\ &\leq +\mu_i^{(+)}(\Lambda_{NPi} + \varepsilon_{Ni}\Lambda_{NCi}) + \varepsilon_{Ti}^{(+)}\varepsilon_{Ni}\Lambda_{TCi}\,; \end{aligned} \tag{8.38}$$

$$\begin{cases} -\mu_i^{(-)}(\Lambda_{NPi} + \varepsilon_{Ni}\Lambda_{NCi}) - \varepsilon_{Ti}^{(-)}\varepsilon_{Ni}\Lambda_{TCi} < \Lambda_{TPi} & \\ < +\mu_i^{(+)}(\Lambda_{NPi} + \varepsilon_{Ni}\Lambda_{NCi}) + \varepsilon_{Ti}^{(+)}\varepsilon_{Ni}\Lambda_{TCi} & \Rightarrow \quad \dot{g}_{TEi} = 0 \\ \Lambda_{TPi} = +\mu_i^{(+)}(\Lambda_{NPi} + \varepsilon_{Ni}\Lambda_{NCi}) + \varepsilon_{Ti}^{(+)}\varepsilon_{Ni}\Lambda_{TCi} & \Rightarrow \quad \dot{g}_{TEi} \leq 0 \\ \Lambda_{TPi} = -\mu_i^{(-)}(\Lambda_{NPi} + \varepsilon_{Ni}\Lambda_{NCi}) - \varepsilon_{Ti}^{(-)}\varepsilon_{Ni}\Lambda_{TCi} & \Rightarrow \quad \dot{g}_{TEi} \geq 0 \end{cases}$$

where

$$\begin{aligned} \mu_i^{(-)} &:= \left[1 - \nu_i\left(\tfrac{1}{2} + \tfrac{\sigma_i}{2}\right)\right]\mu_i \geq 0 \\ \mu_i^{(+)} &:= \left[1 - \nu_i\left(\tfrac{1}{2} - \tfrac{\sigma_i}{2}\right)\right]\mu_i \geq 0 \\ \varepsilon_{Ti}^{(-)} &:= \left[+1 - \varepsilon_{Ti}\left(\tfrac{1}{2} + \tfrac{\sigma_i}{2}\right)\right] \geq 0 \\ \varepsilon_{Ti}^{(+)} &:= \left[-1 + \varepsilon_{Ti}\left(\tfrac{1}{2} - \tfrac{\sigma_i}{2}\right)\right] \leq 0 \\ \sigma_i &= \mathrm{sign}(\Lambda_{TCi})\,. \end{aligned} \tag{8.39}$$

Note that the representation (8.38) of the tangential impact law is no longer symmetric. However, the decomposition in Section 6.3 is still working and will be applied to nonsymmetric characteristics in Section 8.7.

With eqs. (8.25), (8.29), (8.30), (8.35), (8.38) and (8.39) a complete description of the phase of expansion is achieved. The characteristics in Fig. 8.4 differ from those of the compression phase (Fig. 8.3) by the shifting amounts $\varepsilon_{Ni}\Lambda_{NCi}$ and Λ_{TSi}. The evaluation of the expansion phase is done by solving a Linear Complementarity Problem similarly to that of compression, which will be presented in Section 8.9.

8.4 Energy Considerations

In this section it will be shown that the impact laws described above are dissipative or energy preserving for equal coefficients of restitution [21]. In the first part, only a shortened representation of eqs. (8.5), (8.6), (8.13), (8.24), (8.29), (8.30), (8.35) and (8.36) is given in order to derive the energy loss during collisions in the second part. First, the transformations (8.29) and (8.35) together with the inequalities (8.13), (8.24) and (8.30), (8.36) are rewritten in matrix notation:

$$\begin{pmatrix} \boldsymbol{\Lambda}_{NP} \\ \boldsymbol{\Lambda}_{TP} \end{pmatrix} = \begin{pmatrix} \boldsymbol{\Lambda}_{NE} \\ \boldsymbol{\Lambda}_{TE} \end{pmatrix} - \begin{pmatrix} \bar{\bar{\boldsymbol{\epsilon}}}_N & 0 \\ 0 & \bar{\bar{\boldsymbol{\epsilon}}}_N \end{pmatrix} \begin{pmatrix} \boldsymbol{\Lambda}_{NC} \\ \boldsymbol{\Lambda}_{TC} \end{pmatrix} \tag{8.40}$$

$$\begin{pmatrix} \boldsymbol{\Lambda}_{NC} \\ \boldsymbol{\Lambda}_{TC} \end{pmatrix}^T \begin{pmatrix} \dot{\boldsymbol{g}}_{NC} \\ \dot{\boldsymbol{g}}_{TC} \end{pmatrix} \leq 0\,; \quad \begin{pmatrix} \boldsymbol{\Lambda}_{NP} \\ \boldsymbol{\Lambda}_{TP} \end{pmatrix}^T \begin{pmatrix} \dot{\boldsymbol{g}}_{NE} \\ \dot{\boldsymbol{g}}_{TE} \end{pmatrix} \leq 0\,, \tag{8.41}$$

where $\bar{\bar{\boldsymbol{\epsilon}}}_N = \text{diag}\{\varepsilon_{Ni}\}$ is the diagonal matrix which contains the coefficients of restitution in the normal direction, $0 \leq \varepsilon_{Ni} \leq 1$. With that property it is obvious that even

$$\begin{pmatrix} \boldsymbol{\Lambda}_{NC} \\ \boldsymbol{\Lambda}_{TC} \end{pmatrix}^T \begin{pmatrix} \bar{\bar{\boldsymbol{\epsilon}}}_N & 0 \\ 0 & \bar{\bar{\boldsymbol{\epsilon}}}_N \end{pmatrix} \begin{pmatrix} \dot{\boldsymbol{g}}_{NC} \\ \dot{\boldsymbol{g}}_{TC} \end{pmatrix} \leq 0 \tag{8.42}$$

holds. Next, we introduce the abbreviations

$$\begin{aligned}
&\dot{\boldsymbol{g}}_A = \begin{pmatrix} \dot{\boldsymbol{g}}_{NA} \\ \dot{\boldsymbol{g}}_{TA} \end{pmatrix}; \quad \dot{\boldsymbol{g}}_C = \begin{pmatrix} \dot{\boldsymbol{g}}_{NC} \\ \dot{\boldsymbol{g}}_{TC} \end{pmatrix}; \quad \dot{\boldsymbol{g}}_E = \begin{pmatrix} \dot{\boldsymbol{g}}_{NE} \\ \dot{\boldsymbol{g}}_{TE} \end{pmatrix}; \\
&\boldsymbol{\Lambda}_C = \begin{pmatrix} \boldsymbol{\Lambda}_{NC} \\ \boldsymbol{\Lambda}_{TC} \end{pmatrix}; \quad \boldsymbol{\Lambda}_E = \begin{pmatrix} \boldsymbol{\Lambda}_{NE} \\ \boldsymbol{\Lambda}_{TE} \end{pmatrix}; \quad \boldsymbol{\Lambda}_P = \begin{pmatrix} \boldsymbol{\Lambda}_{NP} \\ \boldsymbol{\Lambda}_{TP} \end{pmatrix}; \\
&\tilde{\boldsymbol{w}} = \begin{pmatrix} \tilde{\boldsymbol{w}}_N \\ \tilde{\boldsymbol{w}}_T \end{pmatrix}; \quad \bar{\bar{\boldsymbol{\epsilon}}} = \begin{pmatrix} \bar{\bar{\boldsymbol{\epsilon}}}_N & 0 \\ 0 & \bar{\bar{\boldsymbol{\epsilon}}}_N \end{pmatrix}; \quad \begin{array}{l} \boldsymbol{W} = (\boldsymbol{W}_N\ \boldsymbol{W}_T)\,; \\ \boldsymbol{G} = \boldsymbol{W}^T\boldsymbol{M}^{-1}\boldsymbol{W}\,, \end{array}
\end{aligned} \tag{8.43}$$

and rewrite the dynamics equations (8.5), the kinematic relations (8.6), the transformation (8.40) and the inequalities (8.41), (8.42), which yield

$$
\begin{aligned}
&\boldsymbol{M}(\dot{\boldsymbol{q}}_C - \dot{\boldsymbol{q}}_A) = \boldsymbol{W}\boldsymbol{\Lambda}_C && \boldsymbol{M}(\dot{\boldsymbol{q}}_E - \dot{\boldsymbol{q}}_C) = \boldsymbol{W}\boldsymbol{\Lambda}_E && (8.44)\\
&\dot{\boldsymbol{g}}_C = \boldsymbol{W}^T\dot{\boldsymbol{q}}_C + \tilde{\boldsymbol{w}} && \dot{\boldsymbol{g}}_E = \boldsymbol{W}^T\dot{\boldsymbol{q}}_E + \tilde{\boldsymbol{w}} && (8.45)\\
&\dot{\boldsymbol{g}}_A = \boldsymbol{W}^T\dot{\boldsymbol{q}}_A + \tilde{\boldsymbol{w}} && \dot{\boldsymbol{g}}_C = \boldsymbol{W}^T\dot{\boldsymbol{q}}_C + \tilde{\boldsymbol{w}} && (8.46)\\
& && \boldsymbol{\Lambda}_E = \boldsymbol{\Lambda}_P + \bar{\bar{\boldsymbol{\epsilon}}}\boldsymbol{\Lambda}_C && (8.47)\\
&\boldsymbol{\Lambda}_C^T\dot{\boldsymbol{g}}_C \le 0 && \boldsymbol{\Lambda}_P^T\dot{\boldsymbol{g}}_E \le 0 && (8.48)\\
&\boldsymbol{\Lambda}_C^T\bar{\bar{\boldsymbol{\epsilon}}}\dot{\boldsymbol{g}}_C \le 0 && && (8.49)
\end{aligned}
$$

where the left column belongs to compression and the right column to expansion. Finally, we write eqs. (8.45) and (8.46) as a sum and difference by using (8.44):

$$
\begin{aligned}
\dot{\boldsymbol{g}}_C + \dot{\boldsymbol{g}}_A &= \boldsymbol{W}^T(\dot{\boldsymbol{q}}_C + \dot{\boldsymbol{q}}_A) + 2\tilde{\boldsymbol{w}}\\
\dot{\boldsymbol{g}}_E + \dot{\boldsymbol{g}}_C &= \boldsymbol{W}^T(\dot{\boldsymbol{q}}_E + \dot{\boldsymbol{q}}_C) + 2\tilde{\boldsymbol{w}}
\end{aligned} \tag{8.50}
$$

$$
\begin{aligned}
\dot{\boldsymbol{g}}_C - \dot{\boldsymbol{g}}_A &= \boldsymbol{G}\boldsymbol{\Lambda}_C\\
\dot{\boldsymbol{g}}_E - \dot{\boldsymbol{g}}_C &= \boldsymbol{G}\boldsymbol{\Lambda}_E
\end{aligned} \tag{8.51}
$$

and get with (8.44) and (8.47)–(8.51) a set of ten equations which will be used in the following to prove the dissipative character of the impact law.

Let T_A and T_C denote the kinetic energy at the beginning of the impact and at the end of compression, respectively. Then the loss of energy related to compression is given by

$$
\begin{aligned}
T_C - T_A &= \tfrac{1}{2}\dot{\boldsymbol{q}}_C^T\boldsymbol{M}\dot{\boldsymbol{q}}_C - \tfrac{1}{2}\dot{\boldsymbol{q}}_A^T\boldsymbol{M}\dot{\boldsymbol{q}}_A\\
&= \tfrac{1}{2}(\dot{\boldsymbol{q}}_C + \dot{\boldsymbol{q}}_A)^T\boldsymbol{M}(\dot{\boldsymbol{q}}_C - \dot{\boldsymbol{q}}_A)\\
&\overset{(8.44)}{=} \tfrac{1}{2}(\dot{\boldsymbol{q}}_C + \dot{\boldsymbol{q}}_A)^T\boldsymbol{W}\boldsymbol{\Lambda}_C\\
&\overset{(8.50)}{=} \tfrac{1}{2}\boldsymbol{\Lambda}_C^T(\dot{\boldsymbol{g}}_C + \dot{\boldsymbol{g}}_A) - \boldsymbol{\Lambda}_C^T\tilde{\boldsymbol{w}}\\
&\overset{(8.51)}{=} -\tfrac{1}{2}\boldsymbol{\Lambda}_C^T\boldsymbol{G}\boldsymbol{\Lambda}_C + \boldsymbol{\Lambda}_C^T\dot{\boldsymbol{g}}_C - \boldsymbol{\Lambda}_C^T\tilde{\boldsymbol{w}}\,.
\end{aligned} \tag{8.52}
$$

Similarly, the term $T_E - T_C$ describing expansion is achieved by replacing C with E and A with C. If no external kinematic excitation is considered ($\tilde{\boldsymbol{w}} = 0$), the complete energy difference is

$$
\begin{aligned}
T_E - T_A &= (T_E - T_C) + (T_C - T_A)\\
&= -\tfrac{1}{2}\boldsymbol{\Lambda}_C^T\boldsymbol{G}\boldsymbol{\Lambda}_C - \tfrac{1}{2}\boldsymbol{\Lambda}_E^T\boldsymbol{G}\boldsymbol{\Lambda}_E + \boldsymbol{\Lambda}_C^T\dot{\boldsymbol{g}}_C + \boldsymbol{\Lambda}_E^T\dot{\boldsymbol{g}}_E\,.
\end{aligned} \tag{8.53}
$$

Finally, with the help of (8.47) we express $\boldsymbol{\Lambda}_E$ in terms of $\boldsymbol{\Lambda}_P$ and $\boldsymbol{\Lambda}_C$:

$$\tfrac{1}{2}\boldsymbol{\Lambda}_E^T\boldsymbol{G}\boldsymbol{\Lambda}_E \overset{(8.47)}{=} \tfrac{1}{2}\boldsymbol{\Lambda}_P^T\boldsymbol{G}\boldsymbol{\Lambda}_P + \tfrac{1}{2}\boldsymbol{\Lambda}_C^T\overline{\overline{\boldsymbol{\epsilon}}}\boldsymbol{G}\overline{\overline{\boldsymbol{\epsilon}}}\boldsymbol{\Lambda}_C + \boldsymbol{\Lambda}_C^T\overline{\overline{\boldsymbol{\epsilon}}}\boldsymbol{G}\boldsymbol{\Lambda}_P \tag{8.54}$$

$$\begin{aligned}\boldsymbol{\Lambda}_E^T\dot{\boldsymbol{g}}_E &\overset{(8.47)}{=} \boldsymbol{\Lambda}_P^T\dot{\boldsymbol{g}}_E + \boldsymbol{\Lambda}_C^T\overline{\overline{\boldsymbol{\epsilon}}}\dot{\boldsymbol{g}}_E \\ &\overset{(8.51)}{=} \boldsymbol{\Lambda}_P^T\dot{\boldsymbol{g}}_E + \boldsymbol{\Lambda}_C^T\overline{\overline{\boldsymbol{\epsilon}}}(\dot{\boldsymbol{g}}_C + \boldsymbol{G}\boldsymbol{\Lambda}_E) \\ &\overset{(8.47)}{=} \boldsymbol{\Lambda}_P^T\dot{\boldsymbol{g}}_E + \boldsymbol{\Lambda}_C^T\overline{\overline{\boldsymbol{\epsilon}}}\dot{\boldsymbol{g}}_C + \boldsymbol{\Lambda}_C^T\overline{\overline{\boldsymbol{\epsilon}}}\boldsymbol{G}\boldsymbol{\Lambda}_P + \boldsymbol{\Lambda}_C^T\overline{\overline{\boldsymbol{\epsilon}}}\boldsymbol{G}\overline{\overline{\boldsymbol{\epsilon}}}\boldsymbol{\Lambda}_C \,.\end{aligned} \tag{8.55}$$

Insertion of (8.54) and (8.55) into (8.53) yields

$$\begin{aligned}T_E - T_A = &-\tfrac{1}{2}\boldsymbol{\Lambda}_C^T\boldsymbol{G}\boldsymbol{\Lambda}_C - \tfrac{1}{2}\boldsymbol{\Lambda}_P^T\boldsymbol{G}\boldsymbol{\Lambda}_P + \tfrac{1}{2}\boldsymbol{\Lambda}_C^T\overline{\overline{\boldsymbol{\epsilon}}}\boldsymbol{G}\overline{\overline{\boldsymbol{\epsilon}}}\boldsymbol{\Lambda}_C \\ &+\boldsymbol{\Lambda}_C^T\dot{\boldsymbol{g}}_C + \boldsymbol{\Lambda}_P^T\dot{\boldsymbol{g}}_E + \boldsymbol{\Lambda}_C^T\overline{\overline{\boldsymbol{\epsilon}}}\dot{\boldsymbol{g}}_C \,.\end{aligned} \tag{8.56}$$

Since $\boldsymbol{G}$ is at least symmetric and positive semidefinite (compare eq. 8.43), and $\overline{\overline{\boldsymbol{\epsilon}}}$ is a diagonal matrix consisting of elements $0 \le \varepsilon_{Ni} \le 1$, it follows at least for $\varepsilon_{Ni} = \varepsilon_{Nj}$ that $\frac{1}{2}\boldsymbol{\Lambda}_C^T\overline{\overline{\boldsymbol{\epsilon}}}\boldsymbol{G}\overline{\overline{\boldsymbol{\epsilon}}}\boldsymbol{\Lambda}_C - \frac{1}{2}\boldsymbol{\Lambda}_C^T\boldsymbol{G}\boldsymbol{\Lambda}_C \le 0$. Furthermore, the last three terms in (8.56) are the inequalities (8.48) and (8.49). Thus, the entire expression (8.56) is not greater than zero:

$$T_E - T_A \le 0\,; \tag{8.57}$$

hence, it has been proven that the presented impact law is dissipative or energy preserving at least for equal restitution coefficients.

8.5 Conservation of Energy

Conditions are derived under which the impact laws are necessarily energy preserving. From eq. (8.56) and the remarks below a conservation of energy $T_E - T_A = 0$ is achieved if the conditions

$$\tfrac{1}{2}\boldsymbol{\Lambda}_C^T\boldsymbol{G}\boldsymbol{\Lambda}_C = \tfrac{1}{2}\boldsymbol{\Lambda}_C^T\overline{\overline{\boldsymbol{\epsilon}}}\boldsymbol{G}\overline{\overline{\boldsymbol{\epsilon}}}\boldsymbol{\Lambda}_C \tag{8.58}$$

$$\tfrac{1}{2}\boldsymbol{\Lambda}_P^T\boldsymbol{G}\boldsymbol{\Lambda}_P = 0 \tag{8.59}$$

$$\boldsymbol{\Lambda}_C^T\dot{\boldsymbol{g}}_C = 0 \tag{8.60}$$

$$\boldsymbol{\Lambda}_P^T\dot{\boldsymbol{g}}_E = 0 \tag{8.61}$$

$$\boldsymbol{\Lambda}_C^T\overline{\overline{\boldsymbol{\epsilon}}}\dot{\boldsymbol{g}}_C = 0 \tag{8.62}$$

hold. Since $\boldsymbol{G}$ is symmetric and positive (semi-) definite, it follows from (8.58) and (8.59) that $\boldsymbol{\Lambda}_C = \overline{\overline{\boldsymbol{\epsilon}}}\boldsymbol{\Lambda}_C$ and $\boldsymbol{\Lambda}_P = 0 \overset{(8.47)}{\Rightarrow} \boldsymbol{\Lambda}_E = \overline{\overline{\boldsymbol{\epsilon}}}\boldsymbol{\Lambda}_C$, respectively. Thus

(8.61) is fulfilled, and (8.60) and (8.62) are equivalent. Therefore, energy conservation is possible if

$$\mathbf{\Lambda}_C^T \dot{\mathbf{g}}_C = 0\,; \quad \mathbf{\Lambda}_C = \mathbf{\Lambda}_E = \overline{\overline{\boldsymbol{\epsilon}}} \mathbf{\Lambda}_C\,. \tag{8.63}$$

Insertion of (8.63) into (8.51) yields the relationship of the relative velocities:

$$\dot{\mathbf{g}}_C - \dot{\mathbf{g}}_A = \dot{\mathbf{g}}_E - \dot{\mathbf{g}}_C\,. \tag{8.64}$$

For further discussion we state (8.63) and (8.64) in terms of the normal and tangential directions by resubstituting the magnitudes in (8.43):

$$\mathbf{\Lambda}_{NC}^T \dot{\mathbf{g}}_{NC} + \mathbf{\Lambda}_{TC}^T \dot{\mathbf{g}}_{TC} = 0 \tag{8.65}$$

$$\mathbf{\Lambda}_{NC} = \mathbf{\Lambda}_{NE} = \overline{\overline{\boldsymbol{\epsilon}}}_N \mathbf{\Lambda}_{NC}\,; \quad \mathbf{\Lambda}_{TC} = \mathbf{\Lambda}_{TE} = \overline{\overline{\boldsymbol{\epsilon}}}_N \mathbf{\Lambda}_{TC} \tag{8.66}$$

$$\dot{\mathbf{g}}_{NC} - \dot{\mathbf{g}}_{NA} = \dot{\mathbf{g}}_{NE} - \dot{\mathbf{g}}_{NC}\,; \quad \dot{\mathbf{g}}_{TC} - \dot{\mathbf{g}}_{TA} = \dot{\mathbf{g}}_{TE} - \dot{\mathbf{g}}_{TC}\,, \tag{8.67}$$

where $\mathbf{\Lambda}_{NC}^T \dot{\mathbf{g}}_{NC} = 0$ due to (8.13). We distinguish three cases:

Case 1: $\varepsilon_{Ni} < 1$

The equality in (8.66) holds if and only if $\Lambda_{NCi} = \Lambda_{NEi} = \Lambda_{TCi} = \Lambda_{TEi} = 0$ for contact point i.

If the coefficient of restitution in the normal direction ε_{Ni} is chosen to be less than one, then energy conservation is only possible if no impulse is transferred by this contact. This contact then does not participate in the impact process.

Case 2: $\varepsilon_{Ni} = 1 \wedge \Lambda_{TCi} = 0$

With that assumption the conditions (8.65) and (8.66) are fulfilled for contact i, and we get $\Lambda_{NCi} = \Lambda_{NEi}$, $\Lambda_{TCi} = \Lambda_{TEi} = 0$. If an impulse is transferred by this contact, $\Lambda_{NCi} > 0$, then $\dot{g}_{NCi} = 0$ due to the complementarity condition in (8.13). Generally, for arbitrary initial velocities $\dot{g}_{TAi}$, a tangential impulse $\Lambda_{TCi} = 0$ demands a coefficient of friction $\mu_i = 0$; see Fig. 8.3.

If the impact parameters of a contact which transfers normal impulses are chosen as

$$\varepsilon_{Ni} = 1\,; \quad \mu_i = 0 \tag{8.68}$$

then energy conservation is possible. If energy conservation is ensured, this choice corresponds to a frictionless elastic impact with velocities $\dot{g}_{NEi} = -\dot{g}_{NAi}$.

Case 3: $\varepsilon_{Ni} = 1 \wedge \dot{g}_{TCi} = 0$

With that assumption the conditions (8.65) and (8.66) are fulfilled for contact i, and we get $\Lambda_{NCi} = \Lambda_{NEi} > 0$, $\Lambda_{TCi} = \Lambda_{TEi} \neq 0$ when impulses in both directions, normal and tangential, are transferred. The tangential velocity $\dot{g}_{TCi}$ vanishes for arbitrary $\dot{g}_{TAi}$ generally only if the coefficient of friction μ_i is strong enough; see Fig. 8.3. From the characteristic in Fig. 8.4 we see that the tangential impulse during expansion, Λ_{TEi}, obviously must lie on the boundary to enable velocities $\dot{g}_{TEi} \neq 0$. Thus the shifting condition is $-\mu_i \Lambda_{NEi} \operatorname{sgn}(\Lambda_{TCi}) + 2\Lambda_{TSi} = \Lambda_{TEi} = \Lambda_{TCi}$. With Λ_{TSi} from (8.37) the impact parameters are then $\varepsilon_{Ni} = \varepsilon_{Ti} = \nu_i = 1$.

If the impact parameters of a contact which transfers normal and tangential impulses are chosen as

$$\varepsilon_{Ni} = \varepsilon_{Ti} = \nu_i = 1\,; \quad \mu_i \text{ strong enough} \tag{8.69}$$

then energy conservation is possible. If energy conservation is ensured, this choice corresponds to a completely elastic impact in the normal and tangential directions with velocities $\dot{g}_{NEi} = -\dot{g}_{NAi}$, $\dot{g}_{TEi} = -\dot{g}_{TAi}$.

8.6 Comparison of Newton's and Poisson's Laws

In Section 7 Newton's law was used to describe frictionless impacts where impulses were transferred at each of the impact contacts. Now we will investigate under which conditions the results given by the impact laws of Newton and Poisson are the same. First we have to restrict our kinematics and kinetics equations (8.7), (8.25) to the frictionless case, $\boldsymbol{\Lambda}_{TC} = \boldsymbol{\Lambda}_{TE} = 0$. Thus we can neglect the constraints in the tangential directions and get for compression and expansion that

$$\begin{aligned} \boldsymbol{M}(\dot{\boldsymbol{q}}_C - \dot{\boldsymbol{q}}_A) &= \boldsymbol{W}_N \boldsymbol{\Lambda}_{NC}\,; \quad & \dot{\boldsymbol{g}}_{NC} &= \boldsymbol{W}_N^T(\dot{\boldsymbol{q}}_C - \dot{\boldsymbol{q}}_A) + \dot{\boldsymbol{g}}_{NA} \\ \boldsymbol{M}(\dot{\boldsymbol{q}}_E - \dot{\boldsymbol{q}}_C) &= \boldsymbol{W}_N \boldsymbol{\Lambda}_{NE}\,; \quad & \dot{\boldsymbol{g}}_{NE} &= \boldsymbol{W}_N^T(\dot{\boldsymbol{q}}_E - \dot{\boldsymbol{q}}_C) + \dot{\boldsymbol{g}}_{NC}\,. \end{aligned} \tag{8.70}$$

After the elimination of the f-vectors $\dot{\boldsymbol{q}}_E - \dot{\boldsymbol{q}}_C$ and $\dot{\boldsymbol{q}}_C - \dot{\boldsymbol{q}}_A$, eq. (8.70) is reduced, by using the abbreviation (eq. 7.8)

$$\boldsymbol{G}_N = \boldsymbol{W}_N^T \boldsymbol{M}^{-1} \boldsymbol{W}_N\,, \tag{8.71}$$

to two sets of equations which describe compression and expansion:

$$\dot{\boldsymbol{g}}_{NC} = \boldsymbol{G}_N \boldsymbol{\Lambda}_{NC} + \dot{\boldsymbol{g}}_{NA}\,; \quad \dot{\boldsymbol{g}}_{NE} = \boldsymbol{G}_N \boldsymbol{\Lambda}_{NE} + \dot{\boldsymbol{g}}_{NC}\,. \tag{8.72}$$

Now we assume that each of the impact contacts transfers an impulse during compression, which means by the complementarity condition (8.13) that

$$\boldsymbol{\Lambda}_{NC} \geq \boldsymbol{0}\,; \quad \dot{\boldsymbol{g}}_{NC} = \boldsymbol{0}\,. \tag{8.73}$$

During expansion the Poisson impulses are assumed to be strong enough to admit separation; thus from (8.29), (8.30) we get

$$\boldsymbol{\Lambda}_{NE} = \bar{\bar{\boldsymbol{\epsilon}}}_N \boldsymbol{\Lambda}_{NC}\,; \quad \dot{\boldsymbol{g}}_{NE} \geq \boldsymbol{0}\,. \tag{8.74}$$

Insertion of (8.73), (8.74) into (8.72) yields

$$\dot{\boldsymbol{g}}_{NA} = -\boldsymbol{G}_N \boldsymbol{\Lambda}_{NC}\,; \quad \dot{\boldsymbol{g}}_{NE} = \boldsymbol{G}_N \bar{\bar{\boldsymbol{\epsilon}}}_N \boldsymbol{\Lambda}_{NC} \geq \boldsymbol{0}\,. \tag{8.75}$$

On the other hand, the connection between the relative velocities at the beginning and at the end of the impact, using Newton's law, is given by the relation (7.9):

$$\dot{\boldsymbol{g}}_{NE} = -\bar{\bar{\boldsymbol{\epsilon}}}_N \dot{\boldsymbol{g}}_{NA}\,. \tag{8.76}$$

Inserting eq. (8.75) into eq. (8.76),

$$\boldsymbol{G}_N \bar{\bar{\boldsymbol{\epsilon}}}_N \boldsymbol{\Lambda}_{NC} = \bar{\bar{\boldsymbol{\epsilon}}}_N \boldsymbol{G}_N \boldsymbol{\Lambda}_{NC} \tag{8.77}$$

which is true in the general case only if

$$\boldsymbol{G}_N \bar{\bar{\boldsymbol{\epsilon}}}_N = \bar{\bar{\boldsymbol{\epsilon}}}_N \boldsymbol{G}_N\,. \tag{8.78}$$

Since $\bar{\bar{\boldsymbol{\epsilon}}}_N = \operatorname{diag}\{\varepsilon_{Ni}\}$, we see from eq. (8.78) that the results obtained by Newton's and Poisson's laws coincide if all of the coefficients of restitution are chosen to be equal. This particularly contains the case of *completely inelastic impacts*, $\varepsilon_{Ni} = 0$.

8.7 Decomposition of an Asymmetric Characteristic

The Coulomb friction characteristic (Fig. 6.2) as well as the tangential impact characteristic for compression (Fig. 8.3) have a point symmetry with respect to the origin and can therefore be decomposed by the method shown in Fig. 6.4 and eqs. (6.20)–(6.22). In the case of expansion, however, the impact law in eq. (8.38) misses this symmetry due to different positive and negative maximal absolute values of the impulse Λ_{TPi}. The structure of such an impact characteristic is depicted in Fig. 8.6 and corresponds to an impact law

$$\begin{gathered} -\Lambda_{Gi}^{(-)} \leq \Lambda_{Ti} \leq +\Lambda_{Gi}^{(+)}\,; \quad \Lambda_{Gi}^{(+)} \geq 0\,; \quad \Lambda_{Gi}^{(-)} \geq 0 \\ \begin{cases} -\Lambda_{Gi}^{(-)} < \Lambda_{Ti} < +\Lambda_{Gi}^{(+)} & \Rightarrow \quad \dot{g}_{Ti} = 0 \\ \Lambda_{Ti} = +\Lambda_{Gi}^{(+)} & \Rightarrow \quad \dot{g}_{Ti} \leq 0 \\ \Lambda_{Ti} = -\Lambda_{Gi}^{(-)} & \Rightarrow \quad \dot{g}_{Ti} \geq 0\,. \end{cases} \end{gathered} \tag{8.79}$$

Even this characteristic can be decomposed according to eqs. (6.20)–(6.22) (as long as 0 is contained in the admissible values of Λ_{Ti}) and leads to [21]

$$\Lambda_{Ti} = \Lambda_{Ti}^{(+)} - \Lambda_{Ti}^{(-)} \tag{8.80}$$

$$\begin{aligned} \dot{g}_{Ti} &= z_i^+ - \dot{g}_{Ti}^-\,; \quad & \Lambda_{T0i}^{(+)} &= \Lambda_{Gi}^{(-)} - \Lambda_{Ti}^{(-)} \\ \dot{g}_{Ti} &= \dot{g}_{Ti}^+ - z_i^-\,; \quad & \Lambda_{T0i}^{(-)} &= \Lambda_{Gi}^{(+)} - \Lambda_{Ti}^{(+)} \end{aligned} \tag{8.81}$$

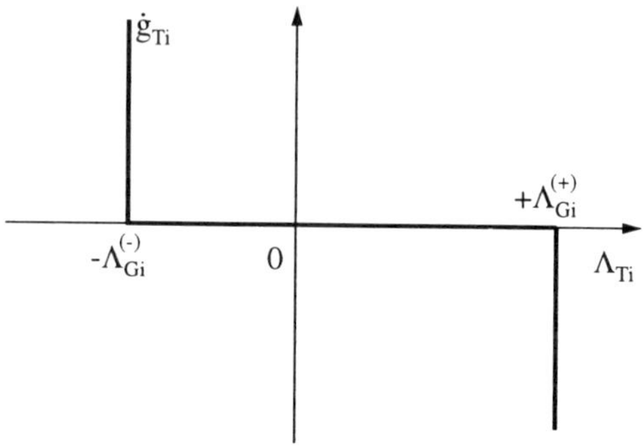

Figure 8.6: Asymmetric Tangential Characteristic

$$\begin{array}{lll}
\dot{g}_{Ti}^{-} \geq 0\,; & \Lambda_{Ti}^{(-)} \geq 0\,; & \dot{g}_{Ti}^{-}\Lambda_{Ti}^{(-)} = 0 \\
\dot{g}_{Ti}^{+} \geq 0\,; & \Lambda_{Ti}^{(+)} \geq 0\,; & \dot{g}_{Ti}^{+}\Lambda_{Ti}^{(+)} = 0 \\
\Lambda_{T0i}^{(+)} \geq 0\,; & z_i^{+} \geq 0\,; & \Lambda_{T0i}^{(+)} z_i^{+} = 0 \\
\Lambda_{T0i}^{(-)} \geq 0\,; & z_i^{-} \geq 0\,; & \Lambda_{T0i}^{(-)} z_i^{-} = 0\,.
\end{array} \tag{8.82}$$

As in Chapter 6, the superscripts $(+)$ and $(-)$ are only chosen for distinctive reasons, whereas $+$ and $-$ denote the positive and negative parts of a term: $a^+ = \frac{1}{2}(|a| + a)$; $a^- = \frac{1}{2}(|a| - a)$; $a = a^+ - a^-$.

The unknown tangential impulse Λ_{Ti} is expressed by the difference in eq. (8.80), which corresponds on force level to eq. (6.20). The four equations in (8.81) consist of the tangential velocities $\dot{g}_{Ti}$, which are split into their positive and negative parts, and the friction saturations on impulse level $\Lambda_{T0i}^{(\pm)}$, which are defined as the difference of the maximal transferable impulses $\Lambda_{Gi}^{(\mp)}$ and the actual tangential impulses $\Lambda_{Ti}^{(\mp)}$. In (8.82) the resulting inequalities and complementarity conditions are summarized, which completes the mathematical description of the impact law.

The tangential characteristics during compression and expansion fit into the scheme of Fig. 8.6 and eqs. (8.80)–(8.82). Thus, the whole impact process can be stated by complementarity conditions, which is done in the following two sections.

8.8 An LCP Formulation for Compression

The compression phase is completely determined by the kinetics and kinematics equations (8.7), the impact law in the normal direction (8.13), and the tangential impact characteristic (8.23). For establishing the equations for the LCP we first consider each of the tangential relative velocities in (8.7) as a pair of simultaneously acting constraints in order to express all of the tangential impulses $\boldsymbol{\Lambda}_{TC}$ of the impact contacts $i \in I_S$ as a difference similar to (8.80):

$$\boldsymbol{\Lambda}_{TC} = \boldsymbol{\Lambda}_{TC}^{(+)} - \boldsymbol{\Lambda}_{TC}^{(-)} . \tag{8.83}$$

Thus the basic set of equations resulting from (8.7) and (8.83) is now

$$\begin{aligned} \boldsymbol{M}(\dot{\boldsymbol{q}}_C - \dot{\boldsymbol{q}}_A) &= \underbrace{\begin{pmatrix} \boldsymbol{W}_N & +\boldsymbol{W}_T & -\boldsymbol{W}_T \end{pmatrix}}_{\boldsymbol{W}_S} \underbrace{\begin{pmatrix} \boldsymbol{\Lambda}_{NC} \\ \boldsymbol{\Lambda}_{TC}^{(+)} \\ \boldsymbol{\Lambda}_{TC}^{(-)} \end{pmatrix}}_{\boldsymbol{\Lambda}_C} \\ \underbrace{\begin{pmatrix} \dot{\boldsymbol{g}}_{NC} \\ +\dot{\boldsymbol{g}}_{TC} \\ -\dot{\boldsymbol{g}}_{TC} \end{pmatrix}}_{\dot{\boldsymbol{g}}_{MC}} &= \underbrace{\begin{pmatrix} \boldsymbol{W}_N^T \\ +\boldsymbol{W}_T^T \\ -\boldsymbol{W}_T^T \end{pmatrix}}_{\boldsymbol{W}_S^T} (\dot{\boldsymbol{q}}_C - \dot{\boldsymbol{q}}_A) + \underbrace{\begin{pmatrix} \dot{\boldsymbol{g}}_{NA} \\ +\dot{\boldsymbol{g}}_{TA} \\ -\dot{\boldsymbol{g}}_{TA} \end{pmatrix}}_{\dot{\boldsymbol{g}}_{MA}} , \end{aligned} \tag{8.84}$$

where the terms below the braces are used as abbreviations in the following. Next, the tangential velocities $\dot{\boldsymbol{g}}_{TC}$ have to be split and the friction saturations $\boldsymbol{\Lambda}_{T0C}$ have to be formulated according to eq. (8.81). Comparing the right part of Fig. 8.3 with Fig. 8.6, we identify $\dot{g}_{TCi}$ as $\dot{g}_{Ti}$ and $\mu_i \Lambda_{NCi}$ as $\Lambda_{Gi}^{(+)} = \Lambda_{Gi}^{(-)}$. Inserting these values into (8.81) and writing the resulting equations for all impact contacts $i \in I_S$ in matrix notation yields

$$\begin{aligned} \underbrace{\begin{pmatrix} \dot{\boldsymbol{g}}_{NC} \\ +\dot{\boldsymbol{g}}_{TC} \\ -\dot{\boldsymbol{g}}_{TC} \end{pmatrix}}_{\dot{\boldsymbol{g}}_{MC}} &= \underbrace{\begin{pmatrix} \dot{\boldsymbol{g}}_{NC} \\ \dot{\boldsymbol{g}}_{TC}^{+} \\ \dot{\boldsymbol{g}}_{TC}^{-} \end{pmatrix}}_{\dot{\boldsymbol{g}}_C} - \underbrace{\begin{pmatrix} 0 & 0 \\ \boldsymbol{E} & 0 \\ 0 & \boldsymbol{E} \end{pmatrix}}_{\boldsymbol{I}_S^T} \underbrace{\begin{pmatrix} \boldsymbol{z}_C^{-} \\ \boldsymbol{z}_C^{+} \end{pmatrix}}_{\boldsymbol{z}_C} \\ \underbrace{\begin{pmatrix} \boldsymbol{\Lambda}_{T0C}^{(-)} \\ \boldsymbol{\Lambda}_{T0C}^{(+)} \end{pmatrix}}_{\boldsymbol{\Lambda}_{T0C}} &= \underbrace{\begin{pmatrix} \overline{\overline{\boldsymbol{\mu}}}_S & -\boldsymbol{E} & 0 \\ \overline{\overline{\boldsymbol{\mu}}}_S & 0 & -\boldsymbol{E} \end{pmatrix}}_{\boldsymbol{N}_S - \boldsymbol{I}_S} \underbrace{\begin{pmatrix} \boldsymbol{\Lambda}_{NC} \\ \boldsymbol{\Lambda}_{TC}^{(+)} \\ \boldsymbol{\Lambda}_{TC}^{(-)} \end{pmatrix}}_{\boldsymbol{\Lambda}_C} , \end{aligned} \tag{8.85}$$

where $\overline{\overline{\boldsymbol{\mu}}}_S$ is a diagonal matrix consisting of the coefficients of friction μ_i from (8.23). After the elimination of $\dot{\boldsymbol{q}}_C - \dot{\boldsymbol{q}}_A$ and $\dot{\boldsymbol{g}}_{MC}$ from (8.84) and (8.85), the remaining two equations are now rewritten as

$$\begin{aligned} \dot{\boldsymbol{g}}_C &= \boldsymbol{W}_S^T \boldsymbol{M}^{-1} \boldsymbol{W}_S \boldsymbol{\Lambda}_C + \boldsymbol{I}_S^T \boldsymbol{z}_C + \dot{\boldsymbol{g}}_{MA} \\ \boldsymbol{\Lambda}_{T0C} &= (\boldsymbol{N}_S - \boldsymbol{I}_S) \boldsymbol{\Lambda}_C \end{aligned} \tag{8.86}$$

and then stated in matrix notation together with the complementarity conditions from (8.13) and (8.82):

$$\begin{aligned}
&\begin{pmatrix} \dot{g}_C \\ \Lambda_{T0C} \end{pmatrix} = \begin{pmatrix} W_S^T M^{-1} W_S & I_S^T \\ N_S - I_S & 0 \end{pmatrix} \begin{pmatrix} \Lambda_C \\ z_C \end{pmatrix} + \begin{pmatrix} \dot{g}_{MA} \\ 0 \end{pmatrix} \\
&\begin{pmatrix} \dot{g}_C \\ \Lambda_{T0C} \end{pmatrix} \geq 0; \quad \begin{pmatrix} \Lambda_C \\ z_C \end{pmatrix} \geq 0; \quad \begin{pmatrix} \dot{g}_C \\ \Lambda_{T0C} \end{pmatrix}^T \begin{pmatrix} \Lambda_C \\ z_C \end{pmatrix} = 0
\end{aligned} \tag{8.87}$$

which is the standard LCP formulation for the compression phase of the impact:

$$\begin{aligned}
&y = Ax + b; \quad y \geq 0; \quad x \geq 0; \quad y^T x = 0 \\
&y,\ x \in \mathbb{R}^{5n_S}.
\end{aligned} \tag{8.88}$$

Its solution $y \in \mathbb{R}^{5n_S}$, $x \in \mathbb{R}^{5n_S}$ contains all unknown contact impulses and velocities during compression. As in eq. (6.35), even here unique solutions with respect to $\dot{q}_C$ can be expected for the frictionless case $N_S = 0$, because the matrix A is then positive semidefinite.

8.9 An LCP Formulation for Expansion

In order to formulate the LCP for expansion we proceed similarly to the case of compression. The expansion phase is determined by the kinetics and kinematics equations (8.25), the transformation onto the new variables $(\Lambda_{NPi}, \Lambda_{TPi})$ in (8.29), (8.35), and the normal and tangential impact law (8.30), (8.38) together with the abbreviations (8.39). In the first step the expansion impulses $(\Lambda_{NEi}, \Lambda_{TEi})$ in (8.25) are expressed in terms of $(\Lambda_{NPi}, \Lambda_{TPi})$, which is done by using eqs. (8.29) and (8.35):

$$\begin{aligned}
&M(\dot{q}_E - \dot{q}_C) = (W_N W_T) \begin{pmatrix} \Lambda_{NP} \\ \Lambda_{TP} \end{pmatrix} + (W_N W_T) \begin{pmatrix} \bar{\bar{\epsilon}}_N \Lambda_{NC} \\ \bar{\bar{\epsilon}}_N \Lambda_{TC} \end{pmatrix} \\
&\begin{pmatrix} \dot{g}_{NE} \\ \dot{g}_{TE} \end{pmatrix} = \begin{pmatrix} W_N^T \\ W_T^T \end{pmatrix} (\dot{q}_E - \dot{q}_C) + \begin{pmatrix} \dot{g}_{NC} \\ \dot{g}_{TC} \end{pmatrix},
\end{aligned} \tag{8.89}$$

where the diagonal matrix $\bar{\bar{\epsilon}}_N$ consists of the coefficients of restitution ε_{Ni} in the normal directions. In a second step we split the tangential impulses similarly as in (8.83),

$$\Lambda_{TP} = \Lambda_{TP}^{(+)} - \Lambda_{TP}^{(-)}; \quad \Lambda_{TC} = \Lambda_{TC}^{+} - \Lambda_{TC}^{-}, \tag{8.90}$$

insert them into eq. (8.89) and write the tangential kinematic condition twice:

$$
\begin{aligned}
\boldsymbol{M}(\dot{\boldsymbol{q}}_E - \dot{\boldsymbol{q}}_C) &= \underbrace{\begin{pmatrix} \boldsymbol{W}_N^T \\ +\boldsymbol{W}_T^T \\ -\boldsymbol{W}_T^T \end{pmatrix}^T}_{\boldsymbol{W}_S} \underbrace{\begin{pmatrix} \boldsymbol{\Lambda}_{NP} \\ \boldsymbol{\Lambda}_{TP}^{(+)} \\ \boldsymbol{\Lambda}_{TP}^{(-)} \end{pmatrix}}_{\boldsymbol{\Lambda}_P} + \underbrace{\begin{pmatrix} \boldsymbol{W}_N^T \\ +\boldsymbol{W}_T^T \\ -\boldsymbol{W}_T^T \end{pmatrix}^T}_{\boldsymbol{W}_S} \underbrace{\begin{pmatrix} \bar{\bar{\boldsymbol{\epsilon}}}_N \boldsymbol{\Lambda}_{NC} \\ \bar{\bar{\boldsymbol{\epsilon}}}_N \boldsymbol{\Lambda}_{TC}^{+} \\ \bar{\bar{\boldsymbol{\epsilon}}}_N \boldsymbol{\Lambda}_{TC}^{-} \end{pmatrix}}_{\boldsymbol{\Lambda}_R} \\
\underbrace{\begin{pmatrix} \dot{\boldsymbol{g}}_{NE} \\ +\dot{\boldsymbol{g}}_{TE} \\ -\dot{\boldsymbol{g}}_{TE} \end{pmatrix}}_{\dot{\boldsymbol{g}}_{ME}} &= \underbrace{\begin{pmatrix} \boldsymbol{W}_N^T \\ +\boldsymbol{W}_T^T \\ -\boldsymbol{W}_T^T \end{pmatrix}}_{\boldsymbol{W}_S^T} (\dot{\boldsymbol{q}}_E - \dot{\boldsymbol{q}}_C) + \underbrace{\begin{pmatrix} \dot{\boldsymbol{g}}_{NC} \\ +\dot{\boldsymbol{g}}_{TC} \\ -\dot{\boldsymbol{g}}_{TC} \end{pmatrix}}_{\dot{\boldsymbol{g}}_{MC}} .
\end{aligned} \tag{8.91}
$$

The terms $\boldsymbol{\Lambda}_{TC}^{+}$ and $\boldsymbol{\Lambda}_{TC}^{-}$ denote the positive and negative parts of the known tangential compression impulses $\boldsymbol{\Lambda}_{TC}$. For the evaluation of the impact law the tangential relative velocities after expansion have to be split with respect to the left two equations in (8.81). From eqs. (8.38) and (8.79) we see that the velocities $\dot{g}_{TEi}$ and $\dot{g}_{Ti}$ correspond to each other; thus we can write (8.81) in matrix notation applied to the terms $\dot{g}_{TEi}$:

$$
\underbrace{\begin{pmatrix} \dot{\boldsymbol{g}}_{NE} \\ +\dot{\boldsymbol{g}}_{TE} \\ -\dot{\boldsymbol{g}}_{TE} \end{pmatrix}}_{\dot{\boldsymbol{g}}_{ME}} = \underbrace{\begin{pmatrix} \dot{\boldsymbol{g}}_{NE} \\ \dot{\boldsymbol{g}}_{TE}^{+} \\ \dot{\boldsymbol{g}}_{TE}^{-} \end{pmatrix}}_{\dot{\boldsymbol{g}}_{E}} - \underbrace{\begin{pmatrix} 0 & 0 \\ \boldsymbol{E} & 0 \\ 0 & \boldsymbol{E} \end{pmatrix}}_{\boldsymbol{I}_S^T} \underbrace{\begin{pmatrix} \boldsymbol{z}_E^{-} \\ \boldsymbol{z}_E^{+} \end{pmatrix}}_{\boldsymbol{z}_E} . \tag{8.92}
$$

Finally, the friction saturations in the right two equations of (8.81) have to be stated. Comparing expressions (8.38) and (8.79), the correspondences

$$
\begin{aligned}
\Lambda_{Ti} &\triangleq \Lambda_{TPi} \\
+\Lambda_{Gi}^{(+)} &\triangleq +\mu_i^{(+)}(\Lambda_{NPi} + \varepsilon_{Ni}\Lambda_{NCi}) + \varepsilon_{Ti}^{(+)}\varepsilon_{Ni}\Lambda_{TCi} \\
-\Lambda_{Gi}^{(-)} &\triangleq -\mu_i^{(-)}(\Lambda_{NPi} + \varepsilon_{Ni}\Lambda_{NCi}) - \varepsilon_{Ti}^{(-)}\varepsilon_{Ni}\Lambda_{TCi}
\end{aligned} \tag{8.93}
$$

can be achieved. After inserting eq. (8.90), we obtain

$$
\begin{aligned}
\Lambda_{Ti}^{(+)} &\triangleq \Lambda_{TPi}^{(+)} \\
\Lambda_{Ti}^{(-)} &\triangleq \Lambda_{TPi}^{(-)} \\
+\Lambda_{Gi}^{(+)} &\triangleq +\mu_i^{(+)}(\Lambda_{NPi} + \varepsilon_{Ni}\Lambda_{NCi}) + (1 - \varepsilon_{Ti})\varepsilon_{Ni}\Lambda_{TCi}^{-} - \varepsilon_{Ni}\Lambda_{TCi}^{+} \\
-\Lambda_{Gi}^{(-)} &\triangleq -\mu_i^{(-)}(\Lambda_{NPi} + \varepsilon_{Ni}\Lambda_{NCi}) - (1 - \varepsilon_{Ti})\varepsilon_{Ni}\Lambda_{TCi}^{+} + \varepsilon_{Ni}\Lambda_{TCi}^{-}
\end{aligned} \tag{8.94}
$$

where the terms $\varepsilon_{Ti}^{(\pm)}\varepsilon_{Ni}\Lambda_{TCi}$ have been simplified by using (8.39) and $\sigma_i\Lambda_{TCi} = |\Lambda_{TCi}| = \Lambda_{TCi}^{+} + \Lambda_{TCi}^{-}$:

$$\begin{aligned}\varepsilon_{Ni}\varepsilon_{Ti}^{(+)}\Lambda_{TCi} &= (1-\varepsilon_{Ti})\varepsilon_{Ni}\Lambda_{TCi}^{-} - \varepsilon_{Ni}\Lambda_{TCi}^{+}\\ \varepsilon_{Ni}\varepsilon_{Ti}^{(-)}\Lambda_{TCi} &= (1-\varepsilon_{Ti})\varepsilon_{Ni}\Lambda_{TCi}^{+} - \varepsilon_{Ni}\Lambda_{TCi}^{-}\,.\end{aligned}$$

With the help of (8.94) the friction saturations in the right two equations of (8.81) can now be stated in matrix notation as

$$\begin{aligned}\underbrace{\begin{pmatrix}\boldsymbol{\Lambda}_{T0P}^{(-)}\\ \boldsymbol{\Lambda}_{T0P}^{(+)}\end{pmatrix}}_{\boldsymbol{\Lambda}_{T0P}} &= \underbrace{\begin{pmatrix}\overline{\overline{\boldsymbol{\mu}}}_S^{(+)} & -\boldsymbol{E} & 0\\ \overline{\overline{\boldsymbol{\mu}}}_S^{(-)} & 0 & -\boldsymbol{E}\end{pmatrix}}_{\widetilde{\boldsymbol{N}}_S-\boldsymbol{I}_S}\underbrace{\begin{pmatrix}\boldsymbol{\Lambda}_{NP}\\ \boldsymbol{\Lambda}_{TP}^{(+)}\\ \boldsymbol{\Lambda}_{TP}^{(-)}\end{pmatrix}}_{\boldsymbol{\Lambda}_P}\\ &+ \underbrace{\begin{pmatrix}\overline{\overline{\boldsymbol{\mu}}}_S^{(+)} & -\boldsymbol{E} & \boldsymbol{E}-\overline{\overline{\boldsymbol{\epsilon}}}_T\\ \overline{\overline{\boldsymbol{\mu}}}_S^{(-)} & \boldsymbol{E}-\overline{\overline{\boldsymbol{\epsilon}}}_T & -\boldsymbol{E}\end{pmatrix}}_{\widetilde{\boldsymbol{N}}_S-\boldsymbol{I}_B}\underbrace{\begin{pmatrix}\overline{\overline{\boldsymbol{\epsilon}}}_N\boldsymbol{\Lambda}_{NC}\\ \overline{\overline{\boldsymbol{\epsilon}}}_N\boldsymbol{\Lambda}_{TC}^{+}\\ \overline{\overline{\boldsymbol{\epsilon}}}_N\boldsymbol{\Lambda}_{TC}^{-}\end{pmatrix}}_{\boldsymbol{\Lambda}_R}\,.\end{aligned} \tag{8.95}$$

The coefficients of the diagonal matrices $\overline{\overline{\boldsymbol{\mu}}}_S^{(\pm)}$ and $\overline{\overline{\boldsymbol{\epsilon}}}_T$ consist of the terms $\mu_i^{(\pm)} = [1-\nu(\frac{1}{2}\mp\frac{\sigma_i}{2})]\mu_i$ from (8.39) and ε_{Ti} for each of the contacts i, where ν_i and ε_{Ti} are the tangential impact parameter from (8.37), and $\sigma_i = \text{sgn}(\Lambda_{TC})$ takes into account the direction of the reversible tangential impulse portions. After the elimination of $\dot{\boldsymbol{q}}_E - \dot{\boldsymbol{q}}_C$ and $\dot{\boldsymbol{g}}_{ME}$ from (8.91) and (8.92), the resulting equation together with (8.95) is

$$\begin{aligned}\dot{\boldsymbol{g}}_E &= \boldsymbol{W}_S^T\boldsymbol{M}^{-1}\boldsymbol{W}_S\boldsymbol{\Lambda}_P + \boldsymbol{I}_S^T\boldsymbol{z}_E + \boldsymbol{W}_S^T\boldsymbol{M}^{-1}\boldsymbol{W}_S\boldsymbol{\Lambda}_R + \dot{\boldsymbol{g}}_{MC}\\ \boldsymbol{\Lambda}_{T0P} &= (\widetilde{\boldsymbol{N}}_S - \boldsymbol{I}_S)\boldsymbol{\Lambda}_P + (\widetilde{\boldsymbol{N}}_S - \boldsymbol{I}_B)\boldsymbol{\Lambda}_R,\end{aligned} \tag{8.96}$$

which yield a representation in matrix notation together with the complementarity conditions from (8.30) and (8.82):

$$\begin{aligned}\begin{pmatrix}\dot{\boldsymbol{g}}_E\\ \boldsymbol{\Lambda}_{T0P}\end{pmatrix} &= \begin{pmatrix}\boldsymbol{W}_S^T\boldsymbol{M}^{-1}\boldsymbol{W}_S & \boldsymbol{I}_S^T\\ \widetilde{\boldsymbol{N}}_S - \boldsymbol{I}_S & 0\end{pmatrix}\begin{pmatrix}\boldsymbol{\Lambda}_P\\ \boldsymbol{z}_E\end{pmatrix}\\ &+ \begin{pmatrix}\boldsymbol{W}_S^T\boldsymbol{M}^{-1}\boldsymbol{W}_S\boldsymbol{\Lambda}_R + \dot{\boldsymbol{g}}_{MC}\\ (\widetilde{\boldsymbol{N}}_S - \boldsymbol{I}_B)\boldsymbol{\Lambda}_R\end{pmatrix}\\ \begin{pmatrix}\dot{\boldsymbol{g}}_E\\ \boldsymbol{\Lambda}_{T0P}\end{pmatrix} \geq 0;\quad &\begin{pmatrix}\boldsymbol{\Lambda}_P\\ \boldsymbol{z}_E\end{pmatrix} \geq 0;\quad \begin{pmatrix}\dot{\boldsymbol{g}}_E\\ \boldsymbol{\Lambda}_{T0P}\end{pmatrix}^T\begin{pmatrix}\boldsymbol{\Lambda}_P\\ \boldsymbol{z}_E\end{pmatrix} = 0\,.\end{aligned} \tag{8.97}$$

As in the case of compression, this structure forms a Linear Complementarity Problem in standard form:

$$\begin{aligned} &\boldsymbol{y} = \boldsymbol{A}\boldsymbol{x} + \boldsymbol{b}\,; \quad \boldsymbol{y} \geq \boldsymbol{0}\,; \quad \boldsymbol{x} \geq \boldsymbol{0}\,; \quad \boldsymbol{y}^T\boldsymbol{x} = 0 \\ &\boldsymbol{y},\, \boldsymbol{x} \in \mathbb{R}^{5n_S}\,. \end{aligned} \tag{8.98}$$

Its solution $\boldsymbol{y} \in \mathbb{R}^{5n_S}$, $\boldsymbol{x} \in \mathbb{R}^{5n_S}$ contains all the contact impulses during expansion and the corresponding relative velocities. Thus the complete process of simultaneously acting impulsive forces can be handled by solving the two complementarity problems from (8.88) and (8.98), where the system state after the impact (sliding, stiction, separation) is determined by the unilateral formulation of the impact laws.

8.10 Remarks on Impacts with Friction

In many classical textbooks impacts are classified by their locations and by the manner of impulse transfer: They are called direct or oblique, central or eccentric, and rough or smooth [54]. All these cases are contained in the impact theory presented above. The decision whether impacts are direct or oblique depends exclusively on the kinematics. An impact with a vanishing tangential approaching velocity ($\dot{g}_{TAi} = 0$) is usually referred to as direct, in the other case ($\dot{g}_{TAi} \neq 0$) as oblique . The classification rough or smooth takes into account the properties of the colliding surfaces. At a smooth collision no tangential impulse can be transferred, which corresponds to a value of $\mu_i = 0$. Rough collisions allow tangential acting impulses ($\mu_i \neq 0$) which might be (partly) reversible or completely irreversible due to dry friction. In our theory both effects are taken into account by the formulation of the expansion phase presented in Section 8.3. The terminology "central or excentric" is originally related to a collision of two single bodies. Such an impact is called central if both centers of masses lie on the line generated by the normal vectors at the contact points, otherwise it is eccentric. Even this case is clearly contained in the theory but in a much more general manner: By dealing with multibody systems generally more than the two colliding bodies are affected by the impact due to the bilateral constraints within the system. The main property of a central impact consists of the fact that an acting tangential impulse does not influence the normal direction. Thus we define an impact to be "generalized central" if the constraint matrices $\boldsymbol{W}_N$ and $\boldsymbol{W}_T$ of the impact contacts show the property $\boldsymbol{W}_N^T\boldsymbol{M}^{-1}\boldsymbol{W}_T = 0$. This definition clearly includes the central impact between two single bodies described above and provides a special structure of the kinematics and kinetics equations.

During compression, for example, the elimination of $\dot{\boldsymbol{q}}_C - \dot{\boldsymbol{q}}_A$ from (8.7) leads to the decoupled equations $\dot{\boldsymbol{g}}_{NC} = \boldsymbol{W}_N^T\boldsymbol{M}^{-1}\boldsymbol{W}_N\boldsymbol{\Lambda}_N + \dot{\boldsymbol{g}}_{NA}$ and $\dot{\boldsymbol{g}}_{TC} = \boldsymbol{W}_T^T\boldsymbol{M}^{-1}\boldsymbol{W}_T\boldsymbol{\Lambda}_T + \dot{\boldsymbol{g}}_{TA}$. Furthermore, the impact law in the normal direction (8.13) does not depend on terms corresponding to the tangential direction.

The normal directions are therefore completely decoupled from the tangential behavior and can be solved separately. After the solution of the normal impacts has been obtained, the tangential behavior is then calculated by using the impact laws (8.23) which contain the known normal impulses. The same procedure can then be applied for the expansion phase. Generalized central impacts therefore lead to a one-sided coupled problem where the normal directions are not affected by the tangential impulses. As we have seen in Section 8.6 the impact laws of Newton and Poisson coincide in the *frictionless* case if all of the coefficients of restitution are chosen to be equal. This result can now be extended to generalized central impacts. A typical system where only generalized central impacts occur is a collection of circular bodies or spheres in the three-dimensional case which is extensively investigated in [34, 62] by a combination of Newton's and Coulomb's laws.

Note that the impact laws described in Sections 8.2 and 8.3 allow any arbitrary combination of simultaneously occurring oblique, eccentric or rough collisions and much more: Even existing contacts which may slide or stick at the moment of a multiple collision at other locations in the system are taken into account and may undergo induced state transitions. Additionally, each contact is checked to see whether impulses are transferred. Superfluous impact constraints are automatically "removed" from the evaluation by the unilateral formulation of the normal impact law. The dividing of the impacts into compression and expansion phases provides various possibilities for curve-fitting methods. Because the four impact parameters are not needed until the expansion phase, they may depend on each of the magnitudes obtained after compression. This enables a realistic modeling of numerous impact processes. However, one should carefully check the assumptions made at the beginning of Chapter 7 in order to decide whether the impact laws may be applied. Wave effects, for example, cannot be taken into account. For some applications, the duration of the impact phases cannot be neglected with respect to the frequencies of the system, and requires the introduction of certain compliance elements. Furthermore, the assumption of rigid bodies and constant displacements during the impact always coincides with a loss of information about the system. One cannot expect to obtain a detailed description of the impact process such as at an evaluation based on a finite element method. On the other hand, such an approach is so time consuming that it cannot be applied on the simulation of the dynamics of large systems. In sum the impact law is clearly only a rough approximation of the processes in the contact zones, but, applied in the right manner, it is a powerful tool for the simulation of complicated dynamical systems.

As a last remark we want to explain why Newton's law should be avoided for impacts with friction. In dynamics, the accelerations always result from the forces which are applied on the system, and *not* vice versa. The same should be valid on the impulse level. At an impact, impulses act in the contacts and lead the system to instantaneous changes in the velocities. Newton's impact law prescribes the velocities at the end of the impact as dependent on the velocities before the impact. This, however, is a kinematic condition which would correspond on the force level with prescribed accelerations. A combination of Newton's law in the normal direction and Coulomb's law in the tangential direction therefore leads to a physically incorrect situation. Due to the prescribed values of the final velocities in the normal directions, no tangential impulse can change them, however strong it may be. In general dynamic systems, however, it is a matter of course that forces which are restricted to certain lines of interaction may influence the relative motions in any directions. The normal and tangential directions are decoupled only in the case of the generalized central impacts and do not interact in the stated manner. Therefore, the concept of Poisson is much more suitable for a combination together with Coulomb's law, because both the normal and tangential impact laws are based on conditions on the transferred impulses.

8.11 Example: Double Impact on a Rod

In this section we continue the discussion of the rod presented in Section 7.5 [21]. We consider the same mechanical model (Fig. 7.2) and the same impact configuration (Fig. 7.3) where the rod hits both obstacles simultaneously. The equations of motion and the kinematics conditions for this situation have been presented in eqs. (7.30) and (7.31) together with the bilateral formulation of a completely inelastic impact at both contacts with respect to Newton's law (eq. 7.32). After the evaluation of the impact equations we derived the transferred impulses (7.38) and saw that one of them may become negative, depending on the location of obstacle 1; compare Fig. 7.4. Instead of eq. (7.32) we now apply a completely inelastic impact ($\varepsilon_{N1} = \varepsilon_{N2} = 0$) with respect to the impact laws presented in Sections 8.2 and 8.3. Due to the vanishing coefficients of restitution the impact process is already completed after the compression phase. Thus we can write the unilateral normal impact equations (8.13) as

$$\begin{array}{lll} \Lambda_{NC1} \geq 0\,; & \dot{g}_{NC1} \geq 0\,; & \Lambda_{NC1}\,\dot{g}_{NC1} = 0 \\ \Lambda_{NC2} \geq 0\,; & \dot{g}_{NC2} \geq 0\,; & \Lambda_{NC2}\,\dot{g}_{NC2} = 0 \end{array} \tag{8.99}$$

$$\begin{array}{lll} \Lambda_{N1} = \Lambda_{NC1} + \Lambda_{NE1}\,; & \dot{g}_{NE1} = \dot{g}_{NC1}\,; & \Lambda_{NE1} = 0 \\ \Lambda_{N2} = \Lambda_{NC2} + \Lambda_{NE2}\,; & \dot{g}_{NE2} = \dot{g}_{NC2}\,; & \Lambda_{NE2} = 0 \end{array} \tag{8.100}$$

A numerical evaluation of the impact law (8.99) together with eqs. (7.30), (7.31) and (8.100) yields the behavior depicted in Fig. 8.7, where the parameter values $m = 1$, $l = 1$, $b = 1$, $-1 \leq a \leq 1$, $\dot{y}_A = -1$ have been used.

Figure 8.7 shows the relative velocities after the impact and the transferred impulses in the normal directions as functions of the distance of the obstacles.

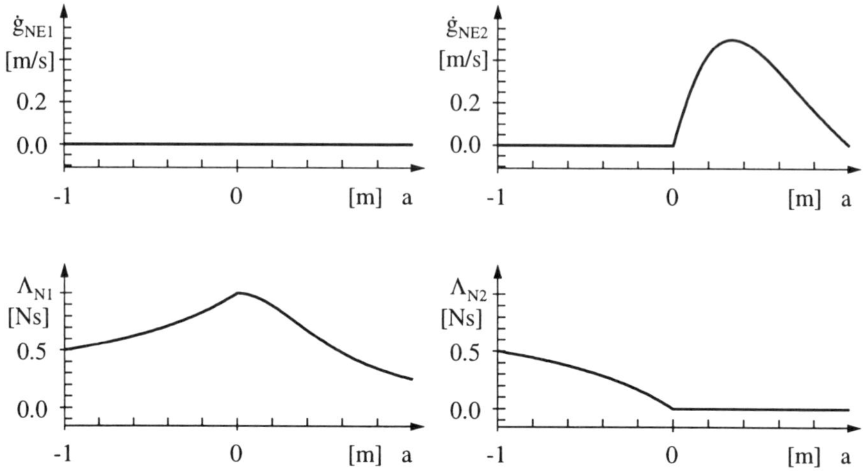

Figure 8.7: Relative Velocities and Impulses

In the case $a = -1$, $b = 1$ both obstacles lie symmetrically at the ends of the rod; thus the transferred impulses are of the same magnitude and, considering an inelastic impact $\varepsilon_{N1} = \varepsilon_{N2} = 0$, the relative velocities after the collision are equal to zero. For $a = 0$, $b = 1$ obstacle 1 rests under the body's center of mass and transfers the whole impulse. The case $0 < a \leq 1$, $b = 1$ corresponds to a configuration where both obstacles lie on the righthand side of the rod. Thus, obstacle 2 does not participate in the impact and could be removed without changing the impact process. No impulse is transferred, and the relative velocity after the impact is greater than zero. By the kinematic condition (7.31) different relative velocities at the end of the impact result in a rotational motion of the rod. Thus the rod remains in contact with obstacle 1 after the impact but rotates counterclockwise and separates from obstacle 2, which is physically correct. The decision whether an impulse is transferred at the obstacles is automatically taken into account by the unilateral formulation of the impact law in the normal direction and may not be explicitly considered by solving this problem. Situations where negative impulses occur, as in the right part of Fig. 7.4, are excluded by the inequalities and complementarity conditions of the impact law (8.99).

One might argue that a double impact as in Fig. 7.3, where both obstacles are hit simultaneously, only happens with a vanishing probability and is of no interest for practical applications. In multibody systems with many impact contacts, however, such situations occur for numerical reasons and must therefore be taken into account. Much more important than the impact configuration in Fig. 7.3 is collisions which affect existing contacts. As an

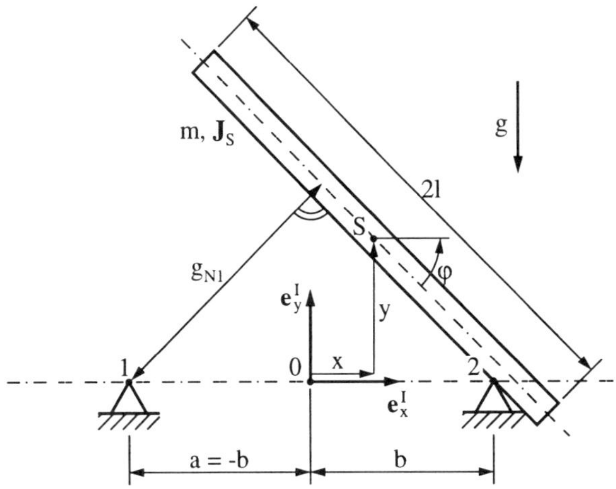

Figure 8.8: The Rocking Rod

example we look once more at the rod and investigate its rocking.

The mechanical model is shown in Fig. 8.8 and coincides with the rod in Fig. 7.2. The obstacles are now arranged symmetrically with respect to the rod's center of mass, which gives of $a = -b$. Now we look at an initial condition corresponding to Fig. 8.8: The rod is inclined and contacts obstruction 2, where the coefficient of friction is chosen to be large enough to enable sticking. From this position we drop the rod. It then turns around contact point 2 until it hits obstacle 1. This collision is assumed to be completely inelastic. After the impact two situations may occur, depending on the distance $2b$ of the obstacles: For small values of b the rod will continue rocking by an alternate turning around the two contact points. A collision with one of the obstacles causes an instantaneous detachment at the other contact. In the second case, the distance between the obstacles is large enough to admit a compressive impulse at *both* contacts, produced by a *single* collision. Due to the completely inelastic behavior, the rod stops turning and rests on both obstacles.

Fig. 8.9 shows the phase space diagrams of the coordinate φ for different distances between the obstacles. The plots were obtained by a numerical evaluation of the equations of motion using the parameters $m = 1$, $l = 1$, $g = 10$, $\varepsilon_{N1} = \varepsilon_{N2} = 0$, $\mu_1 = \mu_2 \gg 1$. In the left diagram we see the rocking of the rod for $b = 0.2$. At each collision the velocity $\dot{\varphi}$ jumps but does *not* change its sign. The turning direction remains the same, but its velocity instantaneously decreases, which corresponds to a loss of energy caused by the completely inelastic impact. After an infinite number of collisions which

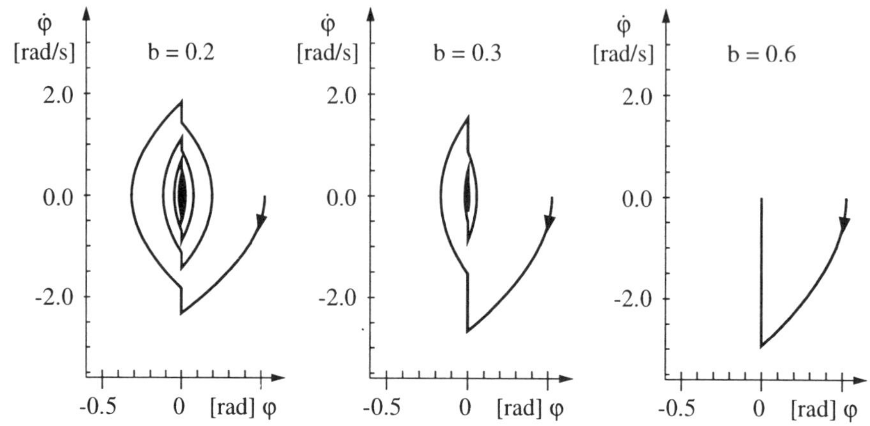

Figure 8.9: Phase Space Diagrams of the Rocking Rod

take place in a finite time interval, the rod stops moving and rests on the obstacles. The middle diagram in Fig. 8.9 shows the same process for an enlarged distance of the obstacles. The velocity jumps are much stronger, but the rod is still rocking. In the right diagram only one impact occurs. After that impact, which transfers compressive impulses at both contacts, the rod rests on both obstacles. Rocking is not possible in that case.

8.12 Example: Poisson's Law in the Frictionless Case

The main difference between Poisson's and Newton's laws in the frictionless case for more than one impact is shown in this example. We consider two balls (see Fig. 8.10) where ball 1 lies on a surface and ball 2 is dropped in order to hit ball 1. The mass matrix and the constraint vectors for this example are

$$\boldsymbol{M} = \begin{pmatrix} m_1 & 0 \\ 0 & m_2 \end{pmatrix} ; \quad \boldsymbol{q} = \begin{pmatrix} y_1 \\ y_2 \end{pmatrix} ; \quad \boldsymbol{w}_{N1} = \begin{pmatrix} 1 \\ 0 \end{pmatrix} ; \quad \boldsymbol{w}_{N2} = \begin{pmatrix} -1 \\ 1 \end{pmatrix} , \tag{8.101}$$

where for numerical evaluation the values $m_1 = m_2 = 1$, $\varepsilon_{N1} = 1$, $0 \leq \varepsilon_{N2} \leq 1$, $\dot{g}_{NA1} = 0$, $\dot{g}_{NA2} = -1$ have been used. The relative velocities at the beginning of the impact are $\dot{g}_{NA1} = 0$, $\dot{g}_{NA2} = -1$. During compression in both contacts, impulses act on the bodies which then are applied at expansion by using Poisson's law. The right part of Fig. 8.10 shows an evaluation of the relative velocities at the end of the impact, where the coefficient of restitution of contact 1 is chosen to be equal to one, $\varepsilon_{N1} = 1$, and ε_{N2} varies from 0 to 1. Two cases are observed by this evaluation. For $0 \leq \varepsilon_{N2} \leq 0.5$, ball 1

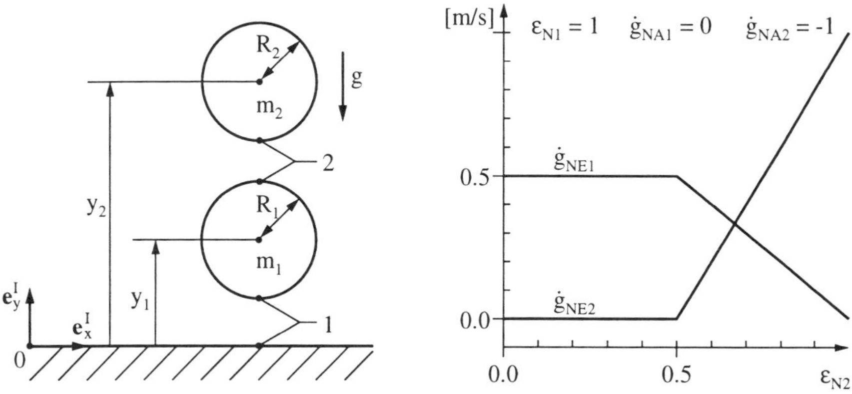

Figure 8.10: Mechanical Model and Relative Velocities

leaves the surface with a velocity $\dot{g}_{NE1} = 0.5$ and remains in contact with ball 2, $\dot{g}_{NE2} = 0$. Due to the completely elastic behavior of contact 1 ($\varepsilon_{N1} = 1$) the Poisson impulse during expansion equals the impulse during compression, $\Lambda_{NC1} = \Lambda_{NE1}$, which enables ball 1 to leave the surface. In contrast, contact 2 is modeled to be only partly reversible. In the range $0 \leq \varepsilon_{N2} \leq 0.5$, the original Poisson impulse applied during expansion would be too small to prevent penetration; thus it is enlarged by the conditions (8.27) such that continual contact is achieved after expansion. For $0.5 \leq \varepsilon_{N2} \leq 1$ the original Poisson impulse is strong enough to enable separation. In this area both balls lose contact after the impact, which is expressed by positive values of their relative velocities. If we would have applied Newton's law to this problem, ball 1 never would leave the surface, since $\dot{g}_{NE1} = -\varepsilon_{N1}\dot{g}_{NA1} = 0$.

8.13 Example: Reversible Tangential Impacts

The mechanical model presented in Fig. 8.11 stands for an example which highlights the different behavior of completely elastic tangential impacts (μ large enough, $\varepsilon_T = \nu = 1$) and irreversible tangential impulses due to Coulomb friction with sticking after compression (μ large enough, $\varepsilon_T = \nu = 0$). When a ball is thrown under a table in such a manner that it bounces alternately between the floor and the bottom of the table top ($\varepsilon_N = 1$), then the horizontal behavior of the ball depends on the tangential impact parameters. The right diagram in Fig. 8.12 shows a trajectory under the influence of only Coulomb friction ($\varepsilon_T = \nu = 0$), which corresponds qualitatively to the behavior of a ping-pong ball. After each impact the tangential velocity decreases and finally tends to zero; thus the ping-pong ball remains under the table. It can be observed that the inclinations of approach and departure are

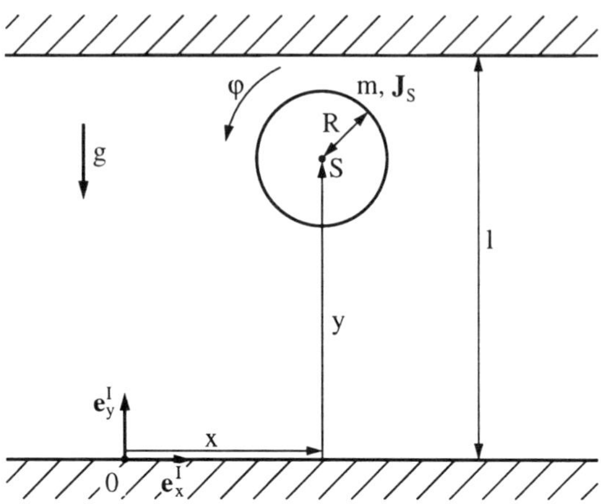

Figure 8.11: Mechanical Model

different in the presence of friction, a fact well known to good billards players. The left diagram in Fig. 8.12 corresponds to perfectly elastic impacts in the tangential direction ($\varepsilon_T = \nu = 1$). The tangential impulses which are stored during compression act in the same direction during expansion. This leads to the amazing effect that a superball thrown under a table bounces back after the second collision.

In order to give a complete mathematical description of the model we state the terms needed for building the equations of motion and the constraints:

$$\boldsymbol{M} = \begin{pmatrix} m & 0 & 0 \\ 0 & m & 0 \\ 0 & 0 & J_S \end{pmatrix} ; \quad \boldsymbol{h} = \begin{pmatrix} 0 \\ -mg \\ 0 \end{pmatrix} ; \quad \boldsymbol{q} = \begin{pmatrix} x \\ y \\ \varphi \end{pmatrix} \tag{8.102}$$

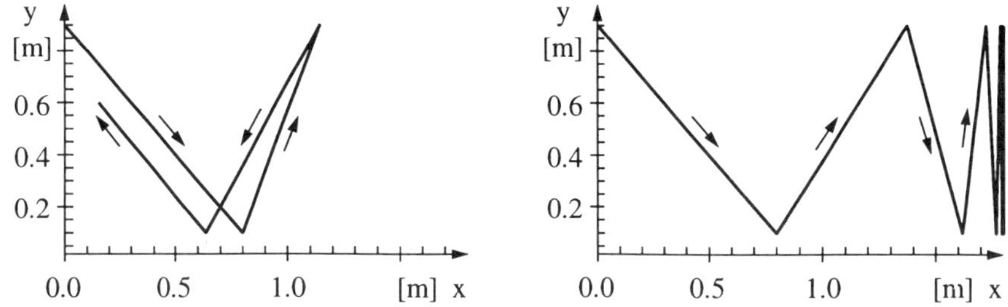

Figure 8.12: Trajectories at Reversible Impact and Coulomb Friction

$$\boldsymbol{w}_{N1} = \begin{pmatrix} 0 \\ 1 \\ 0 \end{pmatrix} ; \quad \boldsymbol{w}_{N2} = \begin{pmatrix} 0 \\ -1 \\ 0 \end{pmatrix} ; \quad \boldsymbol{w}_{T1} = \begin{pmatrix} 1 \\ 0 \\ R \end{pmatrix} ; \quad \boldsymbol{w}_{T2} = \begin{pmatrix} -1 \\ 0 \\ R \end{pmatrix} \tag{8.103}$$

where subscript 1 corresponds to contact of the ball with the floor, and subscript 2 denotes contact with the ceiling. The parameters and initial conditions for deriving the trajectories in Fig. 8.12 were chosen as $m = 1$, $J_S = 0.004$, $R = 0.1$, $g = 0$, $l = 1$, $\mu = 1$, $\varepsilon_N = 1$, $x_0 = 0$, $y_0 = 0.9$, $\varphi_0 = 0$, $\dot{x}_0 = 1$, $\dot{y}_0 = -1$, $\dot{\varphi}_0 = 0$.

8.14 Example: Poisson's Law and Coulomb Friction

The impact pendulum of Fig. 8.13 contains some interesting phenomena with respect to frictional collisions. A completely different behavior is observed by applying either Newton's or Poisson's impact law, where the approach of Newton results in a nearly senseless physical solution due to the kinematic dependence of both the normal and tangential directions. In this example the constraint vectors $\boldsymbol{w}_N$ and $\boldsymbol{w}_T$ are linearly dependent, $\alpha \boldsymbol{w}_N + \beta \boldsymbol{w}_T = 0$, which involves additional complications in the evaluation procedure. Even these cases can be handled partly by applying the pivot algorithm to the formulations of (8.87) and (8.97). The problem becomes more interesting when the surface hit by the pendulum is moved with a constant velocity v_0, because existence problems may arise. This behavior is analyzed for the phase of compression, but first the equations of motion and the constraint vectors have to be stated. For the state of separation the equation of

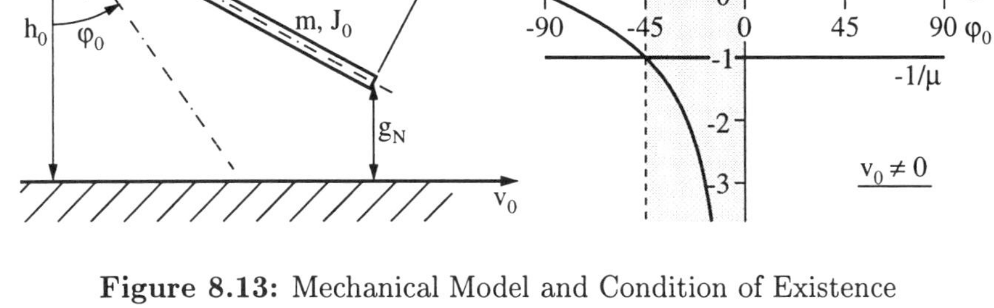

Figure 8.13: Mechanical Model and Condition of Existence

motion with respect to Fig. 8.13 is

$$J_0\ddot{\varphi} + mgs\sin\varphi = 0\,, \tag{8.104}$$

from which we identify the generalized coordinate, the mass matrix and the force vector:

$$\boldsymbol{M} = J_0\,; \quad \boldsymbol{h} = -mgs\sin\varphi\,; \quad \boldsymbol{q} = \varphi\,. \tag{8.105}$$

The relative velocities in the normal and tangential directions between the tip of the pendulum and the moving surface are given by

$$\begin{aligned} \dot{g}_N &= l\dot{\varphi}\sin\varphi \\ \dot{g}_T &= l\dot{\varphi}\cos\varphi - v_0\,, \end{aligned} \tag{8.106}$$

which correspond to the kinematic constraint terms

$$\boldsymbol{w}_N = l\sin\varphi\,; \quad \boldsymbol{w}_T = l\cos\varphi\,; \quad \widetilde{w}_N = 0\,; \quad \widetilde{w}_T = -v_0\,. \tag{8.107}$$

After the elimination of $\dot{\boldsymbol{q}}_C - \dot{\boldsymbol{q}}_A$ from (8.7) the connection of the relative velocities and impulses for this example is given by

$$\begin{pmatrix} \dot{g}_{NC} \\ \dot{g}_{TC} \end{pmatrix} = \frac{l^2}{J_0}\begin{pmatrix} \sin^2\varphi_0 & \sin\varphi_0\cos\varphi_0 \\ \sin\varphi_0\cos\varphi_0 & \cos^2\varphi_0 \end{pmatrix}\begin{pmatrix} \Lambda_{NC} \\ \Lambda_{TC} \end{pmatrix} + \begin{pmatrix} \dot{g}_{NA} \\ \dot{g}_{TA} \end{pmatrix} \tag{8.108}$$

with a singular matrix of coefficients, where φ_0 denotes the angle at which the pendulum hits the surface. Furthermore we have to state the impact law in the normal and tangential directions, which is done according to eqs. (8.13) and (8.23):

$$\Lambda_{NC} \geq 0\,; \quad \dot{g}_{NC} \geq 0\,; \quad \Lambda_{NC}\dot{g}_{NC} = 0 \tag{8.109}$$

$$|\Lambda_{TC}| \leq \mu\Lambda_{NC}\,; \quad \left\{ \begin{array}{lll} |\Lambda_{TC}| < \mu\Lambda_{NC} & \Rightarrow & \dot{g}_{TC} = 0 \\ \Lambda_{TC} = +\mu\Lambda_{NC} & \Rightarrow & \dot{g}_{TC} \leq 0 \\ \Lambda_{TC} = -\mu\Lambda_{NC} & \Rightarrow & \dot{g}_{TC} \geq 0\,. \end{array} \right. \tag{8.110}$$

For a negative approaching velocity $\dot{g}_{NA} < 0$, it follows immediately from the first equation in (8.108) and the impact laws (8.109), (8.110) that the combination $\dot{g}_{NC} \geq 0$, $\Lambda_{NC} = 0$ cannot lead to a solution because, due to (8.110), even the tangential impulse would vanish: $\Lambda_{NC} = 0 \Rightarrow \Lambda_{TC} = 0$. Thus, solutions are only possible for $\dot{g}_{NC} = 0$, $\Lambda_{NC} \geq 0$. Inserting these conditions into the first equation of (8.108) yields

$$0 = \frac{l^2}{J_0}\left(\Lambda_{NC}\sin^2\varphi_0 + \Lambda_{TC}\sin\varphi_0\cos\varphi_0\right) + \dot{g}_{NA}\,. \tag{8.111}$$

With $\dot{g}_{NA} < 0$, eq. (8.111) is only satisfied by impulses $(\Lambda_{NC}, \Lambda_{TC})$ which fulfill

$$\sin^2 \varphi_0 \, (\Lambda_{NC} + \Lambda_{TC} \cot \varphi_0) > 0 \,. \tag{8.112}$$

From the kinematics equations (8.106) we see that for $\dot{g}_{NC} = 0$ the tangential relative velocity is given by $\dot{g}_{TC} = -v_0$, which yields, due to (8.110) and the fact that $v_0 \neq 0$, a tangential impulse $\Lambda_{TC} = -\mu\sigma\Lambda_{NC}$, $\sigma = \text{sgn}(\dot{g}_{TC})$. Inserting this impulse into (8.112), we get

$$\Lambda_{NC} \sin^2 \varphi_0 \, (1 - \mu\sigma \cot \varphi_0) > 0 \,. \tag{8.113}$$

We see that a solution of the impact problem during compression is only achieved if

$$\sigma \cot \varphi_0 < \frac{1}{\mu} \,. \tag{8.114}$$

The condition (8.114) is depicted for $\sigma = -1$ in the right diagram of Fig. 8.13 and leads to the following physical interpretation: The impact process always has a solution if the pendulum is "drawn" by the tangential frictional impulse. If it is "pushed," the solutions are obtained only for large angles. If these angles are too small, the Coulomb friction produces a locking effect and requires sticking, which is not possible by the kinematics. Thus, the problem has no solution. For $v_0 = \dot{g}_{TC} = 0$ we have, from (8.110), $|\Lambda_{TC}| \leq \mu\Lambda_{NC}$. Thus, Λ_{TC} is positive or negative and can always be chosen in such a manner that (8.112) is fulfilled. The problem then always has a solution.

Now we evaluate the problem numerically by using the values $m = 1$, $J_0 = 1/3$, $l = 1$, $s = 1/2$, $g = 10$ and applying a frictional impact with the parameters $\mu = 1$, $\nu = 0$, $\varepsilon_N = 1$, $\varepsilon_T = 0$. Note that the coefficient of restitution in the normal direction is chosen to be equal to one, which means that the *normal* impulses acting during compression and expansion are of the *same* size. As initial conditions we choose $\varphi(t = 0) = 90°$, $\dot{\varphi}(t = 0) = 0$.

Figure 8.14 shows the ratio of the relative velocities in the normal direction at the first occurring impact when the pendulum is dropped from an initial angle of 90°. The six qualitatively different types of solutions in time are plotted for the marked parameters in Fig. 8.15.

We recall that the coefficient of restitution is chosen as $\varepsilon_N = 1$. For $v_0 = 0$ and $\varphi_0 < 45°$ the pendulum hits the surface only once and then remains in contact. Due to the steep angle, separation after the impact is not achieved because the tangential frictional impulses are too strong. Area II corresponds to a dissipative behavior which is well known from partly elastic frictionless impacts. In that case dissipation is introduced by reversed sliding, which means the frictional impulse acts in two different directions during compression and expansion. Case III shows the impact process for a

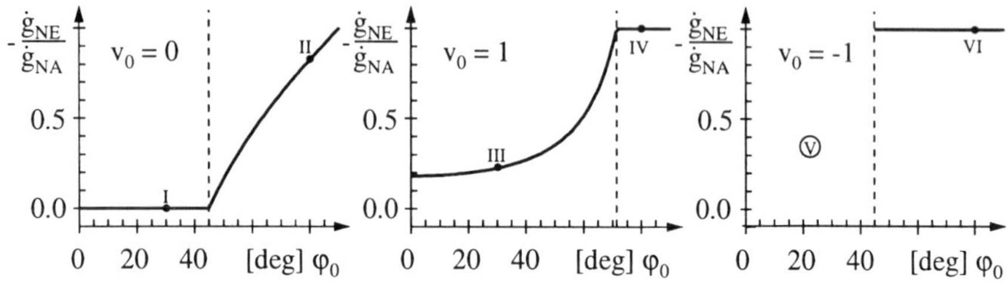

Figure 8.14: Ratio of Relative Velocities

"drawn" pendulum, $v_0 = 1$. For steep angles dissipation will occur only at the first impact, which results from a transition from sliding to stiction at the end of expansion. The pendulum leaves the surface with vanishing tangential relative velocity and remains in the state of an undamped oscillation. In Case IV the impact proceeds without transitions to sticking, so the system is energy preserving. The same behavior is observed in Case VI. Case V corresponds to the parameter range of nonexistence, so the algorithm terminates without a solution. The application of Newton's impact law leads to the solution $\varepsilon_N = -\dot{g}_{NE}/\dot{g}_{NA} = 1$ for any of the parameter values. Due to the fixed proportion of the relative velocities in the normal direction, the tangential impulses would not influence the system behavior which is contradictory to the physics of that example.

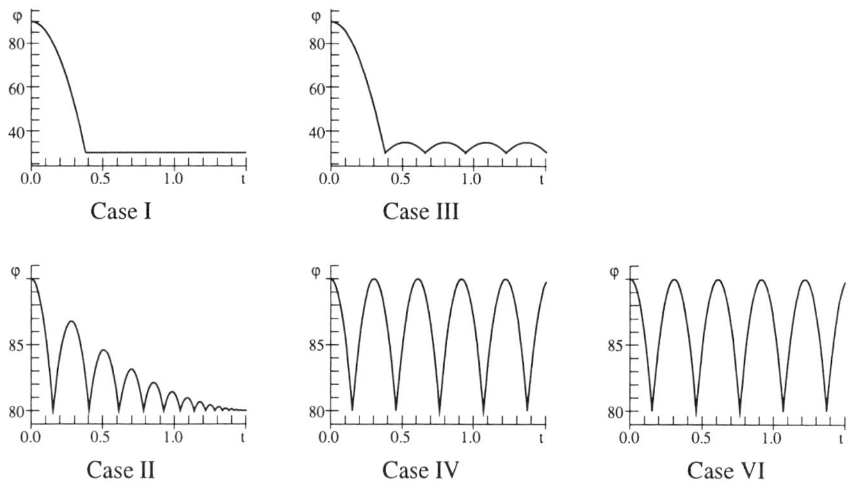

Figure 8.15: Time Histories for Different Cases

9

THE CORNER LAW OF CONTACT DYNAMICS

Kirchhoff's lectures on mathematical physics begin with the statements [39]: "Mechanics is the science of motion; its goal consists in describing the motion *completely* and in the *simplest possible way.* Motion is change of position with time; the object of motion is matter. Imaginations of space, time and matter are necessary and sufficient for an understanding of motion. With these means the science of mechanics must try to reach its goals by developing those concepts necessary for such an effort, for example the concepts of force and mass."

Forces are the only interactions of interest in mechanics inevitably imbedded in the geometry of motion and matter. No matter what we consider — equilibrium, deformation or motion — forces always appear in pairs. They depend on the way we delimit our mechanical system under consideration. If forces can be displaced in their direction, they are called active or applied forces; if they cannot, they are called passive or constraint forces. The concept of active and passive forces is straightforward and was introduced by Lagrange [47]. Active forces generate work or power, passive forces do not. This was first been discussed by d'Alembert [17] and put in modern form by Lagrange and Jourdain. It is the basis of multibody theory and one of the great contributions of French scientists to mechanics.

Before these ideas can be applied to contact dynamics, they need further sophistication. According to the preceding chapters contact dynamics is characterized by changing contact situations. Two bodies might or might not come into contact, and bodies already in contact may have relative motion to each other or they may not. To achieve a general description of such processes we must consider the geometrical or kinematical nature and the kinetic consequences in more detail. First we may remember that passive forces are accompanied by constraints, whereas active forces are not, at least not in a rigorous sense by assigning a constraint force to a kinematical constraint equation. As is well known, sliding friction forces depend on normal constraint forces which follow from a certain contact constraint.

Secondly, from the preceding chapters we know that a passive kinematical constraint equation acts as an indicator for the relative kinematics within the contact location. This might be a relative distance, velocity or acceleration for bodies not in contact, or it might be a relative velocity or acceleration for bodies in sliding contact. For nonzero relative motion the relevant constraint forces or combinations of them are zero. The passive character of the constraint equations ends when an indicator vanishes and then turns to a kinematical constraint equation. This must lead to constraint forces which are not zero and accompanied by a zero and active constraint equation. In this active situation with regard to the kinematic constraint equation, the constraint force or a constraint force combination acts as indicator.

So far so good. But things are more complicated. First, we should keep in mind that we have used the words "active" and "passive" in a double sense, namely for active (applied) and passive (constraint) forces and for active (zero) and passive (nozero) constraint equations. In an idealized way an active constraint equation would result in a passive force, but not always vice versa. In a stick-slip problem a passive stiction force turns to an active sliding friction force. In impact problems the situation active–passive is different. Second, the concept of constraint forces is highly idealized but, nevertheless, quite practical. It allows the evaluation of a passive force due to a constraint without including cumbersome detail. In reality "constraint forces" have a complicated local structure due to difficult local deformation processes, which in most cases are not of particular interest. But considering impacts, for example, we apply the concept of compression and expansion phases of a very local nature disturbing the ideal behavior as discussed above. Moreover, we apply for impulsive processes momenta and not forces, and we work on a velocity, not an acceleration, level. From this it makes sense to extend the classical interaction idea of forces only to physical interactions including forces *and* momenta. In addition, we must keep in mind that the momenta arising in classical impact theory are not passive momenta related to the zero-distance constraint for an impact. Usually energy will be dissipated for $\varepsilon < 1$, which does not go together at all with the idea of a passive force or a passive momentum.

In generalizing these ideas without violating the specific behavior of impulsive and frictional processes as discussed above, we come to the subsequent conclusion. *For dynamical systems of mechanics including multiple or single unilateral contacts, the following complementary situation exists independently of the nature of the specific contact law for each contact: Nonzero relative kinematics (distance, velocity, acceleration) with respect to a contact is accompanied by zero mechanical interactions (forces, momenta) and indicates a passive constraint situation. Zero relative kinematics with respect to a contact is accompanied by nonzero mechanical interactions and indicates an active constraint situation. From this the product of magnitudes* $\boldsymbol{K}_R$ *rep-*

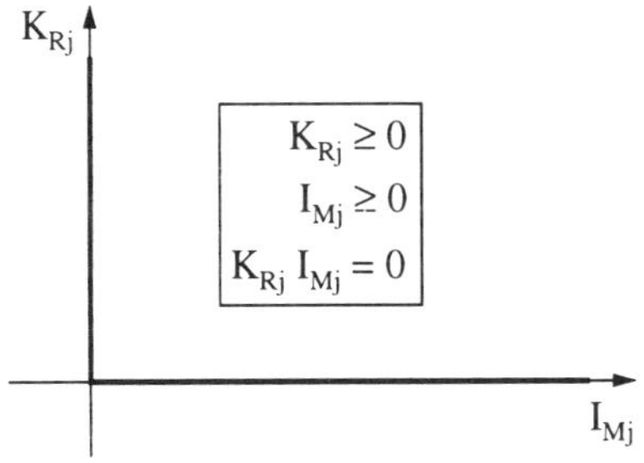

Figure 9.1: Corner Law of Contact Dynamics for a Contact j

resenting relative kinematics and of magnitudes $\boldsymbol{I}_M$ *representing mechanical interactions is always zero.*

Figure 9.1 illustrates this law. For two-dimensional contacts it should be kept in mind that we have two constraint directions within the contact plane. The direction of motion after a contact transition (stiction to sliding for example) follows in these cases from a solution of a nonlinear complementarity problem, which then defines the plane where the corner law applies. The kinetic value of $(I_{MO})_j$ is usually nonzero for impacts with friction where in a compression phase momenta are stored and partially regained during expansion. The corner law is an indispensable basis for a treatment of rigid or elastic multibody systems with multiple contacts. Furthermore, and if we replace the word "mechanical" by the word "physical," it is a very general basis for many physical processes typically characterized by jump phenomena. Examples can be found in rheology, continuum mechanics, heat transfer, electronics, quantum mechanics, optics and possibly biology with its cell membrane problems.

PART 2

APPLICATIONS

10

INTRODUCTION

In the following we present several applications which, with the exception of the woodpecker example, correspond to recent industrial problems. They illustrate the various modeling aspects of the general theory. Impulsive processes under heavy loads can be found in transmission systems of diesel engines and of big plants. In the mesh of the gears the teeth may separate and then bounce back due to external loads, thus generating hammering accompanied by high dynamical overloads. A special case of this phenomenon appears in smaller power plants including a turbine with a spur gearing and a generator. In the case of a short-circuit or a malsynchronization the generator produces impulsively a very large torque leading to a hammering effect in the spur gearing. All these impulsive processes are modeled according to Section 3.2.

An application of the idealized impact theory is possible for rattling processes in switchover gears and for the manipulation of synchronizing units in cars. Rattling in gearboxes is a comfort problem in cars. The gearwheels not under load are vibrationally excited by the oscillations of the drive system, and they move in a free flight within their backlashes, hitting the flanks only briefly, thus generating noise. By proper parameter design this noise can be reduced considerably. Synchronizers are units in changeover gears which adapt the angular speeds of the wheels to be changed in a very short instant of time. They work by a complicated sequence of impacts transmitting torque by friction. We shall present only a plane model.

The operation of turbine blade dampers is governed by stick-slip processes due to dry friction. Turbine blade dampers are manufactured from sheet metal. They are put between the turbine blade fastenings and are supposed to generate damping by sliding friction. Due to the extremely high centrifugal field the relative motion is often destroyed by stiction. A parameter optimization helps to overcome this problem. Another application of pure stick-slip theory concerns clutch cracking, which represents an everlasting problem in the automobile industry. Due to vibrations in the drive train systems the two clutch disks are loaded periodically or quasi-periodically, which leads to the well-known self-excited vibrations of clutch cracking.

The starting point of all our investigations with respect to impacts and friction was a little toy, the woodpecker. Running down a bar the woodpecker performs a self-excited oscillation which can be characterized by a periodic sequence of impacts with friction, by sliding and by vibrating phases. The theory necessary to describe the operation of this toy contains all components discussed in Chapters 5–8. In particular, the concept of the impacts with friction can be verified quite nicely with the woodpecker. An airplane application including aspects of impacts and friction consists of the dynamics of landing gears. Their telescopic structure generates stick-slip processes, and their limited length leads to impacts. The consideration of landing gears very much pushed forward the general theory of the additional constraint matrices. A certain type of drilling machine mainly used in the do-it-yourself business operates impulsively. The removal rate at the cutter head depends greatly on a good layout of the cam disks. This can be investigated by modeling all impacts taking place during operating. The impacts usually are accompanied by sliding friction only between the cams. Professional pneumatic drilling machines are driven by a crankshaft, and they move the drill by a sequence of impulsive and frictional processes. A very careful consideration is necessary to reduce the acceleration loads at the machine's handle, which is part of modern requirements.

A big problem including impacts and friction is connected with assembly processes in manufacturing. Mating parts can lead to various unwanted impulsive and stick-slip phenomena which should be avoided or at least reduced as much as possible. For this purpose the assembly mechanism, in our case a manipulator, must be modeled together with the assembly process itself, which is only possible by application of the theory. The problem of assembly processes very much influenced research on the complementarity problem, both in linear and nonlinear ways.

The following chapters briefly review much of the cooperation with industry in applying the theory. For convenience the designations and abbreviations have not been adapted to those of Part 1. Nevertheless, all chapters are self-explanatory, and the relationship to the formulas derived in Part 1 will be obvious.

11

APPLICATIONS WITH DISCONTINUOUS FORCE LAWS

11.1 Hammering in Gears*

Machines and mechanisms are characterized by rigid or elastic bodies interconnected in such a way that certain functions of the machines can be realized. Couplings in machines are never ideal but may have backlashes or some properties which lead to stick-slip phenomena. Under certain circumstances backlashes generate a dynamical load problem if the corresponding couplings are exposed to loads with a time-variant character. A typical example can be found in gear systems of diesel engines, which usually must be designed with large backlashes due to the operating temperature range of such engines, and which are highly loaded with the oscillating torques of the injection pump shafts and of the camshafts. Therefore, the power transmission from the crankshaft to the camshaft and the injection pump shaft takes place discontinuously by an impulsive hammering process in all transmission elements [82, 87].

Figure 11.1 indicates how the process works. A typical gear unit contains several meshes with backlashes, in the case shown two meshes with backlashes between crankshaft and injection pump shaft and three meshes with backlashes between crankshaft and camshafts.

Due to periodical excitations mainly from the injection pumps and subordinately from the crankshaft and the camshaft, the tooth flanks separate, generating a free-flight period within the backlash which is interrupted by impacts with subsequent penetration. The driven flank (working flank) usually receives more impacts than the nonworking flank (Fig. 11.1). Additionally, in all other backlashes of the gear unit similar processes take place, where the state and the impacts in one mesh with backlash influence considerably

*From [82]. Reprinted by permission of Kluwer Academic Publishers.

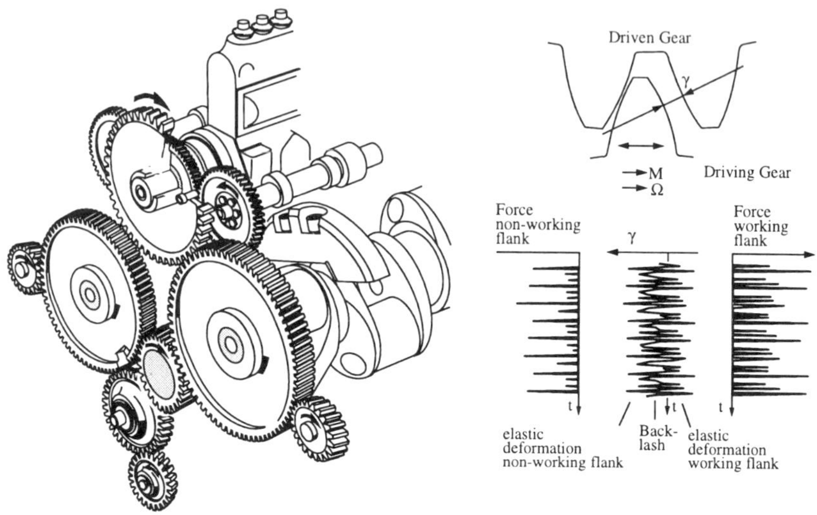

Figure 11.1: Typical Diesel Engine Driveline System and Forces in a Mesh of Gears

the state in all other meshes. This behavior must be accounted for by the mathematical model. As a definition we use the word "hammering" for separation processes within backlashes where high loads cause large impact forces with deformation. Motion within backlashes without loads is called "rattling." This represents a noise problem without load problems. It will be not considered here.

As a rule, such vibrations may be periodic, quasiperiodic or chaotic with a tendency to chaos for large systems. Considering the driveline gear unit as a multibody system with f degrees of freedom and n_p backlashes in the gear meshes, we model the backlash properties by a nonlinear force characteristic with small forces within the backlash and a linear force law in the case of contact of the flanks. The event of a contact is determined by an evaluation of the relative distance in each backlash, which serves as an indicator function. The indicator function for leaving the contact, i.e., flank separation, is given with the normal force in the point of contact, which changes sign in the case of flank separation. These unsteady points (switching points) must be evaluated very carefully to achieve reproducible results. The time series of impact forces will be reduced to load distributions in a last step. They might serve as a basis for lifetime estimates.

Dynamical systems with unsteady behavior have been the subject of increasing investigations during recent years. With regard to impulsive processes the bouncing-ball problem is quite old and famous [43]. With the progress of nonlinear dynamics in the last two decades publications on problems of impact phenomena were in all cases confined to systems with only two or three degrees of freedom [32, 95]. Applications to technical problems were rare and mainly restricted to gear research contributions [103, 102].

The first activities on impulsive processes at the author's institute started in 1982 and led to a series of contributions on rattling and hammering processes in gearboxes and drivelines. The fundamental starting point was a general theoretical approach to mechanical systems with unsteady transitions in 1984 [71], which was very quickly extended to rattling applications [44, 45, 74, 81]. The dissertation [37, 38] deepened the rattling theory and compared one-stage rattling with laboratory tests. From the very beginning all theoretical research focused on general mechanical systems with an arbitrary number of degrees of freedom and with an arbitrary number of backlashes. Application fields are drivelines of large diesel engines, which, due to a large temperature operating range, are usually designed with large backlashes [87].

11.1.1 Modeling

BODY MODELS

Rigid bodies are characterized by six degrees of freedom: three translational and three rotational ones. We combine these magnitudes in an $\mathbb{R}^6$-vector (Fig. 11.2 and [87])

$$\mathbf{p} = \begin{pmatrix} \Phi \\ \mathbf{r}_H \end{pmatrix} \in \mathbb{R}^6 \tag{11.1}$$

with

$$\Phi = (\varphi_x, \varphi_y, \varphi_z)^T \quad , \quad \mathbf{r}_H = (\Delta x_H, \Delta y_H, \Delta z_H)^T \in \mathbb{R}^3 . \tag{11.2}$$

Accordingly, the velocities are

$$\mathbf{v} = \begin{pmatrix} \dot{\Phi} \\ \dot{\mathbf{r}}_H \end{pmatrix} = \begin{pmatrix} \boldsymbol{\omega} \\ \mathbf{v}_H \end{pmatrix} \in \mathbb{R}^6 . \tag{11.3}$$

Elastic bodies in gear or driveline units usually are shafts with torsional and/or flexural elasticity. In the following we consider only torsion by applying a Ritz approach [13] to the torsional deflection φ (Fig. 11.3):

$$\varphi(z,t) = \mathbf{w}(z)^T \boldsymbol{q}_{el}(t) \quad \text{with} \quad \mathbf{w}, \boldsymbol{q}_{el} \in \mathbb{R}^{n_{el}} \tag{11.4}$$

where the subscript *el* stands for "elastic."

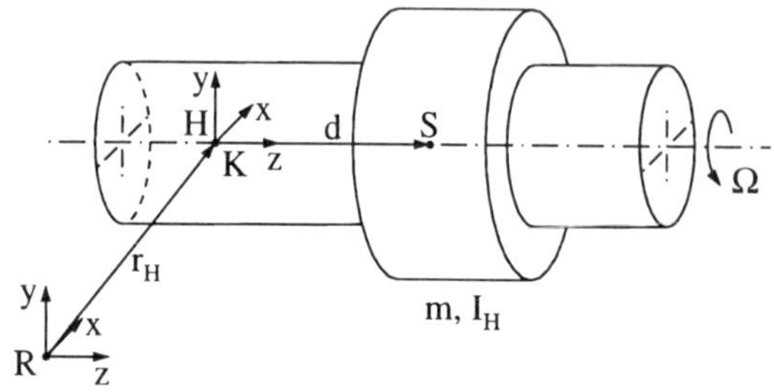

Figure 11.2: Rigid Body Model [87]

COUPLING COMPONENTS

An ideal joint couples two bodies. It can be described kinematically. According to the notation used in multibody theory the free directions f_i of motion of a joint i are given by a matrix $\mathbf{\Phi}_i \in \mathbb{R}^{6,f_i}$. The spatial possible motion of the joint can be described by $(\mathbf{\Phi}_i \boldsymbol{q}_{J_i})$ with relative displacements $\boldsymbol{q}_{J_i}$ in the nonconstrained directions of the joint i. A complementary matrix $\mathbf{\Phi}_i^c \in \mathbb{R}^{6,6-f_i}$ exists for the constrained directions of the joint. Always $\mathbf{\Phi}_i^T \mathbf{\Phi}_i^c = 0$, and the constraint forces are written as

$$\mathbf{f}_{J_i} = \mathbf{\Phi}_i^c \boldsymbol{\lambda}_i \ , \quad \boldsymbol{\lambda}_i \in \mathbb{R}^{6-f_i} \, . \tag{11.5}$$

The Lagrange multipliers $\boldsymbol{\lambda}$ follows from d'Alembert's principle and a Lagrangian treatment of the equations of motion.

Elastic couplings in drivelines are characterized by some force law in a given direction between two bodies (Fig. 11.4). The relative displacement and displacement velocity may be expressed by

$$\gamma_k = \boldsymbol{\psi}_k^T \left(-\mathbf{C}_{ki}\mathbf{p}_i + \mathbf{C}_{kj}\mathbf{p}_j\right) \, , \, \dot{\gamma}_k = \boldsymbol{\psi}_k^T \left(-\mathbf{C}_{ki}\mathbf{v}_i + \mathbf{C}_{kj}\mathbf{v}_j\right) \tag{11.6}$$

$$\mathbf{C}_{ki} = \begin{pmatrix} \mathbf{E} & \mathbf{0} \\ \tilde{\mathbf{c}}_{ki}^T & \mathbf{E} \end{pmatrix} \in \mathbb{R}^{6,6} \, .$$

The vector $\mathbf{c}_{ki}$ follows from Fig. 11.4, and $\tilde{\mathbf{c}}_{ki}$ is the relevant skew-symmetric tensor ($\tilde{\mathbf{a}}\mathbf{b} \equiv \mathbf{a} \times \mathbf{b}$). The vector $\boldsymbol{\psi}_k \in \mathbb{R}^6$ represents a unit vector

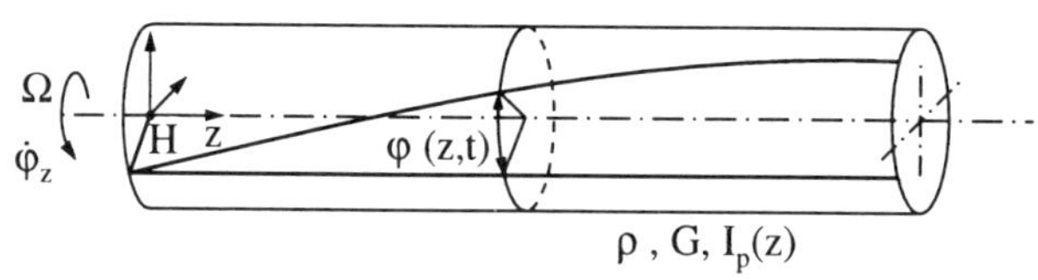

Figure 11.3: Shaft with Torsional Elasticity

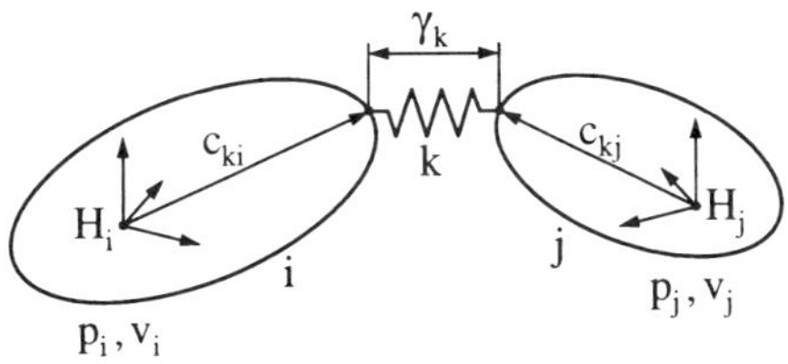

Figure 11.4: Elastic Coupling

in the direction of relative displacement. In that direction we have a scalar force magnitude ζ_k according to the given force law. It can be expressed in the body coordinate frames H_i, H_j (Fig. 11.4) by the generalized forces

$$\mathbf{f}_i = \mathbf{C}_{ki}^T \psi_k \zeta_k \ , \quad \mathbf{f}_j = -\mathbf{C}_{kj}^T \psi_k \zeta_k \, . \tag{11.7}$$

As a simple example a linear force law would be written as

$$\zeta_k = c\gamma_k + d\dot{\gamma}_k \tag{11.8}$$

Of course, any nonlinear relationship might be applied as well, such as the force law with backlash according to Fig. 11.5. Gear meshes with backlash are modeled with the characteristic of Fig. 11.5. In this case care has to be taken with respect to the two possibilities of flank contact on both sides of each tooth. Oil reduces the impact forces of the hammering process. Some application tests have been performed with the nonlinear oil model of Holland [31]. On this basis a simplified model has been derived to approximate the oil influence (Fig. 11.6). The model assumes an exponential damping behavior within the backlash s.

Bearings are very important coupling elements in drivelines. Roller bearings are approximated by the force laws in the (x, y)-directions:

$$\begin{aligned} \zeta_{kx} &= c_x\gamma_x + d_x\dot{\gamma}_x \\ \zeta_{ky} &= c_y\gamma_y + d_y\dot{\gamma}_y \, . \end{aligned} \tag{11.9}$$

For journal bearings with a stationary load we apply the well-known law

$$\begin{pmatrix} \zeta_{kx} \\ \zeta_{ky} \end{pmatrix} = \begin{pmatrix} c_{11} & c_{12} \\ c_{21} & c_{22} \end{pmatrix} \begin{pmatrix} \gamma_{kx} \\ \gamma_{ky} \end{pmatrix} + \begin{pmatrix} d_{11} & d_{12} \\ d_{21} & d_{22} \end{pmatrix} \begin{pmatrix} \dot{\gamma}_{kx} \\ \dot{\gamma}_{ky} \end{pmatrix} . \tag{11.10}$$

For journal bearings with nonstationary loads the Waterstraat solution [101] of the Reynolds equations is applied.

Excitation System

Various excitation sources must be considered in simulating diesel engine driveline vibrations. First of all the crankshaft excites the system with some

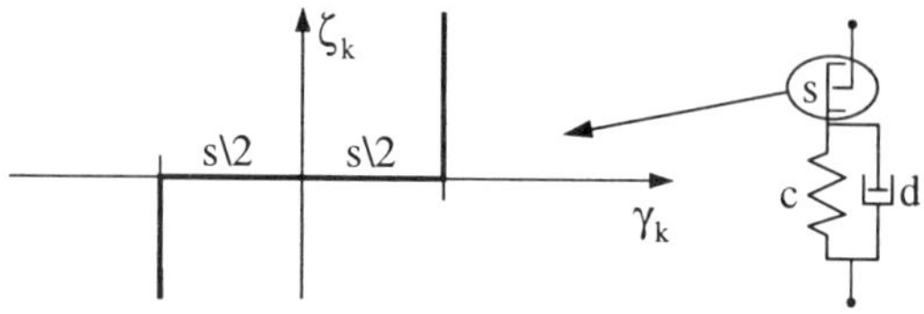

Figure 11.5: Backlash Force Law

harmonics related to the motor speed and depending on the motor design. As a realistic approximation to driveline systems we may assume that the crankshaft motion itself is not influenced by driveline dynamics. Secondly, the valve mechanisms generate a parametric excitation which may be expressed approximately by a time-variant moment of inertia of the camshaft. As a matter of fact, in most applications this influence is fairly small, only a few percent.

A dominant influence comes from the third effect, namely the torques generated by the injection pumps. These torques directly counterbalance the driving torque of the crankshaft, thus inducing the hammering process within the gear meshes. A typical example is given by Fig. 11.7, which represents the situation in a 12-cylinder diesel engine where valves and injection pumps are controlled by one shaft for each of the two cylinder banks.

MATHEMATICAL MODELS

The theory of rigid and elastic multibody systems is applied rigorously. It starts with d'Alembert's principle, which states that passive forces produce no work or, according to Jourdain, generate no power; see Chapter 3. This statement can be used to eliminate passive forces (constraint forces) and to generate a set of differential equations for the coupled machine system under consideration.

We start with the equations of motion for a single rigid body. Combining the momentum and moment of momentum equations and considering the fact

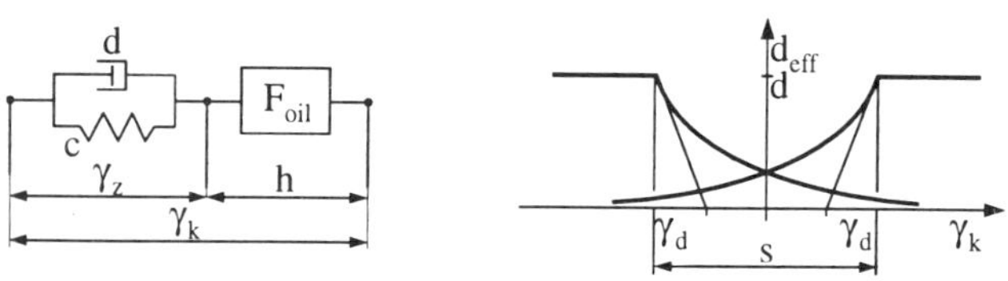

Figure 11.6: Oil Model

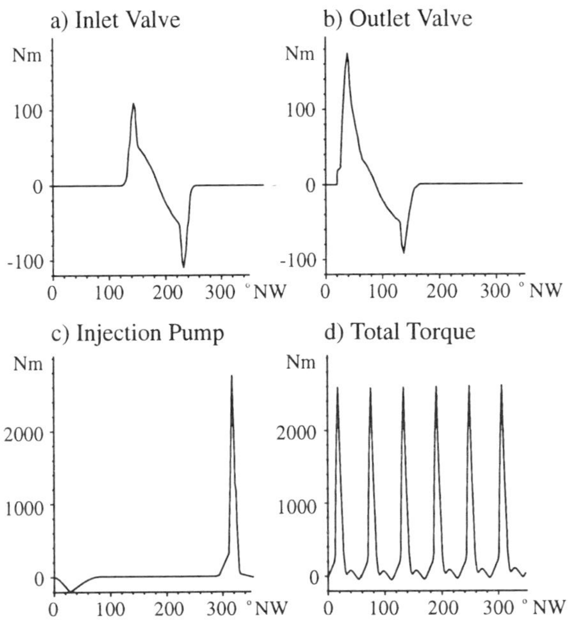

Figure 11.7: Torques at the Camshaft

that the mass center S has a distance d from the body-fixed coordinate frame in H (Fig. 11.2) we obtain

$$\mathbf{I}\dot{\mathbf{v}} = -\mathbf{f}_K + \mathbf{f}_B + \mathbf{f}_E = \mathbf{f}_S(\mathbf{p}, \mathbf{v}, t) \tag{11.11}$$

where

$$\mathbf{I} = \begin{pmatrix} \mathbf{I}_H & m\tilde{\mathbf{d}} \\ -m\tilde{\mathbf{d}} & m\mathbf{E}_3 \end{pmatrix} \in \mathbb{R}^{6,6}$$

$$\mathbf{d} = (0,\, 0,\, d)^T \in \mathbb{R}^3$$

$$\mathbf{I}_H = \operatorname{diag}(A,\, A,\, C) \in \mathbb{R}^{3,3}$$

$$-\mathbf{f}_K = \left[(\tilde{\boldsymbol{\omega}}\mathbf{I}_H\Omega\mathbf{e}_3)^T,\, \mathbf{0}\right]^T \in \mathbb{R}^6$$

$$\mathbf{f}_B = \left[\left(-\mathbf{I}_H\dot{\Omega}\mathbf{e}_3\right)^T, \mathbf{0}\right]^T \in \mathbb{R}^6$$

$$\mathbf{f}_E = \left[\left(\sum \mathbf{M}_{Hk}\right)^T, \left(\sum \mathbf{F}_k\right)^T\right]^T \in \mathrm{I\!R}^6 . \tag{11.12}$$

The magnitudes A, C are moments of inertia; $\mathbf{f}_K$, $\mathbf{f}_B$ and $\mathbf{f}_E$ are gyroscopic, acceleration and applied forces, respectively, and Ω, $\dot{\Omega}$ are prescribed values of angular velocity and acceleration, respectively. The unit vector $\mathbf{e}_3$ is body-fixed in H (Fig. 11.2).

By adding components with torsional elasticity we can influence the rigid body motion only with respect to the third equation of (11.11). Therefore torsional degrees of freedom can be included in a simple way. The equation of motion for a shaft with torsional flexibility is

$$\varrho I_p(z)\frac{\partial^2\overline{\varphi}}{\partial t^2} - \frac{\partial}{\partial z}\left[GI_p(z)\frac{\partial\overline{\varphi}}{\partial z}\right] - M_k\delta(z - z_k) = 0 . \tag{11.13}$$

(ϱ density, I_p area moment of inertia, G shear modulus, M_k torque at location z_k). The total angle $\overline{\varphi}$ has three parts:

$$\overline{\varphi}(z,t) = \int \Omega(t)dt + \varphi_z(t) + \varphi(z,t) , \tag{11.14}$$

where $\Omega(t)$ is the angular velocity program, $\varphi_z(t)$ is the z-component of Φ (eq. 11.2) and $\varphi(z,t)$ is the torsional deflection (Fig. 11.3). Approximating $\varphi(z,t)$ by eq. (11.4) and applying a Galerkin approach to eq. (11.13) result in the set

$$\mathbf{h}(\dot{\Omega} + \ddot{\varphi}_z) + \mathbf{M}_{el}\ddot{\boldsymbol{q}}_{el} + \mathbf{K}_{el}\boldsymbol{q}_{el} = \sum M_k\mathbf{w}(z_k) , \tag{11.15}$$

where

$$\begin{aligned}
\mathbf{M}_{el} &= \textstyle\int_0^l \varrho I_p(z)\mathbf{w}(z)\mathbf{w}(z)^T dz && \in \mathrm{I\!R}^{n_{el},n_{el}} \quad \text{mass matrix} \\
\mathbf{K}_{el} &= \textstyle\int_0^l GI_p(z)\mathbf{w}'(z)\mathbf{w}'(z)^T dz && \in \mathrm{I\!R}^{n_{el},n_{el}} \quad \text{stiffness matrix} \\
\mathbf{h} &= \textstyle\int_0^l \varrho I_p(z)\mathbf{w}(z)dz && \in \mathrm{I\!R}^{n_{el}} \quad \text{coupling vector.}
\end{aligned} \tag{11.16}$$

In agreement with physical arguments eqs. (11.15) are coupled with the rigid body motion only by the term $\mathbf{h}\left(\dot{\Omega} + \ddot{\varphi}_z\right)$. To include torsional coupling in the third equation of (11.11), we only have to complete the angular momentum of the rigid part $C(\Omega + \dot{\varphi}_z)$ by an elastic part $\int_0^l \varrho I_p(z)\left[\partial\varphi(z,t)/\partial t\right]dz$ and evaluate its time derivative. This results in an additional coupling term $\mathbf{h}^T\ddot{\boldsymbol{q}}_{el}$. Equations (11.11) with elastic expansion and eqs. (11.15) can then be combined into

$$\mathbf{I}\dot{\mathbf{v}} + \mathbf{H}^T\ddot{\boldsymbol{q}}_{el} = \mathbf{f}_S \tag{11.17}$$

$$\mathbf{H}\dot{\mathbf{v}} + \mathbf{M}_{el}\ddot{\boldsymbol{q}}_{el} = \mathbf{f}_{S_{el}} \tag{11.18}$$

where all terms not containing accelerations are collected in $\mathbf{f}_S$ and $\mathbf{f}_{S_{el}}$. For formal convenience we define $\mathbf{H} = (\mathbf{0}, \mathbf{0}, \mathbf{h}, \mathbf{0}, \mathbf{0}, \mathbf{0})$.

Proceeding now from a single-body to a multibody system, we consider the constraint (passive) forces $\mathbf{f}_{J_i}$ (eq. 11.5). In a form corresponding to eqs. (11.17), (11.18) we get

$$\mathbf{I}_i \dot{\mathbf{v}}_i + \mathbf{H}_i^T \ddot{\boldsymbol{q}}_{el_i} = \mathbf{f}_{S_i} + \mathbf{f}_{J_i} - \sum_{k \in S(i)} \mathbf{C}_k^T \mathbf{f}_{J_k} \tag{11.19}$$

$$\mathbf{H}_i \dot{\mathbf{v}}_i + \mathbf{M}_{el_i} \ddot{\boldsymbol{q}}_{el_i} = \mathbf{f}_{S_{el_i}} - \sum_{k \in S(i)} \mathbf{D}_k^T \mathbf{f}_{J_k} \,. \tag{11.20}$$

These equations have to be supplemented by a kinematical relation in the form

$$\dot{\mathbf{v}}_i = \mathbf{C}_i \dot{\mathbf{v}}_p(i) + \mathbf{D}_i \ddot{\boldsymbol{q}}_{el_{p(i)}} + \Phi_i \ddot{\boldsymbol{q}}_i \,. \tag{11.21}$$

The first term on the right-hand side represents an acceleration resulting from the absolute acceleration of the predecessor body (index $p(i)$), the second term represents an acceleration resulting from a torsional deformation of the predecessor body (if any) with a matrix $\mathbf{D}_i$ containing the torsional shape functions, and the third term is an acceleration resulting from the relative motion of body i and the predecessor body. $S(i)$ is the set of all bodies following body i.

For computational reasons eqs. (11.19) to (11.21) must be solved recursively by a so-called order-n algorithm (computing time $\sim$ degrees of freedom). This algorithm works as follows [7, 8]: Organize the multibody system under consideration as a treelike structure with a base body connected to the inertial environment and a series of final bodies possessing no successor body. This will be possible for any case, because for closed kinematical loops these may be cut with an additional closing condition as constraint. Start with the last type of bodies and consider their equations of motion (from eqs. 11.19, 11.20):

$$\mathbf{I}_i \dot{\mathbf{v}}_i + \mathbf{H}_i^T \ddot{\boldsymbol{q}}_{el_i} = \mathbf{f}_{S_i} + \mathbf{f}_{J_i} \tag{11.22}$$

$$\mathbf{H}_i \dot{\mathbf{v}}_i + \mathbf{M}_{el_i} \ddot{\boldsymbol{q}}_{el_i} = \mathbf{f}_{S_{el_i}} \,. \tag{11.23}$$

Eliminate from these equations the elastic coordinates to get

$$\hat{\mathbf{I}}_i \dot{\mathbf{v}}_i = \hat{\mathbf{f}}_{S_i} + \mathbf{f}_{J_i} \tag{11.24}$$

with

$$\hat{\mathbf{I}}_i = \mathbf{I}_i - \mathbf{H}_i^T \mathbf{M}_{el_i}^{-1} \mathbf{H}_i$$

$$\hat{\mathbf{f}}_{S_i} = \mathbf{f}_{S_i} - \mathbf{H}_i^T \mathbf{M}_{el_i}^{-1} \mathbf{f}_{S_{el_i}} \,. \tag{11.25}$$

Combine eq. (11.24) with eqs. (11.21) and (11.5) and apply d'Alembert's principle to eliminate the constraint forces (premultiplication with Φ_i^T). We get the generalized relative accelerations

$$\ddot{\boldsymbol{q}}_i = \mathbf{M}_i^{-1}\Phi_i^T\left(\mathbf{f}_{S_i} - \hat{\mathbf{I}}_i\mathbf{C}_i\dot{\mathbf{v}}_{p(i)} - \hat{\mathbf{I}}_i\mathbf{D}_i\ddot{\boldsymbol{q}}_{el_{p(i)}}\right) \tag{11.26}$$

with $\mathbf{M}_i = \Phi_i^T\hat{\mathbf{I}}_i\Phi_i$, which can be used together with eqs. (11.24), (11.21) to obtain the elastic deformation accelerations

$$\ddot{\boldsymbol{q}}_{el_i} = -\mathbf{M}_{el_i}^{-1}\mathbf{H}_i\left(\mathbf{C}_i\dot{\mathbf{v}}_{p(i)} + \mathbf{D}_i\ddot{\boldsymbol{q}}_{el_{p(i)}} + \Phi_i\ddot{\boldsymbol{q}}_i\right) + \mathbf{M}_{el_i}^{-1}\mathbf{f}_{S_{el_i}}\,. \tag{11.27}$$

The accelerations $\ddot{\boldsymbol{q}}_i$ and $\ddot{\boldsymbol{q}}_{el_i}$ still depend on those of the predecessor body. Passing to that body requires determination of the joint forces, which are evaluated from eqs. (11.21, 11.24, 11.26):

$$\mathbf{f}_{J_i} = \mathbf{N}_i\mathbf{C}_i\dot{\mathbf{v}}_{p(i)} - \mathbf{L}_i\mathbf{f}_{S_i} + \mathbf{N}_i\dot{\mathbf{D}}_i\ddot{\boldsymbol{q}}_{el_{p(i)}} \tag{11.28}$$

with

$$\begin{aligned}\mathbf{L}_i &= \mathbf{E} - \hat{\mathbf{I}}_i\Phi_i\mathbf{M}_i^{-1}\Phi_i^T\\ \mathbf{N}_i &= \mathbf{L}_i\hat{\mathbf{I}}_i\,.\end{aligned}$$

With the joint force of eq. (11.28) we enter the equations of motion of the predecessor body and establish a set which corresponds formally to the equations of a final body:

$$\begin{aligned}&\left(\mathbf{I}_{p(i)} + \mathbf{C}_i^T\mathbf{N}_i\mathbf{C}_i\right)\dot{\mathbf{v}}_{p(i)} + \left(\mathbf{H}_{p(i)}^T + \mathbf{C}_i^T\mathbf{N}_i\mathbf{D}_i\right)\ddot{\boldsymbol{q}}_{el_{p(i)}}\\ &\quad = \mathbf{f}_{J_{p(i)}} + \left(\mathbf{f}_{S_{p(i)}} + \mathbf{C}_i^T\mathbf{L}_i\mathbf{f}_{S_i}\right) - \sum_{k\in S(p(i))-\{i\}}\mathbf{C}_k^T\mathbf{f}_{J_k}\end{aligned} \tag{11.29}$$

$$\begin{aligned}&\left(\mathbf{M}_{el_i} + \mathbf{D}_i^T\mathbf{N}_i\mathbf{D}_i\right)\ddot{\boldsymbol{q}}_{el_{p(i)}} + \left(\mathbf{H}_{p(i)} + \mathbf{D}_i^T\mathbf{N}_i\mathbf{C}_i\right)\dot{\mathbf{v}}_{p(i)}\\ &\quad = \left(\mathbf{f}_{S_{el_{p(i)}}} + \mathbf{D}_i^T\mathbf{L}_i\mathbf{f}_{S_i}\right) - \sum_{k\in S(p(i))-\{i\}}\mathbf{D}_k^T\mathbf{f}_{J_k}\end{aligned} \tag{11.30}$$

with magnitudes

$$\begin{aligned}\mathbf{I}_{p(i)} &:= \mathbf{I}_{p(i)} + \mathbf{C}_i^T\mathbf{N}_i\mathbf{C}_i\\ \mathbf{H}_{p(i)} &:= \mathbf{H}_{p(i)} + \mathbf{D}_i^T\mathbf{N}_i\mathbf{C}_i\\ \mathbf{f}_{S_{p(i)}} &:= \mathbf{f}_{S_{p(i)}} + \mathbf{C}_i^T\mathbf{L}_i\mathbf{f}_{S_i}\\ \mathbf{M}_{el_{p(i)}} &:= \mathbf{M}_{el_{p(i)}} + \mathbf{D}_i^T\mathbf{N}_i\mathbf{D}_i\\ \mathbf{f}_{S_{el_{p(i)}}} &:= \mathbf{f}_{S_{el_{p(i)}}} + \mathbf{D}_i^T\mathbf{L}_i\mathbf{f}_{S_i}\,.\end{aligned} \tag{11.31}$$

These equations are analoguous to eqs. (11.22, 11.23). With this procedure we recursively can proceed to the last base body for which eqs. (11.26)

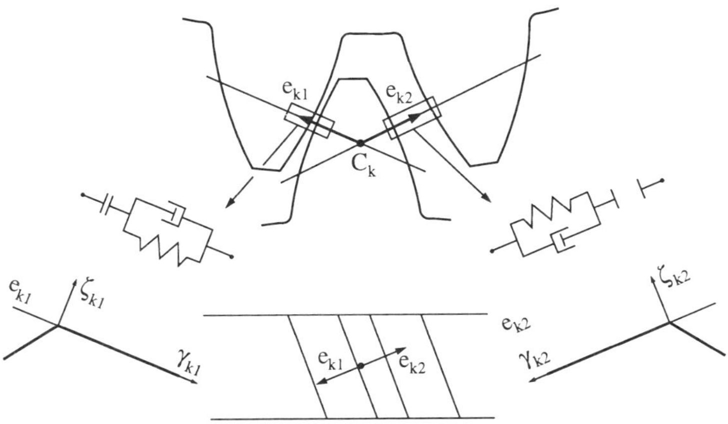

Figure 11.8: Mesh of Gears with Backlash and Relevant Force Laws

are evaluated in an elementary way. Knowing the accelerations of this base body, we again can go forward step by step to the final bodies and determine all accelerations of each body. If necessary we may add recursion for the evaluation of constraint forces.

Equations (11.19) and (11.20) describe a multibody system with n bodies and a maximum number of ($f = 6n$) rigid degrees of freedom and, according to the Ritz Ansatz (eq. 11.4) a certain number of elastic degrees of freedom (number of shape functions per body times number of elastic bodies). In practical applications, however, the number of degrees of freedom might be reduced drastically. For example, driveline units with straight-tooth bevels may be sufficiently modeled by rotational degrees of freedom only.

Backlash Management

To include backlashes we have to implement an algorithm which controls the contact events (Fig. 11.8), by considering contact kinematics and contact forces.

A contact at a tooth flank occurs if the relative distance in contact k becomes zero (eq. 11.6):

$$\gamma_k \left(\mathbf{p}_i, \mathbf{p}_j, \boldsymbol{\psi}_k\right) = 0 \,. \tag{11.32}$$

The subsequent deflection of both teeth follows the force laws of the Figs. 11.5 and 11.6, but the end of the contact is not reached, when we again get $\gamma_k = 0$. The correct condition consists of the requirement that the normal force $(\boldsymbol{\zeta}_k \mathbf{e}_{ki},\ i = 1, 2)$ (Fig. 11.8) vanish.

As we have a unilateral contact problem, separation takes place when the normal force changes sign, which is not the case if $\gamma_k = 0$. Due to the dynamics of the contacting bodies and to the damping influence of the contact oil model (Fig. 11.6), the normal force changes sign before $\gamma_k = 0$, which means the tooth separation takes place when the teeth are still deflected. For separation we therefore must interpolate the force condition (Fig. 11.8)

$$\boldsymbol{\zeta}_{kn} = -\mathbf{e}_{ki}\zeta_k = -\mathbf{e}_{ki}\left(c_k\gamma_k + \zeta_{kD}\left(\gamma_k, \dot{\gamma}_k\right)\right) = 0 \;\; (i = 1, 2)\,, \tag{11.33}$$

where $\boldsymbol{\zeta}_{kn}$ is the normal force vector and ζ_{kD} the damping force law due to oil and structural damping.

Considering several backlashes we need an additional algorithm to determine the shortest time step to the next contact or separation event. This means formally that we have to evaluate in all existing backlashes the following equations:

$$\begin{aligned} \Delta t_{FC} &= \min_{k \in n_p} \left\{\Delta t_k \,|\, \gamma_k(\mathbf{p}_i, \mathbf{p}_j, \boldsymbol{\psi}_k) = 0 \wedge \zeta_{kn}(\gamma_k, \dot{\gamma}_k, \mathbf{e}_n) < 0\right\}, \\ \Delta t_{CF} &= \min_{k \in n_p} \left\{\Delta t_k \,|\, \zeta_{kn}(\gamma_k, \dot{\gamma}_k, \mathbf{e}_n) \geq 0\right\}, \end{aligned} \tag{11.34}$$

where FC means transition from free flight to contact and CF transition from contact to free flight. Equations (11.34) express the situation that during the recursive solution process as presented above at each integration step one must prove the possibility of a contact event or a separation in any of the n_p backlashes. If this proof turns out to be true, the conditions of eqs. (11.34) together with the state of the complete system must be interpolated. The numerical integration process is then started anew at such a switching point. Note that each of the backlash zones of the multibody system might be in a free flight or in a contact state where the transitions are controlled by eqs. (11.34). Note further that the time steps Δt_{FC} or Δt_{CF} are the macrosteps between two events usually taking place in different backlashes; the microsteps for numerical integration are of course considerably smaller.

Numerical Models

The main problem in dealing with unsteady dynamical systems consists of the numerical management of unsteadiness. In the dissertation [87] three possibilities were considered and tested. First, numerical integration routines of first order, such as the Euler method, include no problems with unsteadiness but show stability and convergence difficulties. Second, the numerical method of Shampine-Gordon has been realized with good results for certain cases. But the determination of switching points for an unsteady event may break down abruptly, thus generating numerical instability. Third, a direct

Method	Relative Computing Time
• Direct search of switching points	
– Runge-Kutta 2./3. order	1.0
– Runge-Kutta 5./6. order	1.9
– Gear method (IMSL)	1.4
• Not direct search	
– Adams-Pece integration	1.4

Table 11.1: Comparison of Computing Time

search and interpolation of switching points has been implemented which includes the fewest problems and works quite reliably. Details of this method may be found in [19, 87].

It might be of interest to indicate some computing time aspects (see Table 11.1). From this the best method turned out to be a direct search of switching points together with a Runge-Kutta integration method of order 2/3.

11.1.2 Evaluation of the Simulations

The problem in evaluating the vibrations of unsteady systems consists of a realistic examination of the unsteady events, especially in the further evaluation of the gear forces, which, as a consequence of the hammering process, appear as impulsive forces. In any case we may apply an FFT procedure, and from this we get approximate information about frequencies and amplitudes which represents in steady problems a very powerful tool. In unsteadiness it is different because we do not know with sufficient reliability what force amplitudes will influence lifetime and strength of the gears. If we consider a typical hammering process in the gear mesh (Fig. 11.9) we may conclude

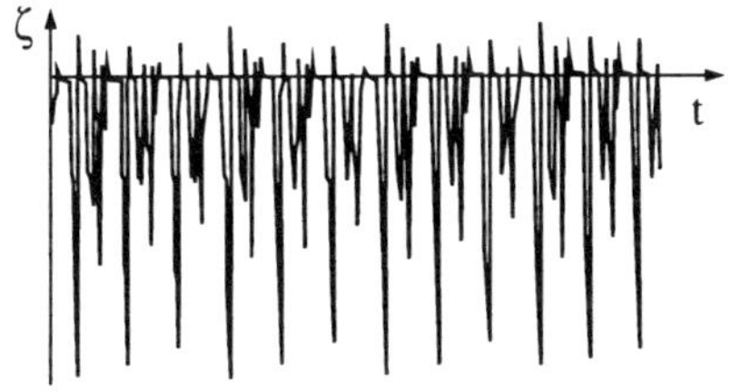

Figure 11.9: Typical Force Sequence of a Hammering Process

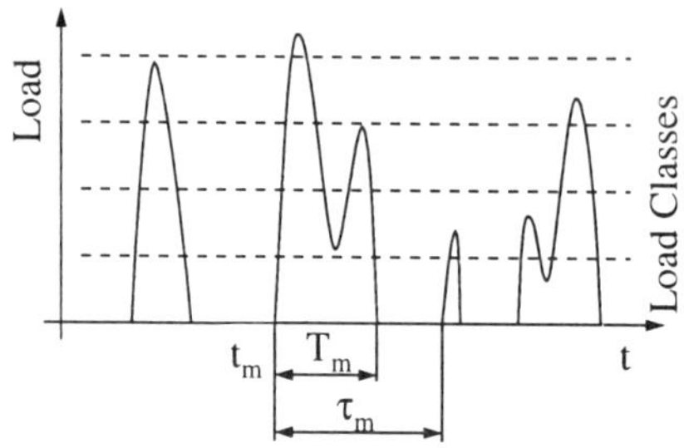

Figure 11.10: Hammering Force Structure

that a statistical consideration would be the most appropriate one. By doing so, some principal ideas of Buxbaum [15, 14] will be applied. A hammering impact can be characterized by the magnitude of the force amplitude and by the time behavior (Fig. 11.10). Analyzing the forces in the gear meshes, we see that the time behavior follows a Gamma distribution quite well (see Fig. 11.11).

Of more interest in practical problems is the load distribution. Again it turns out that the loads in the gear meshes can be well approximated by classical rules valid for different stochastic processes:

$$H(x) = H_0 \ \exp\left[-ax^n\right] , \tag{11.35}$$

where a and n might be determined from the simulation results by using a least square fit. The magnitude H_0 is the number of events of passing the nominal load. For hammering processes the nominal load is zero. Figure 11.12 gives an excellent comparison of formula (11.35) with simulated results.

11.1.3 Results

As an application of the theory presented, a four-stroke diesel engine with 12 cylinders and a power of 3000 kW has been considered. The nonsymmetrical gear system driving the combined camshaft/injection pump shaft is shown in Fig. 11.13 on the left-hand side. On the right-hand side we see the corresponding mechanical model. All gears are spur gears; they will be modeled as rigid bodies according to the section on body models. The camshaft will be described as an elastic body considering torsional elasticity only (see the coupling components section). As an additional option we regard a camshaft damper. The simulations focus on side A of the gear system due to the more complicated dynamical properties. Side A possesses one more stage than side B.

Different models have been established starting from a model with only 5 degrees of freedom and ending with a model with 13 degrees of freedom.

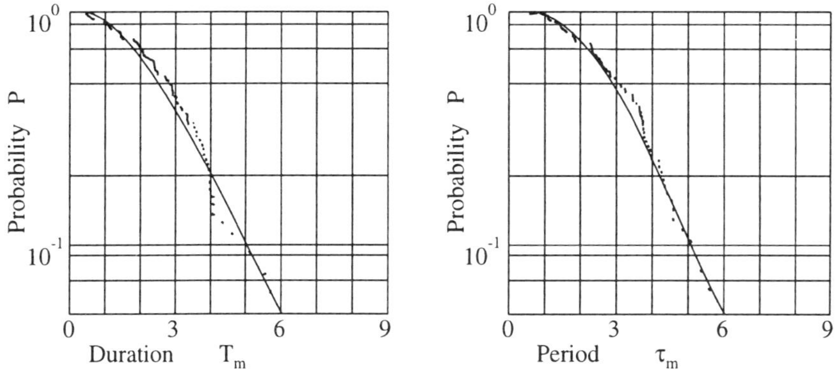

Figure 11.11: Gamma Distribution for the Duration T_m and the Period $\boldsymbol{\tau}_m$ of a Hammering Process

In all cases the torsional elasticity of the camshaft was described by two elastic degrees of freedom only, which turned out to be sufficient. As usual the determination of all stiffnesses proved to be rather difficult. Uncertainties mainly come from unknown flexural influences of the motor housing and from the more or less unknown stiffness reductions in one flanged joint and in a

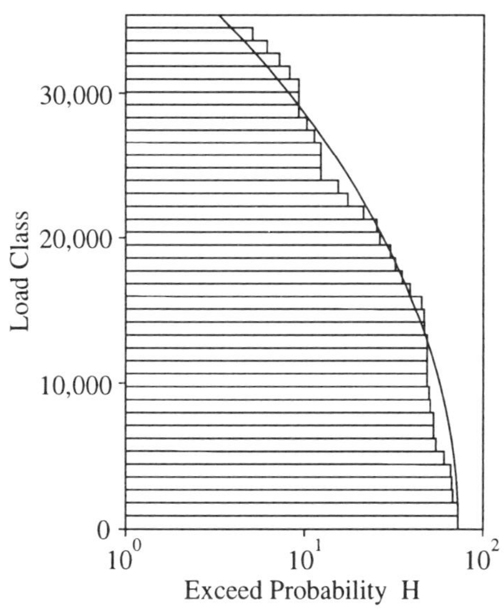

Figure 11.12: Load Distribution for a Typical Hammering Process

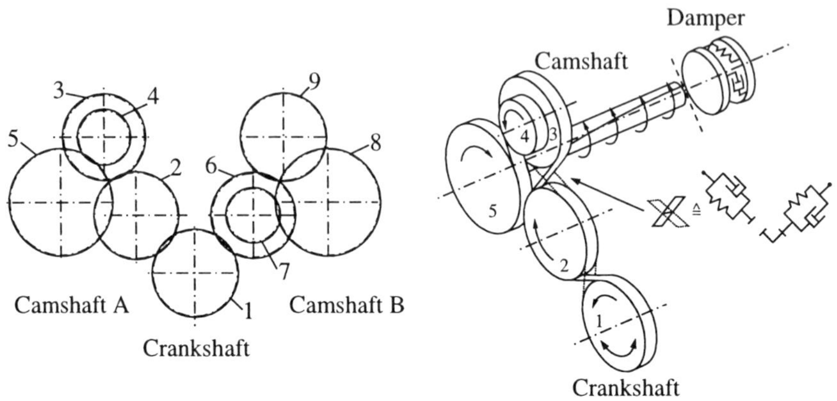

Figure 11.13: Gear System and Equivalent Mechanical Model for a 4-Stroke 12-Cylinder Diesel Engine

press fit. Therefore, stiffnesses had to be adapted to measurements. After this the models compared well with vibration measurements which were performed by a German diesel-engine manufacturer. A comparison is given in Fig. 11.14.

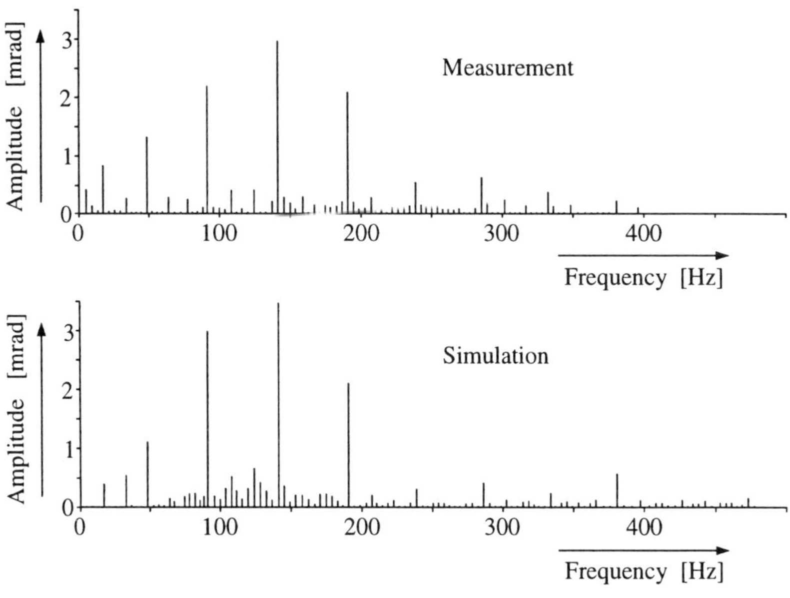

Figure 11.14: Angular Vibrations of the Camshaft, Measurements and Simulation

	Gear Loads [kN]			
Gear Mesh (Fig. 11.13)	Static Mean Load	Quasi-static Maximum Load	Dynamical Loads	
			Backlash = 0	Nom. Backlash
1/2	1.96	11.4	34.5	42.4
2/3	1.96	11.4	31.3	34.8
4/5	3.17	18.5	37.4	38.8

Table 11.2: Gear Loads in Gear Meshes

After the verification of the theory a series of parameter simulations were performed. Although the results obtained relate to the special diesel engine under consideration, they might nevertheless indicate some general parameter tendencies in such drivelines.

First we consider the loads in the gear meshes and compare these loads with statical and quasi-statical cases, which we define in the following way:

- The static load is that one which would be generated by a transmission of the averaged camshaft torque. This mean value would be $\overline{T}_{CS} = 430$ Nm (Fig. 11.7).
- The quasi-static extremum load is that one which would be generated by a static transmission of the maximum torque values for the camshaft (peak values in Fig. 11.7). This maximum torque is $\overline{T}_{\max} = 2520$ Nm.
- The dynamical maximum load is evaluated from the load distribution (see the section on evaluation of the simulations) under the assumption that these loads will be realized with a probability of 99%.

The results for the three gear meshes of Fig. 11.13 are given in Table 11.2. A systematic investigation of parameter tendencies has concluded [87]:

- Excitation Sources
  - Crankshaft excitation is small.
  - Camshaft excitation dominates, especially due to the injection pump loads.

- Gear System
  - Mass parameters show no much influence on the hammering process.
  - Increasing backlashes produce slightly increasing force amplitudes, but significantly increasing camshaft angular vibrations.
  - Increasing gear stiffness and damping gives decreasing force amplitudes.
- Camshaft
  - Increasing stiffness leads to largely decreasing force amplitudes.
  - Damping is of minor influence.
- Bearings
  - Slightly decreasing force amplitudes with increasing damping (all bearings).
  - No large difference exists between journal and roller bearings with respect to vibrations and force amplitudes.
- Camshaft Vibration Damper
  - Damper considerably reduces force amplitudes in the gear meshes.
  - Damper suppresses the hammering between two flanks and supports force transmission at the working flank.

11.2 Overloads in Gears due to Short-Circuit and Malsynchronization in a Generator

11.2.1 Introduction

Small power plants often are composed of a small steam turbine which is connected to a generator by a spur gear transmission with two gearwheels. Figure 11.15 gives an idea of the steam turbine gear combination, and Fig. 11.16 illustrates the mechanical model approximating the dynamics of the power plant, including the generator system [36]. There exist a large number of various power plant systems. The system under consideration represents a typical example.

All mechanical components of the power plant must be designed in such a way that they can survive the large overloads generated in a short-circuit or during malsynchronization, the last case being the more dangerous. The large torques produced by the generator for these cases may lead to hammering in the gears and thus to high dynamical overloads. For a good design of the coupling stiffness, however, hammering can usually be avoided. We consider these parameter influences later.

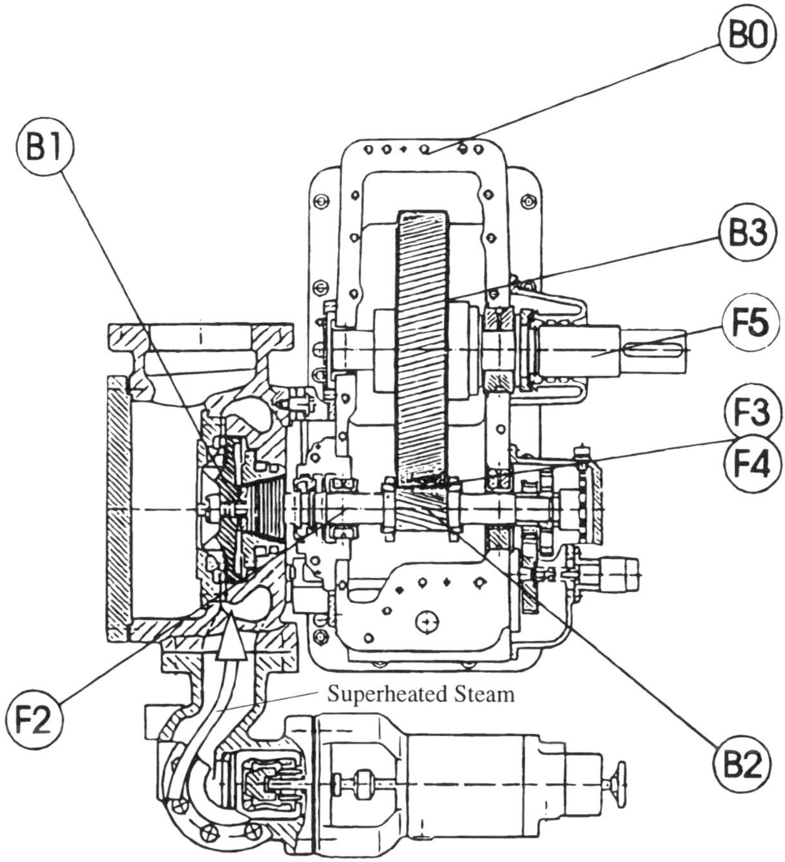

B0: Housing
B1: Turbine Wheel
B2: Bevel Gear
B3: Gearwheel

F2: Turbine Shaft
F3/F4: Helical Gearing
F5: Output Shaft

Figure 11.15: Turbine Gear Combination by KKK-Frankenthal AG

11.2.2 The Equations of Motion

The derivation of the equations of motion can be simplified considerably by the fact that all bodies are rotationally symmetric and that all deviations from a prescribed reference rotation are very small. Defining a body-fixed representation of a vector $\boldsymbol{v}$ by $B_i(\boldsymbol{v})$ and of a tensor $\boldsymbol{T}$ by $B_i(\boldsymbol{T})$, we write

$$B_i(\boldsymbol{w}_i) = {}_i\boldsymbol{w}_i = \begin{pmatrix} {}_i\boldsymbol{r}_i \\ {}_i\boldsymbol{\varphi}_i \end{pmatrix} \in \mathbb{R}^6 \tag{11.36}$$

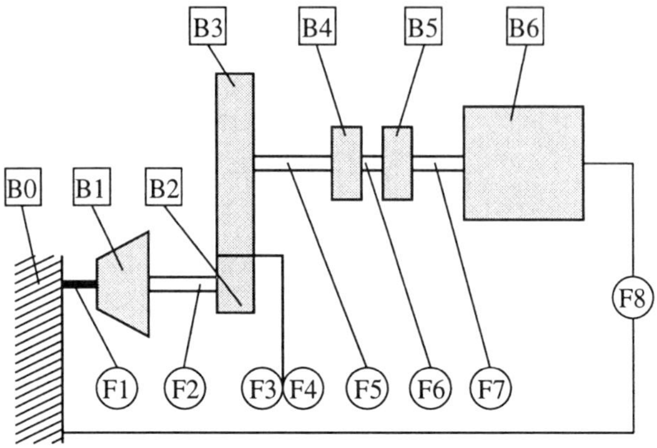

BODIES		FORCE ELEMENTS	
B0:	Housing (Inertial)	F1:	Turbine Torque
B1:	Turbine Wheel	F2:	Turbine Shaft
B2:	Bevel Gear	F3:	Helical Gear, Trailing Flank
B3:	Gearwheel	F4:	Helical Gear, Front Flank
B4/5:	Clutch	F5:	Gear Shaft
B6:	Generator	F6:	Clutch
		F7:	Generator Shaft
		F8:	Generator Torque

Figure 11.16: Mechanical Model of the Turbine Gear Combination

for the deviation vector, and

$$B_i\left(\boldsymbol{I}_{Hi}\right) = {}_i\boldsymbol{I}_i = \begin{pmatrix} A_i & 0 & 0 \\ 0 & A_i & 0 \\ 0 & 0 & C_i \end{pmatrix} \subset \mathbb{R}^{3,3} \tag{11.37}$$

for the moment of inertia tensor of body i.

The assumption $|{}_i\boldsymbol{\varphi}_i| \ll 1$ allows a representation as a vector. With this notation and prescribing a time-dependent rotation $\boldsymbol{\Omega}_{Si} = \boldsymbol{\Omega}_{Si}(t)$, we define the absolute velocities and accelerations in the following form:

$$\begin{aligned}
B_i\left(\boldsymbol{v}_{Hi}\right) &= {}_i\dot{\boldsymbol{r}}_i \;\; ; \;\; B_i\left(\dot{\boldsymbol{v}}_{Hi}\right) = {}_i\ddot{\boldsymbol{r}}_i \\
B_i(\boldsymbol{\Omega}_i) &= {}_i\dot{\boldsymbol{\varphi}}_i + \Omega_{Si}\boldsymbol{e}_3 \\
B_i(\dot{\boldsymbol{\Omega}}_i) &= {}_i\ddot{\boldsymbol{\varphi}}_i + \dot{\Omega}_{Si}\boldsymbol{e}_3 + \Omega_{Si}\,{}_i\dot{\tilde{\boldsymbol{\varphi}}}_i\boldsymbol{e}_3 \\
B_i\begin{pmatrix} \boldsymbol{v}_{Hi} \\ \boldsymbol{\Omega}_i \end{pmatrix} &= {}_i\dot{\boldsymbol{w}}_i + \begin{pmatrix} \boldsymbol{0} \\ \Omega_{Si}\boldsymbol{e}_3 \end{pmatrix} \\
B_i\begin{pmatrix} \boldsymbol{v}_{Hi} \\ \boldsymbol{\Omega}_i \end{pmatrix} &= {}_i\ddot{\boldsymbol{w}}_i + \begin{pmatrix} \boldsymbol{0} \\ \dot{\Omega}_{Si}\boldsymbol{e}_3 + \Omega_{Si}\,{}_i\dot{\tilde{\boldsymbol{\varphi}}}_i\boldsymbol{e}_3 \end{pmatrix} .
\end{aligned} \tag{11.38}$$

Inserting these expression into the general form of the equations of motion (eqs. 3.10), we obtain

$$\sum_{i=1}^{n} \left\{ B_i \left(\boldsymbol{Q}_i^T \right) \left[\begin{pmatrix} m_i \boldsymbol{E} & -m_i B_i(\tilde{\boldsymbol{r}}_{HSi}) \\ m_i B_i(\tilde{\boldsymbol{r}}_{HSi}) & {}_i\boldsymbol{J}_i \end{pmatrix} {}_i\ddot{\boldsymbol{w}}_i \right.\right.$$
$$+ \begin{pmatrix} \boldsymbol{0} \\ D_i \left(\Omega_{Si}^{\bullet} + \Omega_{Si}\, {}_i\tilde{\varphi}_i^{\bullet} \right) \boldsymbol{e}_3 \end{pmatrix}$$
$$\left.\left. - B_i \begin{pmatrix} \boldsymbol{F}_a \\ {}_H\boldsymbol{M}_a + \tilde{\boldsymbol{r}}_{Ha} \boldsymbol{F}_a \end{pmatrix}_i \right] \right\} = \boldsymbol{0} \tag{11.39}$$

$$\sum_{i=1}^{n} \left\{ B_i \left(\boldsymbol{Q}_i^T \right) \left[\boldsymbol{\theta}_i\, {}_i\ddot{\boldsymbol{w}}_i + \boldsymbol{g}_i(t) \right.\right.$$
$$\left.\left. - B_i \begin{pmatrix} \boldsymbol{F}_a \\ {}_H\boldsymbol{M}_a + \tilde{\boldsymbol{r}}_{Ha} \boldsymbol{F}_a \end{pmatrix}_i \right] \right\} = \boldsymbol{0}\,.$$

The active forces and torques $(\boldsymbol{F}_a, \boldsymbol{M}_a, \tilde{\boldsymbol{r}}_{Ha}\boldsymbol{F}_a)$ are given by a force law of the form

$$\zeta_K = \zeta_K \left(\gamma_K, \gamma_K^{\bullet}, t \right) \quad , \quad \zeta_K, \gamma_K \ \in \mathbb{R}^m \,, \tag{11.40}$$

where $\boldsymbol{\zeta}_K$ are the force vectors in the m force laws connecting the bodies of Fig. 11.17. The vectors $\boldsymbol{\gamma}_K$ represent the relative deflections in the direction of the force law under consideration. Figure 11.17 gives a typical configuration of two bodies connected by some force element K. The deflection $\boldsymbol{\gamma}_K$ in the direction of the deflected force element K can be expressed as

$$\gamma_K = \boldsymbol{\psi}_K^T \boldsymbol{C}_{K1\; i1}\boldsymbol{w}_{i1} - \boldsymbol{\psi}_K^T \boldsymbol{C}_{K2\; i2}\boldsymbol{w}_{i2} \,, \tag{11.41}$$

where $\boldsymbol{w}$ is known from Fig. 11.16, $\boldsymbol{C}_{K1}$, $\boldsymbol{C}_{K2}$ are the transformation matrices from the body-fixed reference systems B_{i1}, B_{i2} to the force element system R_K, and $\boldsymbol{\psi}_K$ is the projection matrix from the force element reference system to the deflected force element direction. These matrices are purely geometrical and must be determined once for each force element. They are important not only for the relative displacements $\boldsymbol{\gamma}_K$ but also for transforming the forces ζ_K (eqs. 11.40) into the coordinate space as given by the equations of motion (11.39). We get

$$\begin{aligned} B_{i1} \begin{pmatrix} \boldsymbol{F}_a \\ {}_H\boldsymbol{M}_a + \tilde{\boldsymbol{r}}_{Ha} \boldsymbol{F}_a \end{pmatrix}_{i1} &= \boldsymbol{C}_{K1}^T \boldsymbol{\psi}_K \boldsymbol{\zeta}_K \\ B_{i2} \begin{pmatrix} \boldsymbol{F}_a \\ {}_H\boldsymbol{M}_a + \tilde{\boldsymbol{r}}_{Ha} \boldsymbol{F}_a \end{pmatrix}_{i2} &= \boldsymbol{C}_{K2}^T \boldsymbol{\psi}_K \boldsymbol{\zeta}_K \,. \end{aligned} \tag{11.42}$$

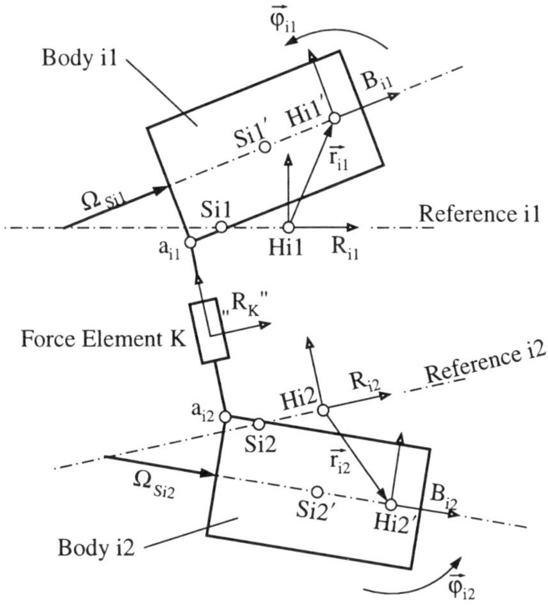

Figure 11.17: Force Element K between Bodies $i1$ and $i2$

The equations of motion (11.39) are then

$$\sum_{i=1}^{n} \left\{ B_i \left(\boldsymbol{Q}_i^T \right) \left[\boldsymbol{\theta}_i \, {}_i\ddot{\boldsymbol{w}}_i + \boldsymbol{g}_i(t) - \boldsymbol{C}_{Ki}^T \boldsymbol{\psi}_K \boldsymbol{\zeta}_K \right] \right\} = \boldsymbol{0}$$

or, briefly,

$$\boldsymbol{Q}^T \left[\boldsymbol{\theta} \ddot{\boldsymbol{w}} - \boldsymbol{h} \right] = \boldsymbol{0} \tag{11.43}$$

with

$$\boldsymbol{Q}^T = \left[B_1 \left(\boldsymbol{Q}_1^T \right), B_2 \left(\boldsymbol{Q}_2^T \right), \cdots, B_n \left(\boldsymbol{Q}_n^T \right) \right]$$

$$\boldsymbol{\theta} = \begin{bmatrix} \boldsymbol{\theta}_1 & \boldsymbol{0} & \cdots & \boldsymbol{0} \\ \boldsymbol{0} & \boldsymbol{\theta}_2 & \ddots & \vdots \\ \vdots & \ddots & \ddots & \boldsymbol{0} \\ \boldsymbol{0} & \cdots & \boldsymbol{0} & \boldsymbol{\theta}_n \end{bmatrix} ; \quad \begin{aligned} \boldsymbol{w}^T &= \left({}_1\boldsymbol{w}_1^T, {}_2\boldsymbol{w}_2^T, \ldots, {}_n\boldsymbol{w}_n^T \right) \\ \boldsymbol{h}^T &= \left(\boldsymbol{h}_1^T, \boldsymbol{h}_2^T, \ldots, \boldsymbol{h}_n^T \right) . \end{aligned}$$

Applying the relationships $\ddot{\boldsymbol{w}} = \boldsymbol{Q}\ddot{\boldsymbol{q}}$ and $\boldsymbol{M} = \boldsymbol{Q}^T \boldsymbol{\theta} \boldsymbol{Q}$, $\overline{\boldsymbol{h}} = \boldsymbol{Q}^T \boldsymbol{h}$ yields the standard form of the equations of motion:

$$\boldsymbol{M}\ddot{\boldsymbol{q}} - \overline{\boldsymbol{h}} = \boldsymbol{0}, \ \boldsymbol{q} \in \mathbb{R}^f, \ \boldsymbol{M} \in \mathbb{R}^{f,f} . \tag{11.44}$$

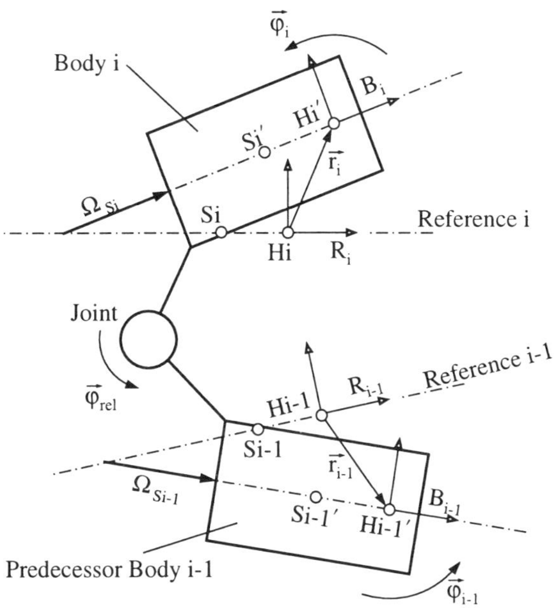

Figure 11.18: Body i with Predecessor Body $i-1$ in a Displaced Situation

11.2.3 Solution Procedure

For machine configurations such as those under consideration, an application of the order-n idea is particularly favorable. The power plant possesses a treelike structure which allows the displacement ${}_i\boldsymbol{w}_i$ of a body i to be represented as a function of the displacement ${}_p\boldsymbol{w}_p$ of its predecessor body and of the relative displacement $\boldsymbol{q}_i$ between them. Such a relationship can easily be written as

$$
{}_i\boldsymbol{w}_i = \boldsymbol{C}_{i\,(i-1)}\boldsymbol{w}_{(i-1)} + \boldsymbol{\phi}_i\boldsymbol{q}_i\,, \tag{11.45}
$$

where $\boldsymbol{C}_i i$ and $\boldsymbol{\phi}_i$ are geometrical matrices. Figure 11.18 indicates the most important correlations.

Applying the eq. (11.45) to the elements of the Jacobian matrix $B_i(\boldsymbol{Q}_i) = (\partial_i\dot{\boldsymbol{w}}_i/\partial\dot{\boldsymbol{q}}_i)$ (eqs. 11.43, 11.44), we obtain a triangular form of $\boldsymbol{Q}$:

$$
\boldsymbol{Q}^T = \begin{pmatrix} \boldsymbol{\phi}_1^T & (\boldsymbol{C}_2\boldsymbol{\phi}_1)^T & (\boldsymbol{C}_3\boldsymbol{C}_2\boldsymbol{\phi}_1)^T & \cdots & (\boldsymbol{C}_n\cdots\boldsymbol{C}_2\boldsymbol{\phi}_1)^T \\ \boldsymbol{0} & \boldsymbol{\phi}_2^T & (\boldsymbol{C}_3\boldsymbol{\phi}_2)^T & \cdots & (\boldsymbol{C}_n\cdots\boldsymbol{C}_3\boldsymbol{\phi}_2)^T \\ \boldsymbol{0} & \boldsymbol{0} & \boldsymbol{\phi}_3^T & \ddots & \vdots \\ \boldsymbol{0} & \boldsymbol{0} & \boldsymbol{0} & \ddots & (\boldsymbol{C}_n\boldsymbol{\phi}_{n-1})^T \\ \boldsymbol{0} & \boldsymbol{0} & \boldsymbol{0} & \boldsymbol{0} & \boldsymbol{\phi}_n^T \end{pmatrix}. \tag{11.46}
$$

Therefore the equation of motion (11.43) will possess the structure

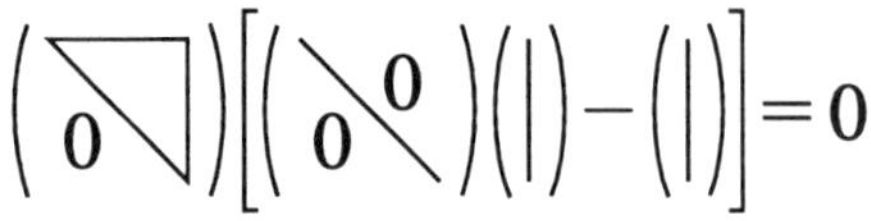

which gives for a last body of the treelike structured plant a particularly simple equation of motion:

$$\begin{aligned} \boldsymbol{\phi}_n^T \boldsymbol{\theta}_n \, {}_n\ddot{\boldsymbol{w}}_n &= \boldsymbol{\theta}_n^T \boldsymbol{h}_n \\ \rightarrow \boldsymbol{M}_n \ddot{\boldsymbol{q}}_n &= \boldsymbol{\theta}_n^T \boldsymbol{h}_n \, . \end{aligned} \tag{11.47}$$

Combining the eqs. (11.47) and (11.45) by eliminating ${}_n\ddot{\boldsymbol{w}}_n$ gives, for $i = n$,

$$\ddot{\boldsymbol{q}}_n = \left(\boldsymbol{\phi}_n^T \boldsymbol{\theta}_n \boldsymbol{\phi}_n\right)^{-1} \left(\boldsymbol{\phi}_n^T \boldsymbol{h}_n - \boldsymbol{\phi}_n^T \boldsymbol{\theta}_n \boldsymbol{C}_n \, {}_{n-1}\ddot{\boldsymbol{w}}_{n-1}\right) . \tag{11.48}$$

Later, according to eqs. (11.43) and (11.46), the predecessor body is described by

$$\begin{aligned} &\boldsymbol{\phi}_{n-1}^T \boldsymbol{\theta}_{n-1} \, {}_{n-1}\ddot{\boldsymbol{w}}_{n-1} + \left(\boldsymbol{C}_n \boldsymbol{\phi}_{n-1}\right)^T \boldsymbol{\theta}_m \, {}_n\ddot{\boldsymbol{w}}_n = \\ &\boldsymbol{\phi}_{n-1}^T \boldsymbol{h}_{n-1} + \left(\boldsymbol{C}_n \boldsymbol{\phi}_{n-1}\right)^T \boldsymbol{h}_n \, . \end{aligned} \tag{11.49}$$

Inserting (11.48) into (11.45) for $i = n$ results in

$${}_n\ddot{\boldsymbol{w}}_n = \left(\boldsymbol{E} - \boldsymbol{\phi}_n \boldsymbol{M}_n^{-1} \boldsymbol{\phi}_n^T \boldsymbol{\theta}_n\right) \boldsymbol{C}_n \, {}_{n-1}\ddot{\boldsymbol{w}}_{n-1} + \boldsymbol{\phi}_n \boldsymbol{M}_n^{-1} \boldsymbol{\phi}_n^T \boldsymbol{h}_n \, , \tag{11.50}$$

which together with eq. (11.49) finally yields

$$\boldsymbol{\phi}_{n-1}^T \boldsymbol{\theta}_{n-1}^* \, {}_{n-1}\ddot{\boldsymbol{w}}_{n-1} = \boldsymbol{\phi}_{n-1}^T \boldsymbol{h}_{n-1}^* \tag{11.51}$$

where

$$\begin{aligned} \boldsymbol{\theta}_{n-1}^* &= \boldsymbol{\theta}_{n-1} + \boldsymbol{C}_n^T \left[\boldsymbol{\theta}_n - \boldsymbol{\theta}_n \left(\boldsymbol{\phi}_n \boldsymbol{M}_n^{-1} \boldsymbol{\phi}_n^T\right) \boldsymbol{\theta}_n\right] \boldsymbol{C}_n \\ \boldsymbol{h}_{n-1}^* &= \boldsymbol{h}_{n-1} + \boldsymbol{C}_n^T \left[\boldsymbol{E} - \boldsymbol{\theta}_n \left(\boldsymbol{\phi}_n \boldsymbol{M}_n^{-1} \boldsymbol{\phi}_n^T\right)\right] \boldsymbol{h}_n \, . \end{aligned} \tag{11.52}$$

This can be carried out from all last bodies to the inertial zeroth body with ${}_0\ddot{\boldsymbol{w}}_0 = \boldsymbol{0}$. In a further forward recursion we determine all unknown accelerations ${}_i\ddot{\boldsymbol{w}}_i$ and $\ddot{\boldsymbol{q}}_i$, and in a second backward recursion we might evaluate any passive forces present.

The computer time of this procedure is proportional to the number of degrees of freedom and not to its third power. In each step we must invert a 6×6 matrix $\boldsymbol{M}_i^{-1}$. The matrices $\boldsymbol{C}_i, \boldsymbol{\phi}_i, \boldsymbol{\Psi}_i, \boldsymbol{\theta}_i$ are determined only once [7].

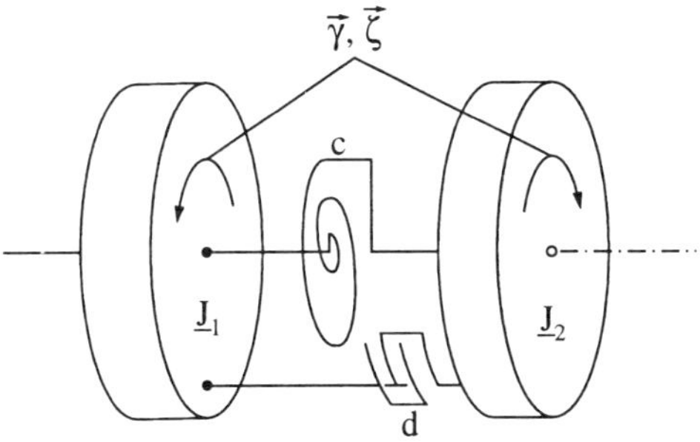

Figure 11.19: Torsional Shaft of a Spring-Damper Element

11.2.4 Force Elements

The same force elements are applied to the power plant problem as to the hammering problem of diesel engines (see Section 11.1). Therefore, we shall only summarize these elements without further discussion. In Section 11.1 the torsional flexibility of shafts was modeled by a RITZ approach. Sometimes it is possible to regard only the first series term of such an approach. This results in a spring-damper approximation with the force law (see Fig. 11.19):

$$\zeta = -c\gamma - d\dot{\gamma}\,. \tag{11.53}$$

The mesh of teeth for a gearwheel pair with helical gearing is illustrated in Fig. 11.20. The matrices $\boldsymbol{C}_K$ and $\boldsymbol{\Psi}$ from the eqs. (11.43, 11.45, 11.52) follow immediately from Fig. 11.20:

$$\begin{aligned}
\boldsymbol{C}_{K1} &= \begin{pmatrix} \boldsymbol{E} & -\begin{pmatrix} 0 & 0 & r_1 \\ 0 & 0 & 0 \\ -r_1 & 0 & 0 \end{pmatrix} \\ \boldsymbol{0} & \boldsymbol{E} \end{pmatrix}, \\
\boldsymbol{C}_{K2} &= \begin{pmatrix} \boldsymbol{E} & -\begin{pmatrix} 0 & 0 & r_2 \\ 0 & 0 & 0 \\ -r_2 & 0 & 0 \end{pmatrix} \\ \boldsymbol{0} & \boldsymbol{E} \end{pmatrix}, \\
\boldsymbol{\Psi}_K &= \begin{pmatrix} {}_K\boldsymbol{e}_K \\ \boldsymbol{0} \end{pmatrix} \in \mathbb{R}^{6\times 1},
\end{aligned} \tag{11.54}$$

with the unit vector in the direction of the pressure angle:

$${}_K\boldsymbol{e}_K = \begin{pmatrix} \cos(\alpha)\cos(\beta) \\ -\sin(\alpha)\cos(\beta) \\ -\sin(\beta) \end{pmatrix} \quad ; \quad \|\,{}_K\boldsymbol{e}_K\,\| = 1\,. \tag{11.55}$$

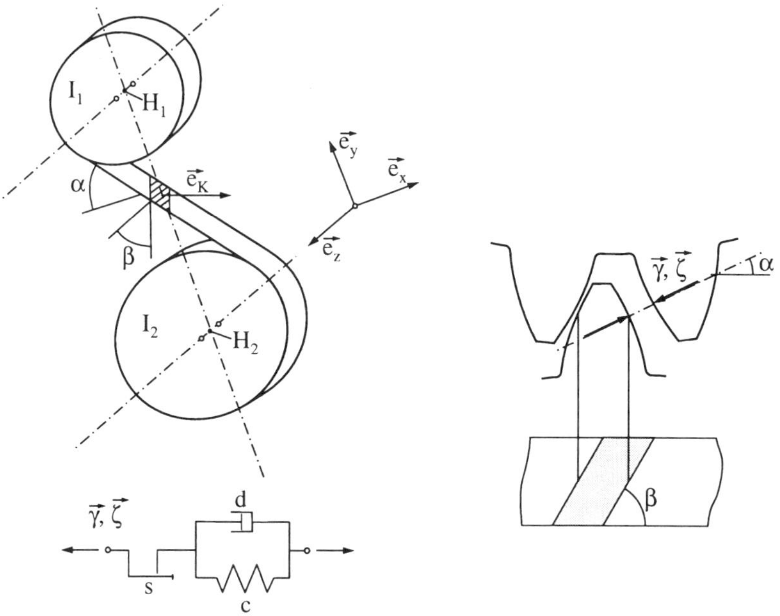

Figure 11.20: The Force Law of Helical Gearing

As a consequence of the backlash in the mesh of gears we must consider a nonlinear force law, because we have three areas of relative motion: a mesh with the front flank, a free-flight phase without contact and a mesh with the trailing flank. This can be written

- $\gamma_1 < 0$ contact front flank
- $\gamma_1 > 0,\ \gamma_2 > 0$ free flight without contact
- $\gamma_2 < 0$ contact trailing flank.

This law is given by Fig. 11.21, where s is the backlash. In this figure we assume a linear spring behavior as soon as contact is established. These force laws depend for gearing on the contact ratio factor, which typically leads to parameter-excited vibrations. For the case under consideration this effect will be neglected because it is of minor influence. We are mainly concerned with the very short time, and thus extremely nonstationary, influence of short-circuits or of malsynchronization.

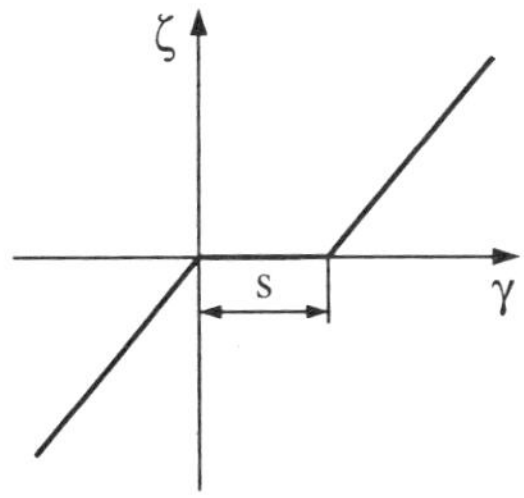

Figure 11.21: Force Law for the Mesh of a Tooth

11.2.5 Synchronous Generator

The turbine gear combination of Fig. 11.15 drives a smooth-core synchronous generator, which is an induction machine. For the stabilization of dynamical transient phenomena the rotor usually has a dampening cage arranged over the field coils. It operates very similar to a cage motor of an asynchronous machine (Fig. 11.22). Synchronous machines move for stationary operation with constant speed

$$\omega_{\text{syn}} = \omega / p \tag{11.56}$$

where ω is the standard frequency and p is the number of pairs of poles.

For the detailed derivation of the equations of motion of a synchronous generator see [36] or standard textbooks in electrical engineering. As a result we get the following set of equations:

$$\boldsymbol{L}\left(\frac{d\boldsymbol{i}}{dt}\right) + \boldsymbol{Z}\boldsymbol{i} = \boldsymbol{u} \tag{11.57}$$

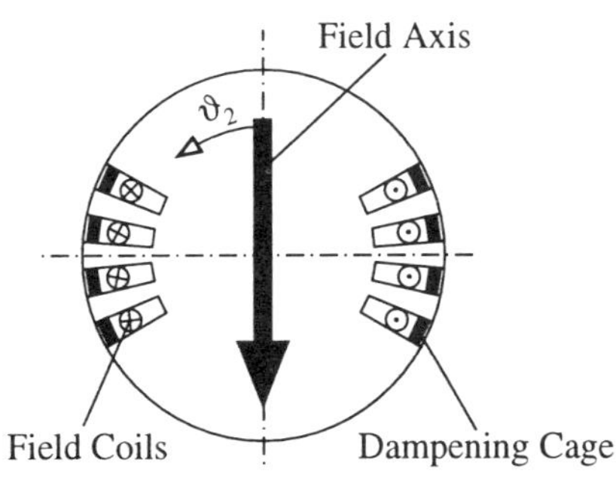

Figure 11.22: Rotor with Field Coils and Dampening Cage

with the vectors and matrices

$$\boldsymbol{i} = (i_{Sd}\ i_{Sq}\ i_{Dd}\ i_{Dq}\ i_E)^T \quad ; \quad \boldsymbol{u} = (u_{Sd}\ u_{Sq}\ 0\ 0\ u_E)^T$$

$$\boldsymbol{L} = \begin{bmatrix} L_{Sd} & 0 & 2M_{DSd} & 0 & 2M_{ES} \\ 0 & L_{Sq} & 0 & 2M_{DSq} & 0 \\ \frac{1}{2}M_{SDd} & 0 & L_{Dd} & 0 & M_{ED} \\ 0 & \frac{1}{2}M_{SDq} & 0 & L_{Dq} & 0 \\ M_{SE} & 0 & 2M_{DE} & 0 & L_E \end{bmatrix} \qquad (11.58)$$

$$\boldsymbol{Z} = \begin{bmatrix} R_S & -p\dot{\vartheta}_m L_{Sq} & 0 & -2p\dot{\vartheta}_m M_{DSq} & 0 \\ p\dot{\vartheta}_m L_{Sd} & R_S & 2p\dot{\vartheta}_m M_{DSd} & 0 & 2p\dot{\vartheta}_m M_{ES} \\ 0 & 0 & R_D & 0 & 0 \\ 0 & 0 & 0 & R_D & 0 \\ 0 & 0 & 0 & 0 & R_E \end{bmatrix} .$$

The magnitudes are labeled as follows: i_S, stator current; i_D, damping current; i_E, exciting current; u_S, stator terminal voltage; u_E, field voltage; subscript d, real parts; subscript q, imaginary parts; subscript S, stator; subscript D, damping; subscript, E, exciting; L, M, inductivities; ϑ_m, mechanical angles from configuration design; p, number of pole pairs; R, resistance; $\boldsymbol{L}$, inductance matrix; $\boldsymbol{Z}$, impedance matrix.

The form of eq. (11.57) is numerically compatible with the form of the mechanical equations of motion (11.44). With known magnitudes from the eqs. (11.57) and (11.58) we are able to compute the rotor torque in the form

$$\begin{aligned} T_G &= \tfrac{3}{2}p\,(\psi_{Sd} i_{Sq} + \psi_{Sq} i_{Sd}) \\ &= \tfrac{3}{2}p\left[i_{Sd} i_{Sq}(L_{Sd} - L_{Sq}) + 2 i_{Dd} i_{Sq} M_{DSd}\right. \qquad (11.59) \\ &\quad \left. -2 i_{Dq} i_{Sd} M_{DSq} + 2 M_{ES} i_E i_{Sq}\right] . \end{aligned}$$

For stationary operation the rotor torque is constant and determined only by inductive expressions. Mechanical dissipation losses are not considered in our

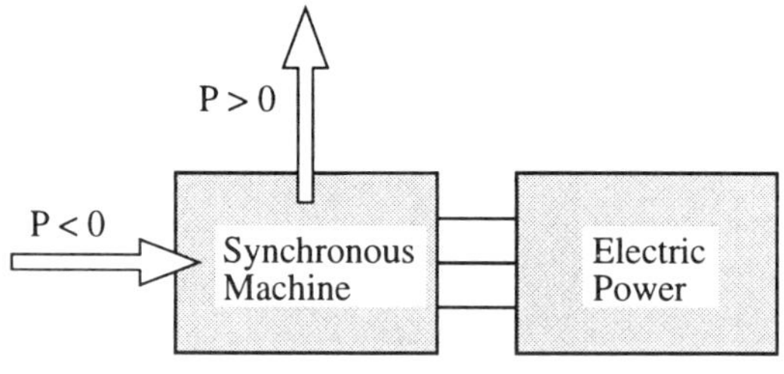

Figure 11.23: Definition of Powers

investigations because they are very small compared with the effects we are interested in.

Knowing the torques we can establish the power budget for the machine. It is assumed that the operational reference speed of the rotor is always positive. Therefore, a negative torque is a braking torque, and a positive torque is a driving one. A definition of the powers follows from Fig. 11.23. From this the power budgets are

$$\begin{aligned} P_W &= \tfrac{3}{2} u_S i_S \cos(\varphi) && \text{Total effective power} \\ P_V &= -\tfrac{1}{2}\left(3R_S i_S^2 + N_D R_D i_D^2\right) && \text{Ohm stray power} \\ P_\delta &= P_W - P_V = \dot{\vartheta}_m T_G && \text{Cyclic power} \\ P_m &= \dot{\vartheta}_m (T_G - T_R) && \text{Mechanical power.} \end{aligned} \tag{11.60}$$

In the following we shall consider first load shedding and its influence on the turbine-gear combination Fig. 11.15. Load shedding or sudden release means that the synchronous generator is decoupled impulsively from the power consumer where this transient situation usually takes place in a very short time, up to 0.02 s depending on the machine.

Second, we shall investigate the cases of short-circuits, which often occur during thunderstorms. The pole braces of a mast might be shorted to ground by lightning, thus generating a short-circuit for the synchronous machine.

In combination with a short-circuit we might have malsynchronization, i.e., a phase error between the induced generator voltage and the consumer voltage.

11.2.6 Simulation and Results

For simulations we consider a synchronous generator driven by a steam turbine with 1040 kW. Data in detail may be found in [36].

The case of load shedding was considered for a time of 0.02 s where all possible dynamics have occurred. It turned out that dangerous gradients of the current might be generated, which is more a critical case for the electrical components than for the mechanical parts. Within a few milliseconds the torque of the synchronous generator decreases to zero. Therefore, the mechanical parts are not touched by this process. The only critical component might be the turbine itself which must be controlled very quickly in the case of load shedding. This aspect has not been investigated here.

Short-circuits have been simulated for 0.1 s, where the control system of generator and turbine show no actions of significance. The results are given by Fig. 11.24. The generator torque exciting the system with 50 Hz and 100 Hz is supercritical with respect to the torque acting at the clutch with 20 Hz eigenfrequency. As a consequence the torque at the clutch is smaller

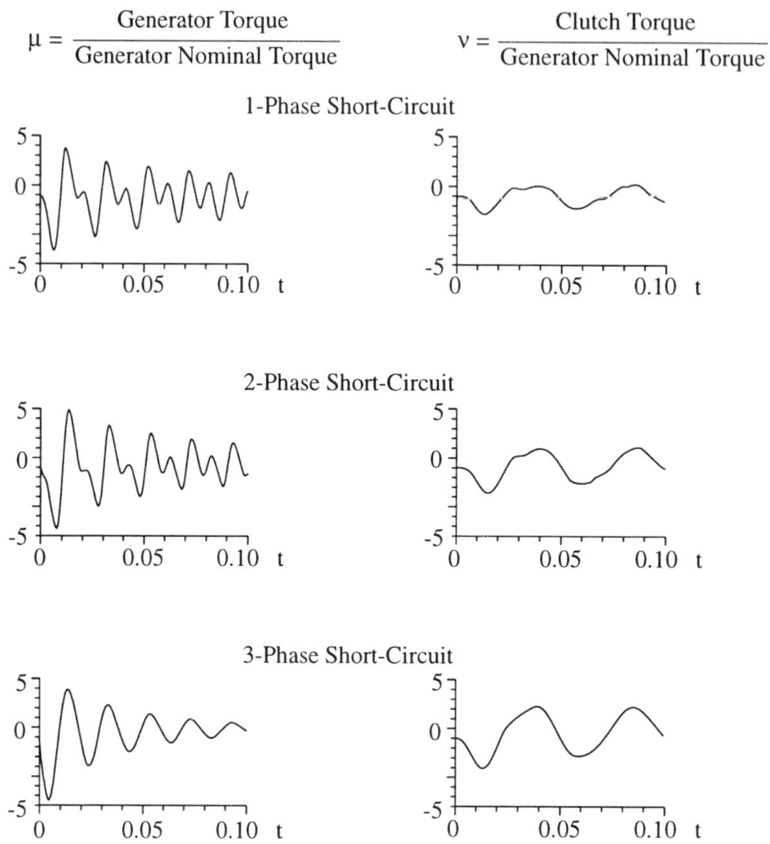

Figure 11.24: Short-Circuits for a 2-Mass Model (Generator Nominal Torque = 6730 Nm)

than the generator torque. With the number of phases with short-circuits the torques increase. On the other hand, the differences, especially for the generator torques, are small. Of importance for an excitation of the mechanical components is the 100-Hz vibration of the short-circuits in the one-phase and two-phase cases. Therefore, for further consideration of parameter influences the two-phase short-circuit case will be taken as standard.

The worst case for malsynchronization occurs with a phase error of 180° (opposition of phases). The dynamical loads are three times as large as for short-circuits (Fig. 11.25).

The mechanical components cannot withstand loads of that magnitude. On the other hand, malsynchronizations usually take place with phase errors up to 30°, which is not as dangerous. They are generated by manual coupling

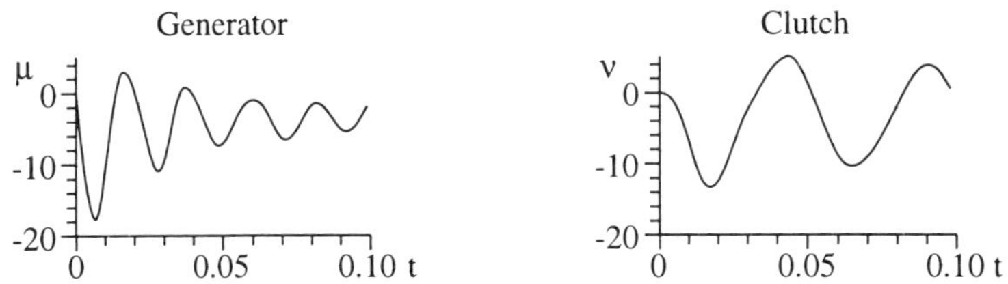

Figure 11.25: Malsynchronization with 180°, 2-Mass Model, Torques μ, ν Related to Nominal Torque

of networks or by short-circuits due to thunderstorms. Figure 11.26 illustrates the maximum torques μ, ν related to the nominal torque for phase errors from 0° to 180°. The difference of the two curves represents the torque loads (shaded area).

In the following we shall investigate some parameter influences with respect to the dynamical loads. We consider the two-phase short-circuit on the basis of six-mass model and vary the clutch stiffness, the backlash, the tooth stiffness and the moment of inertia of the gearwheel (see Figs. 11.15, 11.16). Figure 11.27 shows the torques for the nominal set of data for all mechanical components, the backlash being zero.

First, assuming a constant backlash $s = 0.15$ mm, we vary the clutch stiffness. With increasing stiffness we see a high-frequency vibration. For a value of 3.5 times the nominal stiffness a first short hammering effect occurs which grows with still more increasing stiffness (Fig. 11.28). This leads to dynamical overloads up to a factor of 6. Typically, the hammering process does not occur at the eigenfrequency of a tooth, which is about 2776 Hz, but at a lower eigenfrequency of 255 Hz, which corresponds to an eigenform where the teeth of the gears vibrate against the clutch. It can be shown that the hammering vibration has no direct influence on the generator motion on the one side and on the turbine motion on the other.

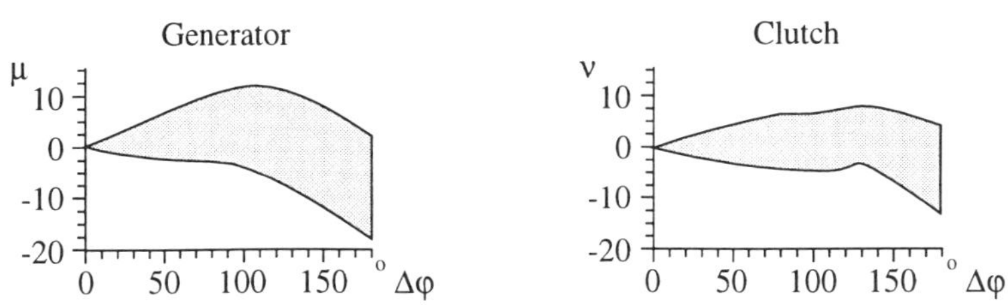

Figure 11.26: Torque Loads by Malsynchronization Dependent on the Phase Error $\Delta\varphi$ (2-Mass Model)

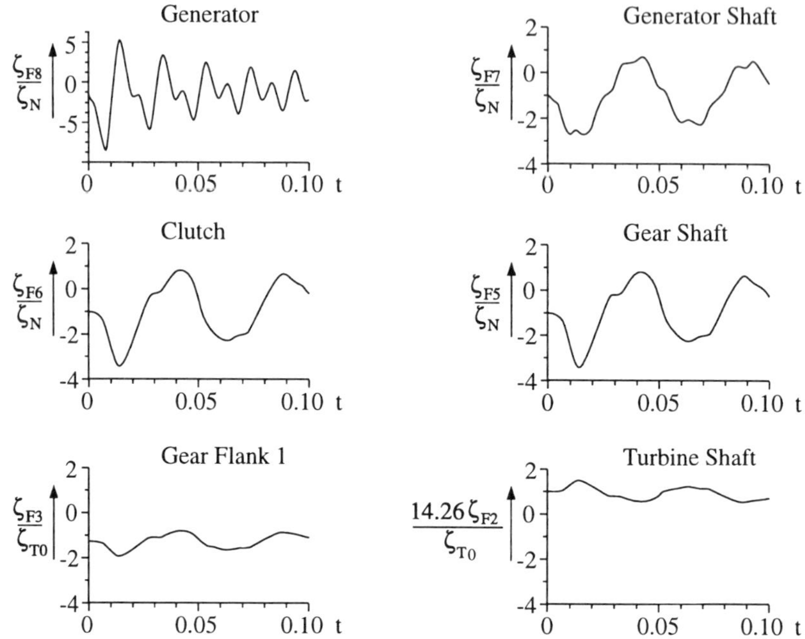

Figure 11.27: Loads for Nominal Data Set (ζ_n = Generator Nominal Torque = 6729 Nm, ζ_{T0} = Gear Tooth Nominal Force = 12025 N)

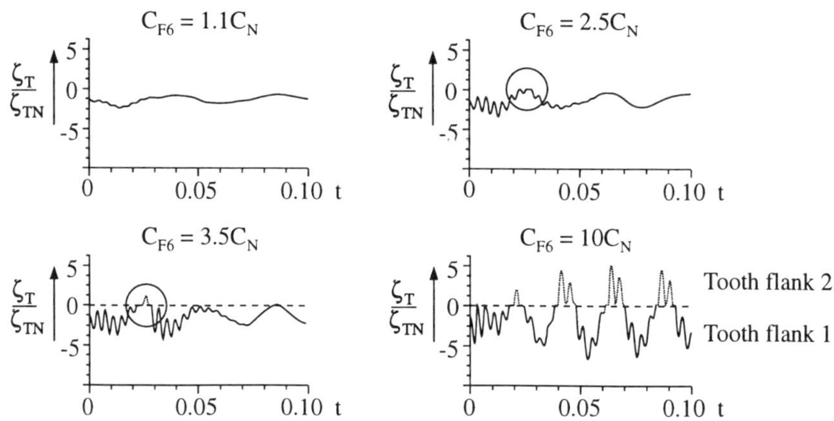

Figure 11.28: Vibration of the Clutch Stiffness (Nominal Value $C_N = 4.75 \cdot 10^5$ Nm/rad, ζ_T Tooth Force, ζ_{T0} = Nominal Tooth Force)

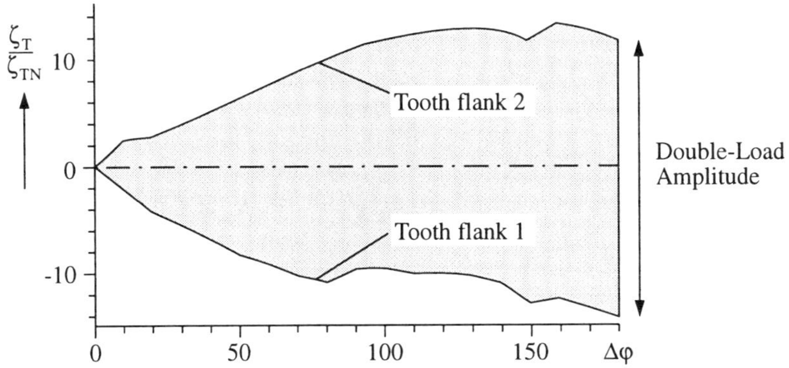

Figure 11.29: Load Amplitudes for the Critical Stiffness Case

The maximum possible loads for malsynchronization can be expected for a clutch stiffness leading to a more or less perfect coupling between the electrical

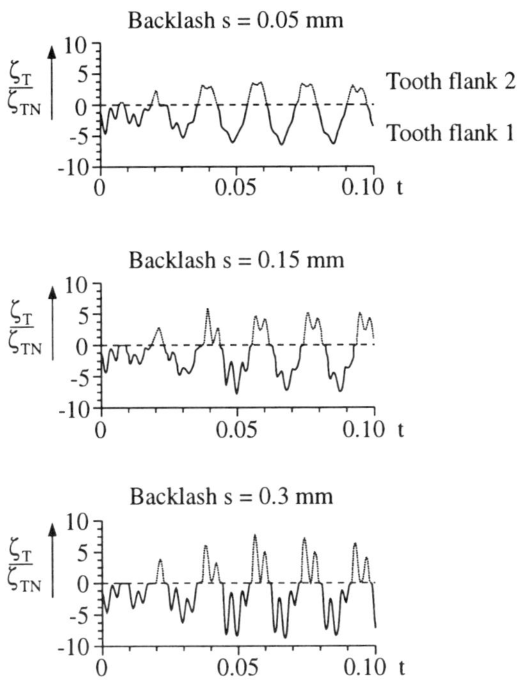

Figure 11.30: Tooth Forces Depending on the Backlash

and mechanical components. This critical stiffness is in our case $C_N = 3.46 \cdot 10^7$ Nm/rad, which corresponds to a value of 72.6 times the nominal stiffness. Figure 11.29 depicts the resulting loads as functions of the phase error. The maximum overloads are 12 times the nominal load, which occurs at a phase error $\Delta\varphi > 100°$. In practice, however, no one would design such large clutch stiffnesses. The nominal stiffness does not endanger the system during malsynchronization. As might be expected by physical arguments, increasing backlashes lead to increasing loads. Depending on the amplitudes of the gear vibrations the teeth flanks might separate and then hammer back with very large relative accelerations. The kinetic energy of this kind of relative motion is transformed into elastic or plastic deformations energy in the contact zone, thus generating high dynamical loads. Figure 11.30 gives an example for three different backlashes for a two-phase short-circuit and critical clutch stiffness.

The moment of inertia of the gearwheel possesses a positive influence on the development of teeth loads. The big mass of the wheel works as a low-pass filter with respect to the vibrations of the drive system. Therefore, increasing moments of inertia of the gearwheel reduces the loads in the mesh of the teeth. Figure 11.31 illustrates this behavior for a two-phase short-circuit and

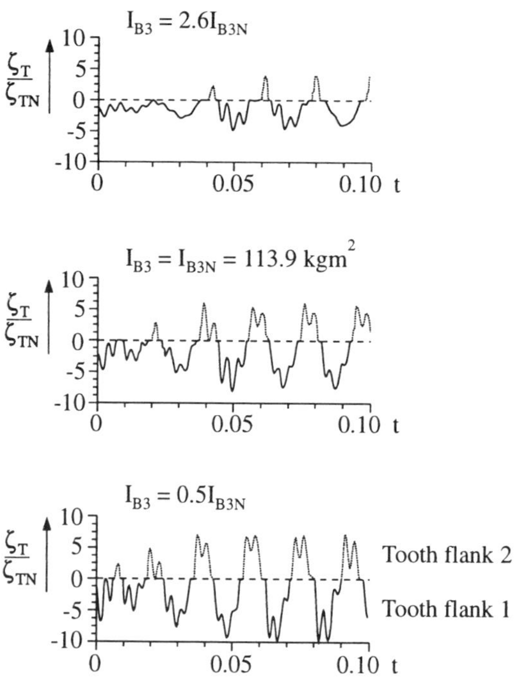

Figure 11.31: Tooth Forces Depending on the Gearwheel Moment of Inertia

critical clutch stiffness.

Tooth stiffness does not significantly influence the loads. The eigenfrequencies of the teeth with $f_T > 2$ kHz are too far away from any important frequency of the mechanical system. In addition, there are no possibilities of varying the tooth stiffness, which is determined by the design according to power transmission requirements.

12

APPLICATIONS WITH CLASSICAL IMPACT THEORY

12.1 Gear Rattling*

12.1.1 Introduction

Couplings with backlash may generate an unsteady motion of the coupled components which generally is composed of free flight and contact phases. The duration of these phases depends on the configuration of the system under consideration, on the contact structure and on the external forces acting on the system components. Considering gearboxes or gear trains, two situations may occur, depending on the exciting torques. In changing speed gears, for example, we always find some gearwheels not under load that are able to rattle. In this case the teeth of the gearwheels come into contact only for a very short time, where contact forces, or better the contact momenta, are small. An impulsive process of this type can be well approximated by a generalized impact theory. Quite another situation occurs if the teeth of some gearwheels separate and bounce back under the influence of large forces, thus generating high dynamical loads in the contact zone. Spur gears in large diesel engines driving the camshafts and the injection pump shafts are an example. In this case the deformation process within the area of contact must be considered very thoroughly (see Chapter 11). In the following we shall focus our attention on the first, the rattling problem.

The basic idea behind the solution process is simple. We consider a multibody system with several couplings including backlashes, which are either meshes of teeth or bearings. One or more of the bodies are externally excited by forces with periodic behavior, or they are excited just kinematically. We assume that in one of these plays an impact has just taken place so that we start with free-flight phases in the various plays. Contact in one of the plays is

*From [38]. Reprinted by permission of Kluwer Academic Publishers.

indicated by zero relative distances which serve as indicator functions for the events "contact". Considering a multibody system such as a shift transmission with many plays, we must evaluate that backlash where a contact occurs at the earliest time. Then we apply a generalized impact theory as derived in Chapter 7, yielding the state shortly after that impact, which serves as the initial state for the next free-flight phase. The results of such a patching method depend greatly on the accuracy of interpolating the impulsive events.

Another representation can be achieved by considering the rattling state before impact $j+1$, say in some backlash m in dependence on the state before impact j in some backlash n. Such a description can be derived analytically, resulting in a set of discrete mappings which give a good basis for more structural investigations of rattling [73]. Moreover, looking at the results for rattling processes and their resemblance to stochastic appearance it makes sense to introduce some mean values to describe rattling properties.

There exists some literature in related fields from which a small selection is given. Theoretical aspects are discussed in the books [43, 44], whereas [27] gives many interesting experimental examples of chaotic vibrations. The very complete and fundamental book on dynamics [70] was one of the first to present impulsive processes more thoroughly. The paper [71] was the starting point for the rattling investigations as presented here. It establishes a Lagrangian approach to impulsive processes and applies it to some toy woodpecker dynamics, which resembles very much a rattling process. These ideas have been extended in two dissertations [37, 46], where the first one [37] investigates rattling in a very fundamental and comprehensive way and the second one [46] gives stochastic approximations to rattling.

12.1.2 Gearbox Model

Equations of Motion

A changeover gear system with five stages is depicted in Fig. 12.1. The driving shaft runs the countershaft, which itself rotates the main shaft. All stages with the exception of the fourth are located on the countershaft/main shaft combination and switched on the main shaft. The fourth stage is switched by connecting the driving and main shafts directly. The mesh of gears between the driving shaft and the countershaft is a constant mesh; it cannot be switched. Therefore the countershaft is rotating continuously even with an active fourth stage. At the gearbox entrance some angular vibrations enter, being part of the vibrations of the complete drive train system which is excited by the engine's unbalances. These vibrations are approximately harmonic; they come in with the driving shaft and may lead to some rattling in the constant mesh of the countershaft (if the fourth stage is in operation). The already slightly irregular motion of the countershaft acts as an excitation to all gear stages not under load, thereby generating rather chaotic vibrations.

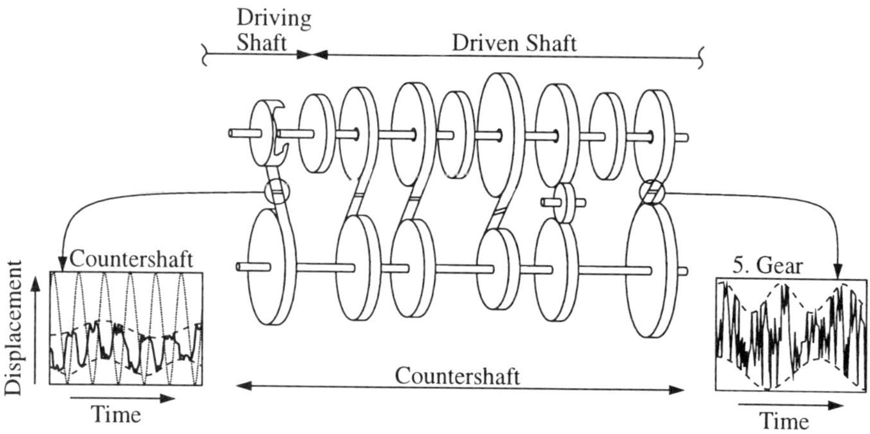

Figure 12.1: Model of n_s Stage Gear Mechanism ($n_s = 5$ and Reverse Stage)

An example is displayed in Fig. 12.1. For the fourth-stage operation we get two-stage rattling (countershaft, stage) for stages 1, 2, 3, 5 and three-stage rattling for the reverse gear. For any other stage in operation the countershaft is under load, and we get one-stage rattling in the forward and two-stage rattling in the backward gears (fourth gear excepted). This situation will be a starting point for later consideration [38].

We shall investigate a gear mechanism with n_s stages, taking into account only rotational motions. The system includes f degrees of freedom and n_p contacts where impulsive motion might take place. We assume the possibility of n impacts taking place at the same instant of time in n contacts. In the specific case of gearbox rattling we might simplify our approach considerably by the fact that rattling takes place in gear stages not under load. Therefore, we may assume that the process of rattling within the load-free stages has no influence on those parts of the changeover gear which are under load and which act as an excitation source for rattling. These loaded stages are part of the complete drive train system and have to be analyzed in this way. Rattling is decoupled, allowing a model where the number of degrees of freedom f is equal to the number of possible contacts n_p.

If there were no exciting fluctuation in the input shaft angular velocity, the gearwheels would rotate with their respective nominal constant speeds. Therefore it is obviously useful to describe the system state using the deviations from these nominal velocities. Mathematically we use the angular coordinates

$$\boldsymbol{\varphi} = [\varphi_1, \ldots, \varphi_f]^T \in \mathbb{R}^f$$

as generalized coordinates. According to the above-mentioned subdivision of the motion, we distinguish free flight, where the wheels have no contact, and the impact phase, when one or more constraints are hit. During free flight only the torques transmitted by the gear lubricant oil appear. They are modeled as linear damping and constant moments. From this the equations of motion can be written as

$$\boldsymbol{I}\ddot{\boldsymbol{\varphi}} + \boldsymbol{D}\dot{\boldsymbol{\varphi}} + \boldsymbol{T}_r = 0 \tag{12.1}$$

if $s_j \in (-v_j, 0)$, $j = 1, \ldots, f$, with

$\boldsymbol{I} = \text{diag}\{I_1, I_2, \ldots, I_f\} \in \mathbb{R}^{f,f}$ the matrix of the moments of inertia,

$\boldsymbol{D} = \text{diag}\{d_1, d_2, \ldots, d_f\} \in \mathbb{R}^{f,f}$ the damping matrix,

$\boldsymbol{T}_r = \{T_{r1}, T_{r2}, \ldots, T_{rf}\} \in \mathbb{R}^f$ the vector of constant moments,

v_j the backlash within the jth gear mesh (given magnitude),

s_j the relative distance in the jth gear mesh.

The vector of relative distances in the backlashes $\boldsymbol{s} = [s_1, s_2, \ldots, s_f]^T \in \mathbb{R}^f$ in the following will be called the vector of play coordinates. These coordinates will have values in the ranges $(-v_j, 0)$, $j = 1, \ldots, f$. They are related to the angle coordinates by the regular transformation

$$\boldsymbol{s} = \boldsymbol{W}^T \boldsymbol{\varphi} + \boldsymbol{v}e \tag{12.2}$$

where the regular matrix $\boldsymbol{W} \in \mathbb{R}^{f,f}$ and the vector $\boldsymbol{v} \in \mathbb{R}^f$ depend on the geometric properties of the gearbox and $e(t) = r_e\varphi_e(t)$ represents the excitation of the input gear; r_e is the basic radius of the input gearwheel, and φ_e is the fluctuation of the motion of the input shaft. Measurements of the excitation φ_e showed that it may be sufficiently approximated by a single harmonic function $\varphi_e(t) = A\sin(\omega t)$, with amplitude A, of the fluctuations of the angular motion.

The solution of the equations of motion (12.1) is

$$\begin{aligned} \boldsymbol{\varphi}(t) &= \boldsymbol{\varphi}(t_0) + \boldsymbol{B}^{-1}\left[\boldsymbol{E} - \exp\left[-\boldsymbol{B}(t-t_0)\right]\right]\left(\dot{\boldsymbol{\varphi}}(t_0) + \boldsymbol{B}^{-1}\boldsymbol{c}\right) \\ &\quad -\boldsymbol{B}^{-1}\boldsymbol{c}(t-t_0) \end{aligned} \tag{12.3}$$

$$\dot{\boldsymbol{\varphi}}(t) = \exp\left[-\boldsymbol{B}(t-t_0)\right]\left(\dot{\boldsymbol{\varphi}}(t_0) + \boldsymbol{B}^{-1}\boldsymbol{c}\right) - \boldsymbol{B}^{-1}\boldsymbol{c} \tag{12.4}$$

if $s_j \in (v_j, 0)$, $j = 1, \ldots, f$, where $\boldsymbol{B} = \boldsymbol{I}^{-1}\boldsymbol{D}$, $\boldsymbol{c} = \boldsymbol{I}^{-1}\boldsymbol{T}_r$, and $\boldsymbol{E}$ is the identity matrix, $\boldsymbol{E} \in \mathbb{R}^{f,f}$. The contact phase is modeled as partly elastic impact since the gearwheels are made of hardened steel and the load is very low. Using impact theory one gets transition equations from the system state before an impact to the state after it.

The classical impact equations for n colliding bodies at time t_i relate the relative velocities before and after an impact by

$$\dot{\boldsymbol{s}}_i^+ = -\bar{\bar{\boldsymbol{\epsilon}}}_i \dot{\boldsymbol{s}}_i^- \quad \in \mathbb{R}^n \tag{12.5}$$

which in generalized coordinates may be written as (see eq. 12.2)

$$\begin{gathered} \boldsymbol{W}_i^T \dot{\boldsymbol{\varphi}}^+ + \bar{\bar{\boldsymbol{\epsilon}}}_i \boldsymbol{W}_i^T \dot{\boldsymbol{\varphi}}^- + (\boldsymbol{E} + \bar{\bar{\boldsymbol{\epsilon}}}_i)\boldsymbol{v}_i \dot{e} = \boldsymbol{0} \\ \boldsymbol{W}_i \in \mathbb{R}^{f,n}, \; \boldsymbol{v}_i \in \mathbb{R}^n . \end{gathered} \tag{12.6}$$

It is assumed that n impacts take place at the same instant of time. Starting with the usual type of equations of motion

$$\boldsymbol{I}(\boldsymbol{\varphi}, t) \cdot \ddot{\boldsymbol{\varphi}} + \boldsymbol{h}(\boldsymbol{\varphi}, \dot{\boldsymbol{\varphi}}, t) = \boldsymbol{0} \tag{12.7}$$

and performing the same steps as in Chapter 7, we get a typical set of transition equations valid for each impacting contact and the type of system as presented in Fig. 12.1 [38]:

$$\begin{aligned} \dot{\boldsymbol{\varphi}}^+ &= \left[\boldsymbol{E} - \boldsymbol{I}^{-1}\boldsymbol{W}_i(\boldsymbol{W}_i^T\boldsymbol{I}^{-1}\boldsymbol{W}_i)^{-1}(\boldsymbol{E} + \bar{\bar{\boldsymbol{\epsilon}}}_i)\boldsymbol{W}_i^T\right]\dot{\boldsymbol{\varphi}}^- \\ &\quad -\boldsymbol{I}^{-1}\boldsymbol{W}_i(\boldsymbol{W}_i^T\boldsymbol{I}^{-1}\boldsymbol{W}_i)^{-1}(\boldsymbol{E} + \bar{\bar{\boldsymbol{\epsilon}}}_i)\boldsymbol{v}_i\dot{e} \\ &= \boldsymbol{U}_i\dot{\boldsymbol{\varphi}}^- + \boldsymbol{\zeta}_i\dot{e} \\ \boldsymbol{\Lambda}_i &= -(\boldsymbol{W}_i^T\boldsymbol{I}^{-1}\boldsymbol{W}_i)^{-1}(\boldsymbol{E} + \bar{\bar{\boldsymbol{\epsilon}}}_i)(\boldsymbol{W}_i^T\dot{\boldsymbol{\varphi}}^- + \boldsymbol{v}_i\dot{e}) \end{aligned} \tag{12.8}$$

The constraint impulses $\boldsymbol{\Lambda}_i \in \mathbb{R}^n$ appearing at each impact or at a sequence of impacts represent an astonishly good measure for rattling noise. Therefore, we shall assume that the rattling noise intensity will be proportional to $\sum |\Lambda_{ij}|$; i.e.,

$$\text{Noise Intensity} \;\sim \frac{1}{N}\sum_i \left(\sum_{j=1}^{n(i)} |\Lambda_{ij}|\right), \tag{12.9}$$

where the first sum gives a summation over all N impacts, which occur during simulation time. The second sum regards all impacts taking place at one time. The above assumption is a result of many efforts to model rattling noise. For example, the loss of energy gives a good measure as well. The physical background is clear: rattling noise energy can only be covered by impact losses. Of course, there are additional losses by transporting noise from the meshes of the gears to the housing walls of the gearbox, but modeling these phenomena is extremely difficult. For parameter studies we need only a relative measure of noise, not an absolute one. Thus, it turns out that the impact losses averaged over all impacts are a really good magnitude to represent rattling noise intensity.

Combining the above considerations, we write the transition equations for backlash i as

$$\begin{aligned} t^+ &= t^- \\ \boldsymbol{\varphi}^+ &= \boldsymbol{\varphi}^- \\ \dot{\boldsymbol{\varphi}}^+ &= \boldsymbol{U}_i\dot{\boldsymbol{\varphi}}^- + \boldsymbol{\zeta}_i\dot{e} \end{aligned} \tag{12.10}$$

where the superscript minus specifies values just before and plus sign specifies those just after the impact. $\boldsymbol{U}_i \in \mathbb{R}^{f,f}$ and $\boldsymbol{\zeta}_i \in \mathbb{R}^f$ are the transition matrix and a transition vector, respectively, that depend on the geometry of the gearbox, the coefficient of restitution and the impacting pair of gearwheels. In order to achieve a more evident mathematical form the equations of motion are transformed into the play coordinates:

$$\ddot{\boldsymbol{s}} + \boldsymbol{W}^T\boldsymbol{B}\boldsymbol{W}^{-T}\dot{\boldsymbol{s}} = \boldsymbol{v}\ddot{e} + \boldsymbol{W}^T\boldsymbol{B}\boldsymbol{W}^{-T}\boldsymbol{v}\dot{e} - \boldsymbol{W}^T\boldsymbol{c} \tag{12.11}$$

if $s_j \in (-v_j, 0)$, $j = 1, \ldots, f$, and

$$\dot{\boldsymbol{s}}^+ = \boldsymbol{W}^T\boldsymbol{U}_i\boldsymbol{W}^{-T}\dot{\boldsymbol{s}}^- + (\boldsymbol{E} - \boldsymbol{W}^T\boldsymbol{U}_i\boldsymbol{W}^{-T})\boldsymbol{v}\dot{e} \tag{12.12}$$

if $s_i \in \{-v_i, 0\}$, $s_j \in (-v_j, 0)$, $j = 1, \ldots, f$, $j \neq i$.

The system (12.1, 12.10) or (12.11, 12.12) may be called nonholonomic, because the constraint during impact appears in the form of a velocity condition (eqs. 12.10 or 12.12). The vibrations possess a self-exciting character due to some typical mechanisms for this kind of motion. The excitation $e(t)$ represents the energy source, which transfers energy to the driven wheel by the switching behavior of the impact itself. The energy is stored in the driven wheel in the form of kinetic energy and dissipated by oil drag and by impact losses. Typically, a balance of transferred energy and dissipation results in a periodic solution represented by a stable limit cycle, at least in the periodic case.

Discrete Model

From eqs. (12.6) and (12.7) it is obvious that time should be introduced into the system state space in order to have the dynamical behavior of the system completely described. Now $\boldsymbol{y} = \left[t, \boldsymbol{\varphi}^T, \dot{\boldsymbol{\varphi}}^T\right]^T \in \left[0, \frac{2\pi}{\omega}\right) \times \mathbb{R}^{2f}$ stands for the state vector of the angle coordinates, whereas $\boldsymbol{z} = \left[t, \boldsymbol{s}^T, \dot{\boldsymbol{s}}^T\right]^T \in \left[0, \frac{2\pi}{\omega}\right) \times \mathbb{R}^{2f}$ is the state vector in play coordinates. As the excitation is periodic the vector field of the dynamical system has the same period, and therefore the time t may be normed within the interval of the excitation period.

The motion is exactly defined if the sequence of system states before or after impacts is known. In addition, the trajectories between impacts have no influence on noise emission caused only by the impacts themselves. Hence, it is preferable to use a time-discrete description instead of the time-continuous one. This is easily done by the definition of point mappings onto Poincaré sections (P.S.). For the system states before impacts the P.S. is ($\vee$ logical OR, $\wedge$ logical AND)

$$\overset{-}{\sum} = \{(\boldsymbol{y} \text{ resp. } \boldsymbol{z}) \,|\, (s_1 = -v_1 \wedge \dot{s}_1 < 0) \vee (s_1 = 0 \wedge \dot{s}_1 > 0) \\ \vee \ldots \vee (s_f = -v_f \wedge \dot{s}_f < 0) \vee (s_f = 0 \wedge \dot{s}_f > 0)\}$$

whereas for states after impacts the P.S. is defined as

$$\overset{+}{\sum} = \{(\boldsymbol{y} \text{ resp. } \boldsymbol{z}) \,|\, (s_1 = -v_1 \wedge \dot{s}_1 \geq 0) \vee (s_1 = 0 \wedge \dot{s}_1 \leq 0) \\ \vee \ldots \vee (s_f = -v_f \wedge \dot{s}_f \geq 0) \vee (s_f = 0 \wedge \dot{s}_f \leq 0)\} \,.$$

Two mappings are defined: the mapping describing the impact

$$\boldsymbol{F}_s : \overset{-}{\sum} \rightarrow \overset{+}{\sum} \quad \in \mathbb{R}^{2f+1}$$

is defined by eq. (12.10) and

$$\boldsymbol{F}_f : \overset{+}{\sum} \rightarrow \overset{-}{\sum} \quad \in \mathbb{R}^{2f+1}$$

represents the free flight. The latter results from the solution of the equations of motion during free flight and an additional equation determining the position at the next impact. Now the complete motion may be written in the form

$$\cdots \boldsymbol{y}_i^- \rightarrow \boldsymbol{y}_i^+ \rightarrow \boldsymbol{y}_{i+1}^- \rightarrow \boldsymbol{y}_{i+1}^+ \rightarrow \boldsymbol{y}_{i+2}^- \cdots .$$

The stability of the process may be judged by the functional matrices of the mappings. The functional matrix of the entire motion is achieved by sequential multiplication of the matrices corresponding to the separate transitions. The form of the functional matrices for the kth impact and the following free flight, respectively, is

$$\boldsymbol{M}_{s_k} = \frac{\partial \boldsymbol{y}_k^+}{\partial \boldsymbol{y}_k^-} \,, \quad \boldsymbol{M}_{f_k} = \frac{\partial \boldsymbol{y}_{k+1}^-}{\partial \boldsymbol{y}_k^+} \,.$$

It is impractical to judge on the stability of periodic solutions by the investigation of eigenvalues of the functional matrix in the neigborhood of fix points since these points, in general, cannot be determined analytically. In this case the motion is calculated by numerical evaluation of the mappings so that the stability of periodic solutions becomes obvious. Nevertheless, we determine the greatest Ljapunov exponent σ as a measure for the regularity of the motion:

$$\sigma = \lim_{n\to\infty} \left[\frac{1}{n} \ln \left(\frac{|\boldsymbol{w}_n|}{|\boldsymbol{w}_0|}\right)\right] \tag{12.13}$$

with

$$\boldsymbol{w}_n = \prod_{k=1}^{n} \boldsymbol{M}_{f_k} \boldsymbol{M}_{s_k} \boldsymbol{w}_0 \,.$$

12.1.3 Results

RESULTS FOR THE ONE-STAGE MODEL

Within shift transmissions the gearwheels not under load show in many cases a one-stage rattling behavior, because the motion is approximately reactionless with respect to the driving wheel. Moreover, the one-degree-of-freedom simplicity gives a good insight into the structure of the vibrations and the parameter influences. Figure 12.2 presents such a single-stage model.

The parameter space of the one-stage gear has small dimension. Even more, using similarity rules it can be further reduced by 3. The equations of motion for the normalized play coordinates are

$$\left.\begin{aligned} \ddot{s} + \beta\dot{s} &= \ddot{e} + \beta\dot{e} + \gamma \text{ if } s \in (-1,0) \\ \dot{s}^+ &= -\varepsilon\dot{s}^- \qquad\quad \text{if } s \in \{-1,0\} \end{aligned}\right\} \tag{12.14}$$

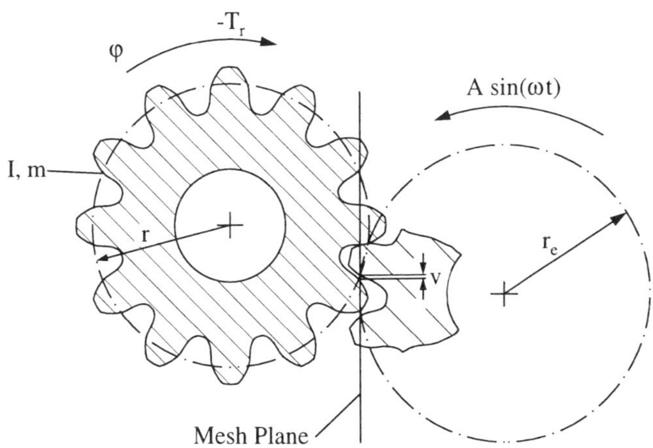

Figure 12.2: Model of One-Stage Rattling

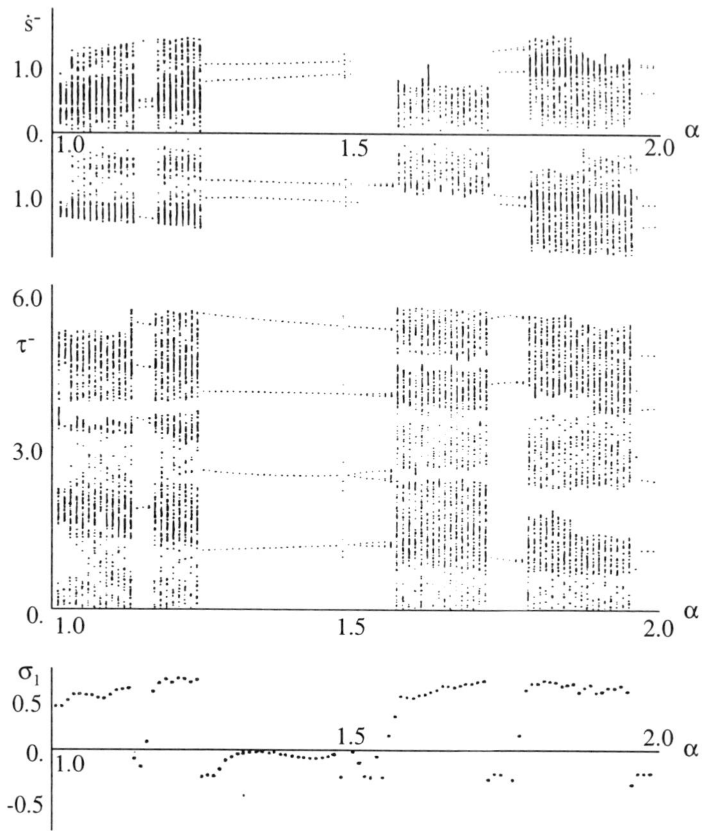

Figure 12.3: Bifurcation of One-Stage Gear; (a) Bifurcations of Relative Velocities within Backlash, (b) Bifurcations of Normalized Time, (c) Greatest Ljapunov Exponent, the Bifurcation Parameter is α and the other Parameters are $\beta = 0.1, \gamma = 0.1, \varepsilon = 0.9$

with $e = \alpha \sin(\tau)$, where $s = \hat{s}/v$, $\tau = \omega t$, $\beta = d/I\omega$, $\gamma = T_r r/I\omega^2 v$, $\alpha = r_e A/v$ are the normalized play coordinates themselves, the time, damping, constant moments and amplitudes, respectively. Derivatives in eq. (12.14) are with regard to the normalized time τ.

As mentioned, the vector $\boldsymbol{z} = (\tau, s, \dot{s})^T \in [0, 2\pi) \times \mathbb{R}^2$ is used within the description of the discrete system. The mapping $\boldsymbol{F}_{s_k}$ of the kth impact is given by

$$\tau_k^+ - \tau_k^- = 0 \ , \ s_k^+ - s_k^+ = 0 \ , \ \dot{s}_k^+ - \varepsilon \dot{s}_k^- = 0 \tag{12.15}$$

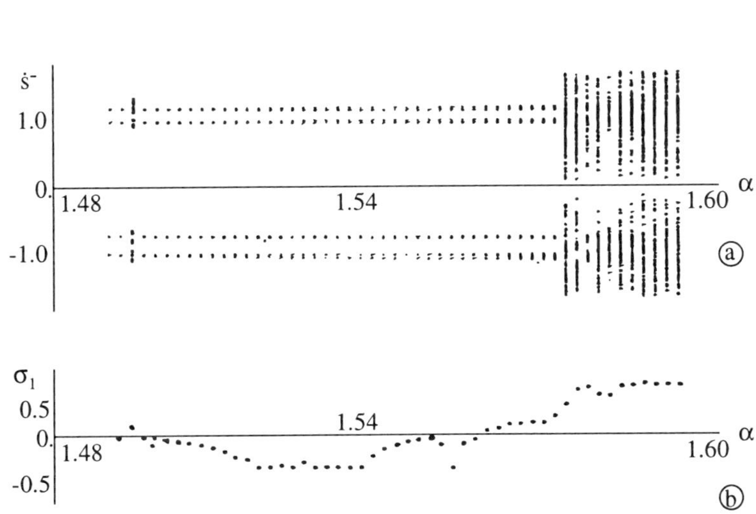

Figure 12.4: Details of Bifurcation Diagram Fig. 12.3: (a) Detail of Fig. 12.3(a), (b) Detail of Fig. 12.3(c)

whereas for the kth free-flight phase the appropriate mapping $\boldsymbol{F}_{f_k}$ is

$$\left.\begin{aligned}
&-s_{k+1}^{-} + s_k^{+} + \alpha\left(\sin(\tau_{k+1}^{-}) - \sin(\tau_k^{+})\right)\\
&+\tfrac{1}{\beta}\left(1 - \exp\left(-\beta\left(\tau_{k+1}^{-} - \tau_k^{+}\right)\right)\right)\left(\dot{s}_k^{+} - \alpha\cos(\tau_k^{+}) - \tfrac{\gamma}{\beta}\right)\\
&+\tfrac{\gamma}{\beta}\left(\tau_{k+1}^{-} - \tau_k^{+}\right) = 0\\
&\alpha\cos(\tau_{k+1}^{-}) - \dot{s}_{k+1}^{-}\\
&+\left(\dot{s}_k^{+} - \alpha\cos(\tau_k^{+}) - \tfrac{\gamma}{\beta}\right)\exp\left(-\beta\left(\tau_{k+1}^{-} - \tau_k^{+}\right)\right)\tfrac{\gamma}{\beta} = 0\\
&s_{k+1}^{-} - s_k^{+} \in \{-1, 1, 0\}\ .
\end{aligned}\right\} \qquad (12.16)$$

The greatest Ljapunov exponent can be calculated from eq. (12.13). Although the mappings are extremely complex, the sum of the Ljapunov exponents may be calculated since the determinant of the functional matrices can be given analytically. The determinant of the mapping from the state before impact k to the state before impact $k+1$ is

$$\det\left(\boldsymbol{M}_{f_k}\boldsymbol{M}_{s_k}\right) = \varepsilon^2 \frac{\dot{s}_k^{-}}{\dot{s}_{k+1}^{-}} \exp\left(-\beta\left(\tau_{k+1}^{-} - \tau_k^{+}\right)\right)\ .$$

Consequently, the determinant of the mapping from before the first impact to before the nth results from the multiplication of the determinants of elementary mappings:

$$\begin{aligned} \det\left(\prod_{k=1}^{n} \boldsymbol{M}_{f_k}\boldsymbol{M}_{s_k}\right) &= \prod_{k=1}^{n} \det\left(\boldsymbol{M}_{f_k}\boldsymbol{M}_{s_k}\right) \\ &= \varepsilon^{2n}\frac{\dot{s}_1^-}{\dot{s}_{n+1}^-}\exp\left(-\beta\sum_{k=1}^{n}\left(\tau_{k+1}^- - \tau_k^+\right)\right). \end{aligned} \tag{12.17}$$

From these equations the sum of the exponents can be calculated:

$$\begin{aligned} \sigma_1+\sigma_2+\sigma_3 &= \lim_{n\to\infty}\frac{1}{n}\ln\left|\det\left(\prod_{k=1}^{n}\boldsymbol{M}_{f_k}\boldsymbol{M}_{s_k}\right)\right| \\ &= 2\ln(\varepsilon)-\beta\langle\tau_{k+1}^- - \tau_k^+\rangle. \end{aligned} \tag{12.18}$$

One of the exponents is identically zero as it corresponds to the defined mapping within the P.S. The sum of the other two is negative, which is not surprising since the mappings have a dissipative character resulting in a shrinking phase space. Note that in eq. (12.18) only dissipative terms appear, i.e.,

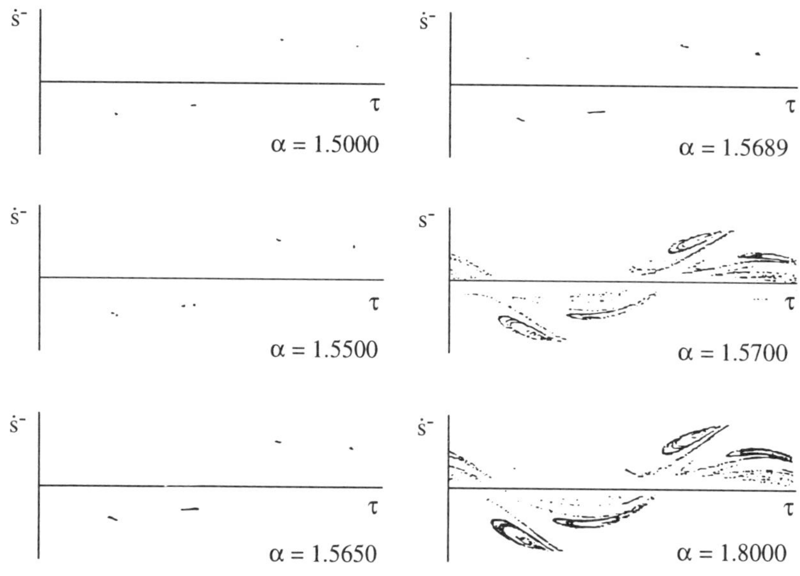

Figure 12.5: Change in the Form of the Attractor in the P.S. Parameters as in Fig. 12.3

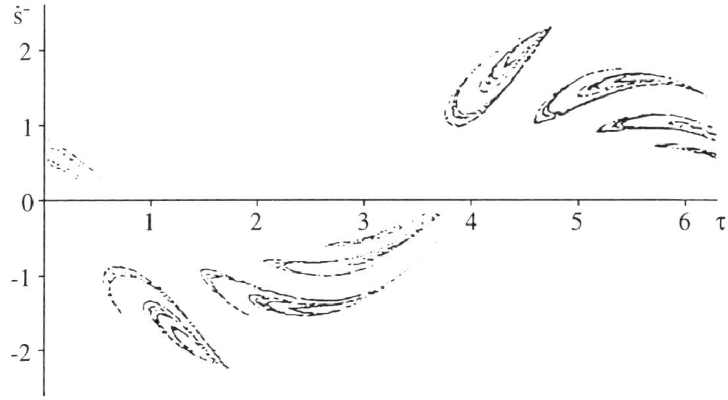

Figure 12.6: Strange Attractor. Parameters are $\alpha = 3$, $\beta = 0.1$, $\gamma = 0.1$, $\varepsilon = 0.9$ and 10.000 iterations

the coefficient of restitution ε and the normalized damping multiplied by the mean flight time, which is a measure for the mean shrinkage during the flight phase.

The numerical evaluation of the mappings and the calculation of the greatest exponent has been performed on a digital computer, the results being shown below. In Fig. 12.3(a) a bifurcation diagram is displayed where the parameter α (amplitude of excitation/backlash) is varied. (In order to get a stationary situation the mapping is performed 5000 times, of which the last 300 results are displayed.) Figure 12.3(a) shows the relative velocities within the backlash, Fig. 12.3(b) shows the time points τ^- in the range of $(0, 2\pi]$, and Fig. 12.3(c) shows the greatest Ljapunov exponent. The qualitative properties of the motion strongly depend on the parameters of the system. In certain parameter regions the solution is periodic and through a sequence of bifurcations becomes chaotic. This feature repeats itself over and over. The bifurcation itself is the destabilization of a periodic solution resulting in another solution with twice the period order. The distance between two bifurcations decreases rapidly so that the bifurcation sequence soon results in chaotic behavior.

This feature is shown in detail in Fig. 12.4, which magnifies parts of Fig. 12.3. From this diagram the parameters α_1, α_2, α_3 are extracted which correspond to the first, second and third bifurcation, respectively. From these a Feigenbaum number as well as the parameter for which chaos appears can be calculated. The values are

$$\delta = \frac{\alpha_2 - \alpha_1}{\alpha_3 - \alpha_2} = 5(4.669)$$

$$\alpha_\infty = \alpha_1 + \frac{1}{1-1/\delta}(\alpha_2 - \alpha_1) = 1.556(1.557)$$

with the exact theoretical results given for comparison in parentheses. The development of the corresponding strange attractor in the P.S. is given in Fig. 12.5, which nicely shows the loss of periodicity and the spreading of the attractor in the direction of the unstable manifold. In Fig. 12.6 a fully developed strange attractor can be seen with its geometric structures displayed quite well.

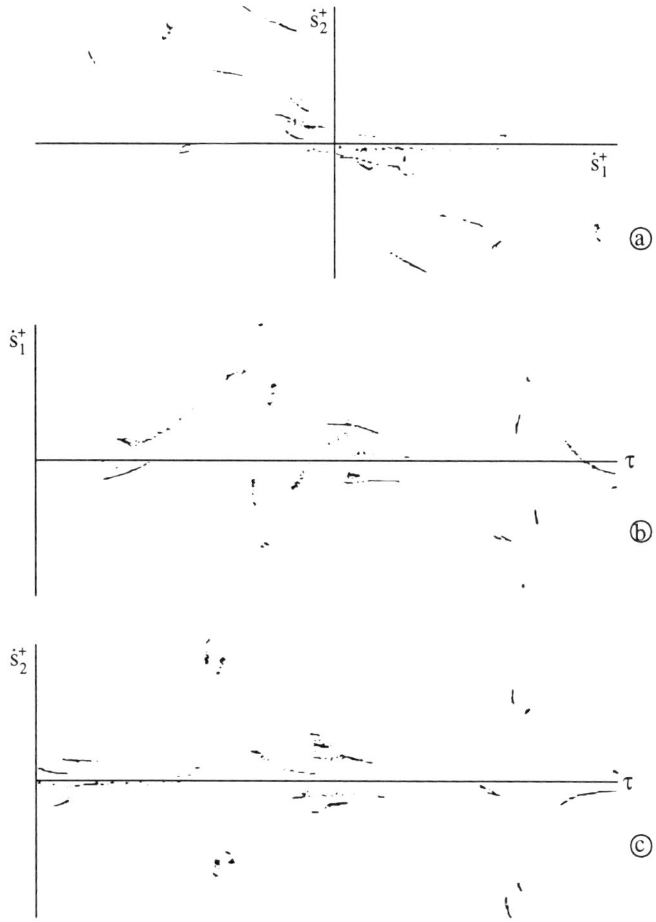

Figure 12.7: Chaotic Motion of the Two-Stage Model; (a) Plane of Velocities, (b/c) Plane of Normalized Time and Velocity in one Backlash

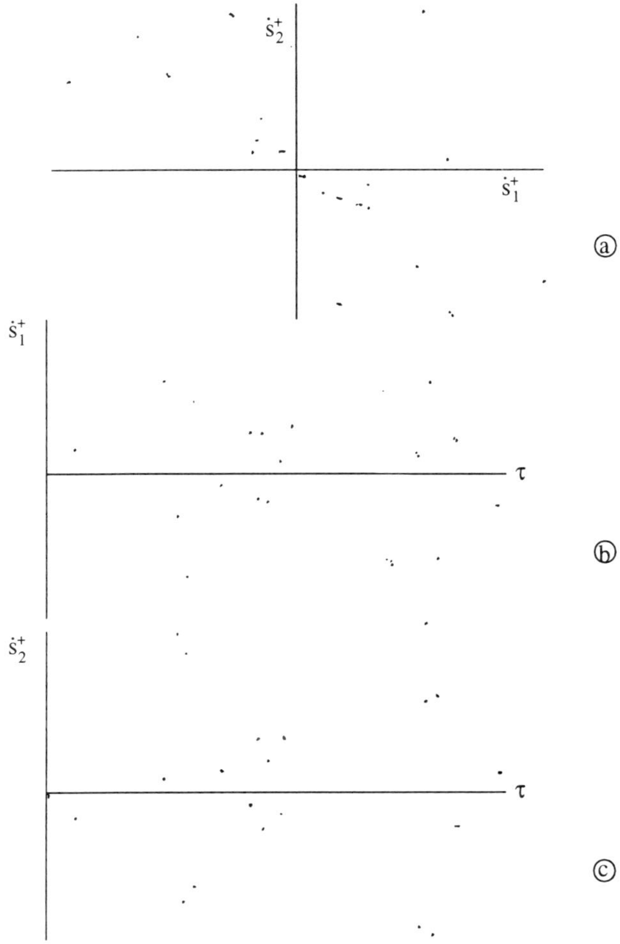

Figure 12.8: Periodic Motion of the Two-Stage Model; (a) Plane of Velocities, (b/c) Plane of Normalized Time and Velocity in one Backlash

Results for the Two-Stage Model

The modeling has already been presented, and the complete description of the motion is possible only in $\mathbb{R}^5$. Since the dimension of the parameter space is very high irrespective of a possible reduction by 3 by means of similarity rules, we will use in the following unnormalized parameters.

Qualitatively the results are the same as in the above case of the one-stage gear, i.e., periodic as well as chaotic motion with a sequence of bifurcations in between. Since the phase space has more dimensions, the representation of results becomes more difficult. Here only projections of this space onto certain planes will be shown.

Figure 12.7, for example, shows different projections for a chaotic motion for a specific parameter combination. The first diagram (a) is the projection into the plane of velocities, the second (b) a projection onto normalized time and the velocities in backlash 1 after impacts, and the third (c) a similar graph for backlash 2. The inner structure of the attractor as compared to the single-stage model has vanished as a property of the projection, and the interpretation of the results becomes more difficult. Figure 12.8 shows the same projections as Fig. 12.7 for a periodic motion of high order.

12.1.4 Parameter Dependence of Mean Values

PARAMETER DEPENDENCE FOR ONE-STAGE MODEL

The results as evaluated so far give some insight into the structure of the impulsive processes connected with rattling. As could be shown [37], periodic rattling corresponds to periodic impact patterns, chaotic rattling to chaotic impact patterns. This is nice to know, but it has little influence on the solution of practical problems. It turns out [45, 81] that rattling noise in real gearboxes does not depend in a significant way on the structural properties of the corresponding nonlinear vibrations. It depends mainly on the parameter influences on the mean values of the impulsive processes. Therefore, in order to optimize the design of the gearbox the influence of the parameters on noise emission has to be investigated. The mechanisms of sound transmission and radiation are very sophisticated and not a topic of this Chapter. It is assumed that the squared means of the relative velocities in the backlashes can be used as a measure for the radiated noise. The derivation of the parameter dependence of the mean values is certainly possible numerically, but at this point we would like to introduce a different modeling approach to achieve analytical approximations.

Here we investigate a one-stage gear with neglect of damping and constant moments. The mappings are written with respect to normalized values and the angle will be used as a generalized coordinate. The resulting form of the mappings is

$$\left.\begin{aligned} \dot{\phi}_{k+1} &= -\varepsilon\dot{\phi}_k + (1+\varepsilon)\alpha\cos(\tau_{k+1}) \\ \dot{\phi}_k(\tau_{k+1}-\tau_k) &- \alpha(\sin\tau_{k+1} - \sin\tau_k) + (s_{k+1} - s_k) = 0 \end{aligned}\right\} \tag{12.19}$$

with the normalized angle $\phi = r\varphi/v$ and the normalized velocity $\dot{\phi}_k$ during the kth flight phase. Superscript + and − need not be used as the angular velocity does not change during the flight. From the analysis of vibrations it is well known that for the system of eq. (12.19) there exist periodic as well as chaotic solutions. Irrespective of the qualitative characteristics of the motion it always appears in the attractor basin. Infinitely repeating the mapping very densely fills the attractor manifold. For a description of the motion of our system (eq. 12.19) it would be sufficient to know the distribution functions for velocity and time in some Poincaré sections. These distributions are continuous functions in the chaotic case and Dirac functions in the periodic case. It corresponds to the invariant distribution in the ergodic theory of dynamical systems. It would even be sufficient to have the marginal distribution of the normalized time points of the impacts to achieve characteristic statistical results. This marginal distribution also depends on the behavior of the system. It becomes continuous for the chaotic and a number of Dirac impulses for the periodic case.

Figure 12.5 shows that the peridoc as well as the chaotic motions appear in the vicinity of the fix points. This results in the fact that the marginal distribution in the neighborhood of the corresponding points also shows significant values compared to the rest of the region. Since the eqs. (12.19) are invariant under the transformation

$$\begin{aligned} &\tau \to \tau + \pi \\ &\dot{\phi} \to -\dot{\phi} \\ &(s_{k+1} - s_k) \to -(s_{k+1} - s_k) \end{aligned} \tag{12.20}$$

every distribution of impacts will be periodic with period π. On the other hand, if a motion is not symmetric within the backlash a transformation of the motion by eq. (12.20) will always yield in a correct solution. Taking the mean of the transformed and the untransformed motion regains the periodicity of the marginal distribution. Now let us assume the marginal distribution is known. Equations (12.19) are replaced by

$$\left.\begin{aligned} &\dot{\phi}_{k+1} = -\varepsilon\dot{\phi}_k + (1+\varepsilon)\alpha\cos(\tau) \\ &\rho(\tau),\ \tau \in [0, 2\pi) \end{aligned}\right\} \tag{12.21}$$

which is a discrete stochastic equation. Time τ is assumed a random variable with some probability density $\rho(\tau)$ corresponding to the distribution found above. From eq. (12.21) the mean values can be determined analytically. Calculating the expected value for eq. (12.21) yields the following recursive formula for the first-order moment:

$$m_{1_{k+1}} = -\varepsilon m_{1_k} + (1+\varepsilon)\alpha J_1$$

where $J_1 = \int_0^{2\pi} \rho(\tau)\cos\tau\, d\tau$ and $m_{1_k} = E[\dot{\phi}_k]$.

Here m_{1_k} is the expected value of the angular velocities after k repetitions. The term J_1 vanishes due to the periodicity in the marginal distribution $\rho(\tau)$. The above iteration converges to zero if $\varepsilon < 1$:

$$m_1 = \lim_{k\to\infty} m_{1_k} = 0\,.$$

The derivation of the second moment is possible by quadrature of eq. (12.21) and determination of the expected values:

$$m_{2_{k+1}} = \varepsilon^2 m_{2_k} + (1+\varepsilon)^2\alpha^2 J_2 - 2\varepsilon(1+\varepsilon)\alpha E[\dot{\phi}_k \cos\tau]$$

where $J_2 = \int_0^{2\pi} \rho(\tau)\cos^2\tau\, d\tau$ and $m_{2_k} = E[\dot{\phi}_k^2]$.

In this formula m_{2_k} stands for the expected value of the squared angular velocities after k iterations. Obviously,

$$E[\dot{\phi}_k \cos\tau] \approx 0$$

holds since the values $\dot{\phi}$ are symmetric and the normalized time points of impacts appear in the vicinity of $\pi/2$ and $3\pi/2$. This approximation results in the following iteration formula for the second moment:

$$m_{2_{k+1}} = \varepsilon^2 m_{2_k} + (1+\varepsilon)^2\alpha^2 J_2\,.$$

m_{2_k} can explicitly be given as a function of the initial conditions:

$$m_{2_k} = \frac{1-\varepsilon^{2k}}{1-\varepsilon}(1+\varepsilon)\alpha^2 J_2 + \varepsilon^{2k} m_{2_0}\,. \tag{12.22}$$

Here m_{2_0} represents the influence of the initial conditions. Equation (12.22) converges very rapidly as $\varepsilon < 1$. In parallel, the influence of the initial conditions, or of the assumed distribution of the initial conditions, diminishes. For $k \to \infty$ eq. (12.22) converges to the second moment m_2:

$$m_2 = \lim_{k\to\infty} m_{2_k} = J_2 \frac{1+\varepsilon}{1-\varepsilon}\alpha^2\,. \tag{12.23}$$

The factor J_2 could be determined by numerical simulation, but is only estimated since as a proportionality factor its influence is small. Let us assume marginal distributions $\rho(\tau)$, one for a periodic and the other one for a chaotic case (see Fig. 12.9).

From these distributions the integral J_2 is calculated and introduced into eq. (12.23). Figure 12.10 shows the second moments resulting from this approximation for both of the assumed marginal distributions, on the one

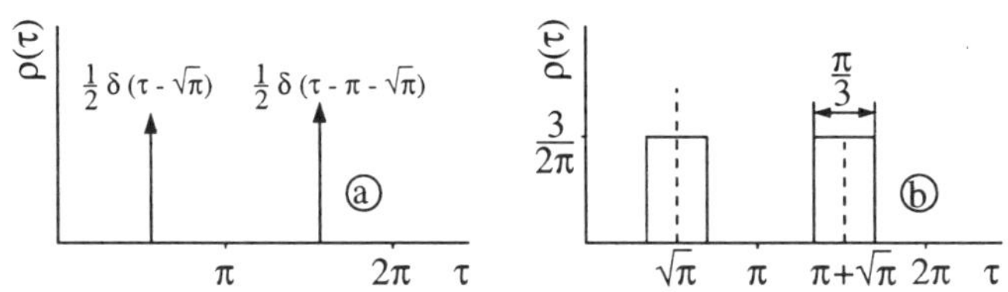

Figure 12.9: Assumed Marginal Distribution $\rho(\tau)$

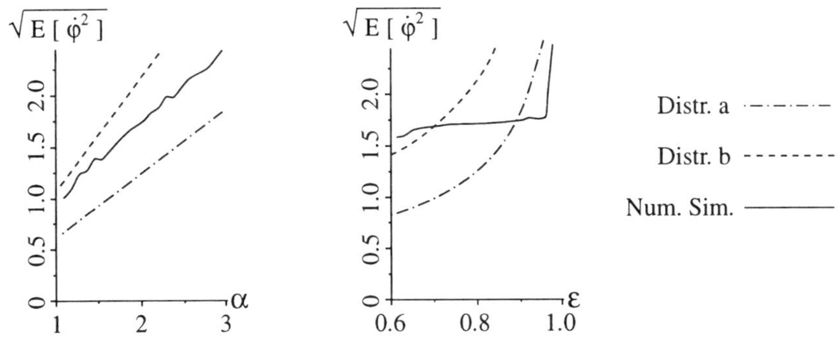

Figure 12.10: Second Moments from Approximation (a and b) and Simulation (c)

hand, and the numerical simulation, on the other. In spite of the simplicity of the applied method the qualitative agreement is reasonable. Resetting the normalized parameters into unnormalized forms yields the following relation between the second moment and the system parameters:

$$E[\dot{\varphi}^2] = J_2 \frac{A^2\omega^2}{(r/r_e)^2} \frac{1+\varepsilon}{1-\varepsilon} . \tag{12.24}$$

This form clearly shows the influence of the single parameters. The intensity of the velocities rises quadratically with the excitation frequency and amplitude, decreases quadratically with the transmission factor, but for $\varepsilon < 0.8$ seems to be independent of the number of restitution.

It is interesting to compare eq. (12.24) with some results obtained from the discrete mappings as derived for one-stage rattling and which can be written as [72, 73]

$$\text{Noise} \sim \dot{\varphi}^2 \sim (1+\varepsilon)^2 I \frac{A^2\omega^2}{(r/r_e)^2} ,$$

which closelyresembles eq. (12.24) with slight modifications. The equation says that noise is proportional to the squares of the coefficient of restitution ε, of the excitation frequency ω and amplitude A. It is further inversely proportional to the square of the base circle of radius r of the driven wheel. It is proportional to the driven wheel moment of inertia I. This formula is quite practical because it has a direct connection with the design of shift transmission systems and of complete drive trains. The amplitude A at the gearbox entrance should be small, which means that within the operating range of a car its drive train should have no eigenforms with a large amplitude at the gearbox entry. This again gives direct recommendations for drive train design. Moreover, a large wheel should not drive a small one due to the term

$1/r^2$, which means that the upper stages should possibly not be shifted on the main shaft but on the countershaft. Most modern gearbox designs take this fact into consideration.

Parameter Dependence for the Two-Stage Model

Also in the two-stage case let us investigate a simplified model, neglecting damping and constant moments. As mentioned for one-stage rattling, we will use nonnormalized parameters and try to find analytical relations between the expected mean values and the system parameters. As in the one-stage case, depending on the parameters, there appear periodic as well as chaotic solutions in the vicinity of the fix points (compare Figs. 12.7 and 12.8). Again the system can be described completely by the invariant distribution. Here we consider only the marginal distribution of the impact times within the excitation period.

Compared to the one-stage case the situation is more complicated as impacts may appear in backlash 1 or 2. The P.S. achieved from the numerical simulation for the two-stage model shows that the impacts in the two backlashes are correlated. This suggests a simple approximate relation between the distributions of the impacts in backlashes 1 and 2. It is assumed that after l_1 impacts in backlash 1 l_2 impacts in backlash 2 appear. The value of the quotient l_1/l_2 depends on the backlashes and moments of inertia and should be identified numerically. For the case of approximately equal moments of inertia of the loose gearwheels and if the quotient of the backlashes is in the range of 0.5–2, then $l_1 \approx l_2 \approx 1$. This means that after an impact in backlash 1 an impact in backlash 2 may be expected. Assuming this to be exactly the case yields an approximation of the real motion, which will sufficiently approximate the attractor. Let us now assume the marginal distribution of impact times in backlash 1 was known a priori. Instead of the mappings we will use the following discrete stochastic equations

$$\left.\begin{aligned} \dot{\boldsymbol{\varphi}}_{2_{k+2}} &= \boldsymbol{U}_2 \dot{\boldsymbol{\varphi}}_{1_{k+1}} \\ \dot{\boldsymbol{\varphi}}_{1_{k+1}} &= \boldsymbol{U}_1 \dot{\boldsymbol{\varphi}}_{2_k} + \boldsymbol{\zeta}\cos(\omega t) \\ \rho(t)\,,\ \ t &\in \left[0,\ \frac{2\pi}{\omega}\right)\,. \end{aligned}\right\} \tag{12.25}$$

$\dot{\boldsymbol{\varphi}}_{2_k} = (\dot{\varphi}_1, \dot{\varphi}_2)^T \in \mathbb{R}^2$ is the vector of angular velocities of the two gearwheels after the kth impact in backlash 2, and $\dot{\boldsymbol{\varphi}}_{1_{k+1}} = (\dot{\varphi}_1, \dot{\varphi}_2)^T \in \mathbb{R}^2$ are the angular velocities after the $(k+1)$st impact in backlash 1. The vector $\boldsymbol{\zeta} = \omega \boldsymbol{\zeta}_1$ is introduced as an abbreviation. The transfer matrices $\boldsymbol{U}_1$, $\boldsymbol{U}_2$ and the vector $\boldsymbol{\zeta}$ have already been defined, and subscripts 1 and 2 represent the stage in which the impact occurs. Equations (12.25) assume a probability density for the impacts in backlash 1. After each of these impacts one in backlash 2 follows. Using the stochastic model of eqs. (12.25) one can calculate the expected values of the velocities. For this purpose we formulate the equations in the following way:

$$\left.\begin{aligned} \dot{\varphi}_{2_{k+2}} &= \boldsymbol{U}_2 \boldsymbol{U}_1 \dot{\varphi}_{2_k} + \boldsymbol{U}_2 \boldsymbol{\zeta} \cos(\omega t) \\ \dot{\varphi}_{1_{k+1}} &= \boldsymbol{U}_2^{-1} \dot{\varphi}_{2_k} \\ \rho(t)\,,\ \ & t \in \left[0,\ \frac{2\pi}{\omega}\right)\,. \end{aligned}\right\} \tag{12.26}$$

The first equation contains only relations between system states in backlash 2 and can easily be processed further. Determination of the expected values yields

$$\left.\begin{aligned} \boldsymbol{\mu}_{2_{k+2}} &= \boldsymbol{U}_2 \boldsymbol{U}_1 \boldsymbol{\mu}_{2_k} + \boldsymbol{U}_2 \boldsymbol{\zeta} J_1 \\ \boldsymbol{\mu}_{1_{k+1}} &= \boldsymbol{U}_2^{-1} \boldsymbol{\mu}_{2_k} \end{aligned}\right\} \tag{12.27}$$

$$\text{with } J_n = \int_0^{2\pi/\omega} \rho(\omega t) \cos^n(\omega t)\, dt\,.$$

$\boldsymbol{\mu}_{2_k} = E(\dot{\varphi}_1, \dot{\varphi}_2)^T$ is the expected value of the velocities after the kth impact in backlash 2, and $\boldsymbol{\mu}_{1_{k+1}} = E(\dot{\varphi}_1, \dot{\varphi}_2)^T$ represents the expected value of the velocities after the $(k+1)$st impact in backlash 1. Equation (12.27) converges if the absolute eigenvalues of the matrix $\boldsymbol{U}_2\boldsymbol{U}_1$ are less than one. For the parameter region of interest this holds. The expected values can explicitly be found as functions of the initial conditions and their limits can be given:

$$\left.\begin{aligned} \boldsymbol{\mu}_2 &= \lim_{k\to\infty} \boldsymbol{\mu}_{2_k} = (\boldsymbol{E} - \boldsymbol{U}_2\boldsymbol{U}_1)^{-1} \boldsymbol{U}_2 \boldsymbol{\zeta} J_1 \\ \boldsymbol{\mu}_1 &= \lim_{k\to\infty} \boldsymbol{\mu}_{1_{k+1}} = \boldsymbol{U}_2^{-1} \boldsymbol{\mu}_2\,. \end{aligned}\right\} \tag{12.28}$$

Evaluating this form with respect to the parameters of the model yields

$$\boldsymbol{\mu}_1 = \boldsymbol{\mu}_2 = J_1 A \omega (1/i_1, 1/i_1 i_2)^T \tag{12.29}$$

with $i_1 = \frac{r_1}{r_e}, i_2 = \frac{r_2}{r_1}$ the transmission factors of the two stages. This result was to be expected, as the excitation is amplified by the transmission factors weighted by the value J_1, which depends on the marginal distribution of the impacts. As in calculating the first-order moments, we will distinguish between the two backlashes for the calculation of expected values of the squares of the velocities. $\boldsymbol{C}_{2_k} = [\dot{\varphi}_1 \dot{\varphi}_2^T] \in \mathbb{R}^{2,2}$ is the correlation matrix of velocities after impacts in backlash 2 and k interations of the process, and

$\boldsymbol{C}_{1_{k+1}} = [\dot{\boldsymbol{\varphi}}_1 \dot{\boldsymbol{\varphi}}_2^T] \in \mathbb{R}^{2,2}$ is the correlation matrix of the velocities after impacts in backlash 1 and $k+1$ iterations of the process. The evaluation of $\boldsymbol{C}_2$ is done by multiplying the first of eqs. (12.26) by its transpose and extracting the expected values:

$$\begin{aligned} \boldsymbol{C}_{2_{k+2}} &= \boldsymbol{U}_2\boldsymbol{U}_1\boldsymbol{C}_{2_k}(\boldsymbol{U}_2\boldsymbol{U}_1)^T \\ &\quad + \left[\boldsymbol{U}_2\boldsymbol{\zeta}\boldsymbol{\mu}_{2_k}^T(\boldsymbol{U}_2\boldsymbol{U}_1)^T + \boldsymbol{U}_2\boldsymbol{U}_1\boldsymbol{\mu}_{2_k}(\boldsymbol{U}_2\boldsymbol{\zeta})^T\right] J_1 \\ &\quad + \boldsymbol{U}_2\boldsymbol{\zeta}(\boldsymbol{U}_2\boldsymbol{\zeta})^T J_2 \,. \end{aligned} \tag{12.30}$$

This equation is a recursive matrix equation which converges under the condition that the absolute eigenvalues of the matrix $\boldsymbol{U}_2\boldsymbol{U}_1$ are less than one. The limit is a solution of the above equation, and so a matrix equation for its determination appears:

$$\begin{aligned} \boldsymbol{C}_2 &= \boldsymbol{U}_2\boldsymbol{U}_1\boldsymbol{C}_2(\boldsymbol{U}_2\boldsymbol{U}_1)^T \\ &\quad + \left[\boldsymbol{U}_2\boldsymbol{\zeta}\boldsymbol{\mu}_2^T(\boldsymbol{U}_2\boldsymbol{U}_1)^T + \boldsymbol{U}_2\boldsymbol{U}_1\boldsymbol{\mu}_2(\boldsymbol{U}_2\boldsymbol{\zeta})^T\right] J_1 \\ &\quad + \boldsymbol{U}_2\boldsymbol{\zeta}(\boldsymbol{U}_2\boldsymbol{\zeta})^T J_2 \quad \text{where } \boldsymbol{C}_2 = \lim_{k\to\infty} \boldsymbol{C}_{2_k} \,. \end{aligned} \tag{12.31}$$

Similarly, a corresponding matrix equation can be derived for the correlation matrix of the velocities in backlash 1:

$$\begin{aligned} \boldsymbol{C}_1 &= \boldsymbol{U}_1\boldsymbol{U}_2\boldsymbol{C}_1(\boldsymbol{U}_1\boldsymbol{U}_2)^T + \left[\boldsymbol{\zeta}\boldsymbol{\mu}_1(\boldsymbol{U}_1\boldsymbol{U}_2)^T + \boldsymbol{U}_1\boldsymbol{U}_2\boldsymbol{\mu}_1\boldsymbol{\zeta}^T\right] J_1 \\ &\quad + \boldsymbol{\zeta}\boldsymbol{\zeta}^T J_2 \quad \text{where } \boldsymbol{C}_1 = \lim_{k\to\infty} \boldsymbol{C}_{1_{k+1}} \,. \end{aligned}$$

The correlation matrices can be determined analytically with the results being a bit lengthy. As in the one-stage case the integral J_1 vanishes due to the symmetry of the attraction region if all possible solutions are summed up. If only a single attractor is considered, the integral should still have a small value. The elements of the correlation matrix are

$$C_{1_{11}} = \frac{A^2\omega^2}{i_1^2(1-\varepsilon)} \frac{g_1(\varepsilon,x)}{g(\varepsilon,x)} J_2 \qquad C_{1_{22}} = \frac{A^2\omega^2}{i_1^2 i_2^2(1-\varepsilon)} \frac{g_2(\varepsilon,x)}{g(\varepsilon,x)} J_2$$

$$C_{1_{12}} = \frac{A^2\omega^2}{i_1^2 i_2} \frac{g_3(\varepsilon,x)}{g(\varepsilon,x)} J_2$$

and

$$C_{2_{11}} = \frac{A^2\omega^2}{i_1^2(1-\varepsilon)} \frac{f_1(\varepsilon,x)}{f(\varepsilon,x)} J_2 \qquad C_{2_{22}} = \frac{A^2\omega^2}{i_1^2 i_2^2(1-\varepsilon)} \frac{f_2(\varepsilon,x)}{f(\varepsilon,x)} J_2$$

$$C_{2_{12}} = \frac{A^2\omega^2}{i_1^2 i_2} \frac{f_3(\varepsilon,x)}{f(\varepsilon,x)} J_2 \qquad \text{with } x = \frac{I_2/I_1}{i_2^2} \,.$$

The functions $f_1, f_2, f_3, f, g_1, g_2, g_3, g$ are high-order polynomials of ε and x. The properties of the proposed formulas agree well with experimental results. From the mentioned expected values other characteristics can be derived such as energy dissipation by the impacts or the intensity of the impacts. Again, these results in terms of mean values are of great significance on the estimation of parameter influences with regard to gearbox design, where detailed structural properties like periodicity, chaos or bifurcation do not play an important role.

12.1.5 Experimental Results

The theoretical methods as given above have been developed primarily for practical purposes. The fundamental insights into parameter influences on rattling are the basis for improvements of gearboxes and or drive train systems, as indicated in Section 12.1.4.

To generate some confidence in the theoretical methods and to be sure of a good correspondence of idea and reality, experimental investigations and comparisons have to be performed. Therefore, a test setup for one-stage rattling has been realized according to Fig. 12.11. A wheel with a single involute tooth is excited by vibrating constraints. In addition, an eddy current

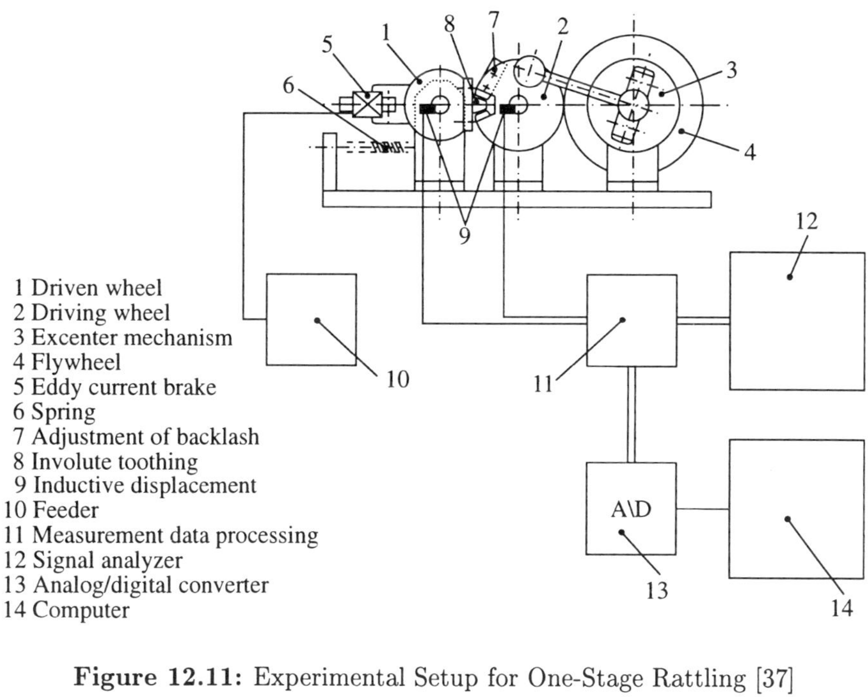

Figure 12.11: Experimental Setup for One-Stage Rattling [37]

brake simulates linear damping forces, whereas a week spring adds an almost constant momentum. The vibrating constraints themselves are driven via an excenter mechanism by a direct current motor. All important parameters can be varied, such as the excitation frequency and amplitude, the backlash, the damping factor and the constant moment. The following ranges are possible:

- excitation amplitude (0 ... 9) mm
- excitation frequency (0 ... 50) Hz
- backlash (0 ... 7) mm
- viscous damping (0 ... 100) Nmm s/rad
- constant damping (0 ... 500) Nmm

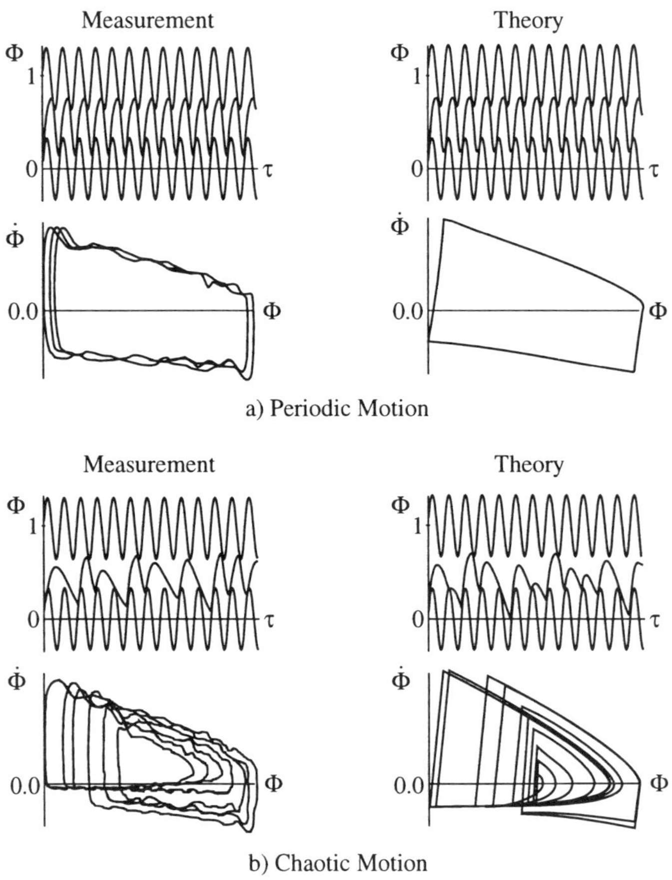

Figure 12.12: Comparison of Theory and Measurements for One-Stage Rattling [37]

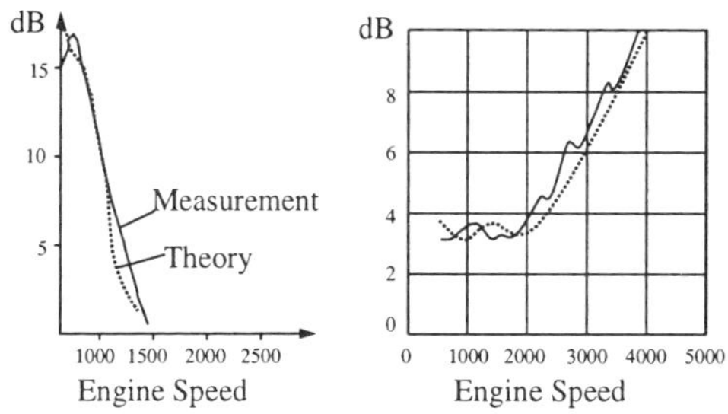

Figure 12.13: Comparison of Theory with Rattling Noise Measurements

Measurement results are the motions of the two wheels: the tooth wheel and the one with the vibrating constraints. The signals are then further processed on a digital computer. Figure 12.12 gives a comparison to theory and measurements for periodic and chaotic motion [37]. As can be seen, periodic rattling motion generates a periodic impact pattern within the backlash of the gear meshes, and chaotic motion generates a chaotic impact pattern. The backlashes in the two figures are represented in a periodic form indicating the external periodic excitation.

Another, more industrial-oriented verification of the above theories was achieved by comparing noise measurements of gear rattling with the approximation of formula (12.9) with the use of eqs. (12.8). Figure 12.13 gives a comparison of the computed noise curves with the airborne noise measured in the surroundings of a gearbox integrated in the drive trains of two German cars with completely different behavior during constant acceleration. Agreement is excellent. In recent years there have been a series of such comparisons with noise measurements of German car producers, which always confirmed the assumptions chosen for the theoretical methods.

12.2 A Ship-Turning Gear

A highly illustrative example for some remarkable properties of rattling was detected in an application connected with noise problems of a pleasure boat [19, 78]. It was equipped with a drive train system combining a 340-hp eight-cylinder engine with a reversal gear which in one direction operates as a one-stage gear system and in the other direction acts as a two-stage gear system (Fig. 12.14). For switching from rotation to counterrotation a clutch is used which possesses backlash.

In addition all meshes of the gears have backlashes. Therefore, and dependent on the switched state, the system can rattle, the rattling process being excited by large engine unbalances. Rattling was so loud that the drive system could not be sold. Several parameter studies were performed to ap-

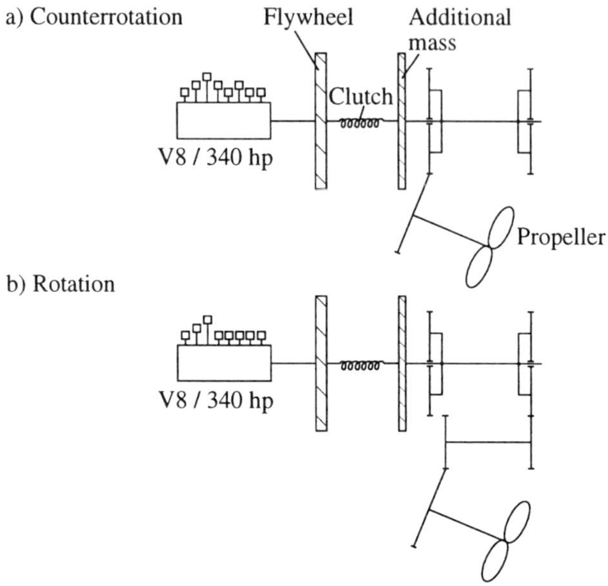

Figure 12.14: Scheme of the Ship-Turning Gear

ply the theory of Chapter 7. These parameter variations improved the noise behavior only marginally. A breakthrough occurred as a result of an idea of the company's design engineering department, which demonstrated a special clutch with a maximum angular backlash up to 35°. The experiments indicated a noise breakdown for an angular play of about 17°. Our simulations confirmed this magnitude. Figure 12.15 displays the principal situation.

An explanation of this somehow strange behavior can be achieved by considering a very simple one-DOF model with backlash. The system is periodically excited and has a nonlinearity by a spring with backlash which corresponds to the mesh of the gear teeth (Fig. 12.16).

The equation of motion in dimensionless form is

$$\xi'' + D\xi' + \varphi(\xi) = \varphi_0 \cos\tau \tag{12.32}$$

($\xi = x/v$, $\tau = \omega t$, $(\,\cdot\,)' = d/d\tau$, $D = d/m\omega$, $\varphi(\xi) = F(x)/m\omega^2 v$, $\varphi_0 = F_o/m\omega^2 v$, $\eta = c/m\omega^2$). The nonlinear force law $F(x)$ is then

$$\begin{aligned} \varphi(\xi) &= \eta(\xi - \tfrac{1}{2}) \text{ for } \xi > \tfrac{1}{2}, \\ \varphi(\xi) &= \eta(\xi + \tfrac{1}{2}) \text{ for } \xi < \tfrac{1}{2}, \\ \varphi(\xi) &= 0 \qquad\quad\; \text{for } -\tfrac{1}{2} \le \xi \le +\tfrac{1}{2}. \end{aligned} \tag{12.33}$$

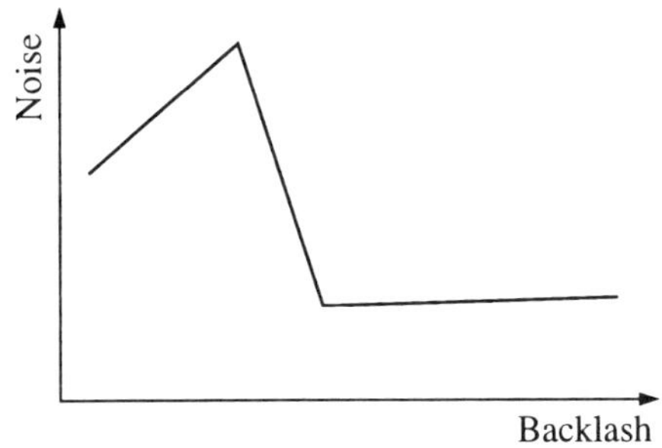

Figure 12.15: Noise over Backlash, Principal Situation

We approximate the nonlinear equation (12.32) by the linear one

$$\xi'' + D\xi' + \eta_0 \xi = \varphi_0 \cos \tau \tag{12.34}$$

and evaluate η_0 by a least square method, taking into account the areas of eq. (12.33). This results in

$$\eta_0 = \left(\frac{2\eta}{\pi \xi_0}\right) \left[\xi_0 \arccos\left(\frac{1}{2\xi_0}\right) - \frac{1}{2}\sqrt{1 - \left(\frac{1}{2\xi_0}\right)^2}\,\right], \tag{12.35}$$

where $\xi_0 = x_0/v$ is the gain of the solution of eq. (12.34).

Gain and phase ψ are then

$$\left(\frac{\xi_0}{\varphi_0}\right) = (m\omega^2)\left(\frac{x_0}{F_0}\right) = \left[(1-\eta_0)^2 + D^2\right]^{-1/2} \tag{12.36}$$

$$\tan \psi = \frac{D}{1-\eta_0}. \tag{12.37}$$

Figure 12.17 depicts the resonance curves for this approximation. They reveal some astonishing properties.

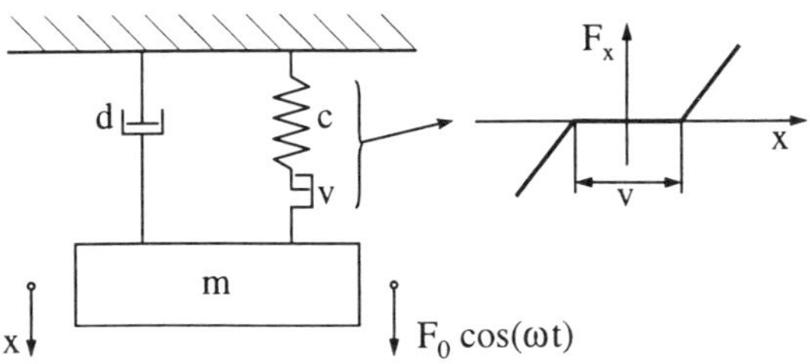

Figure 12.16: One-DOF System with a Backlash Nonlinearity

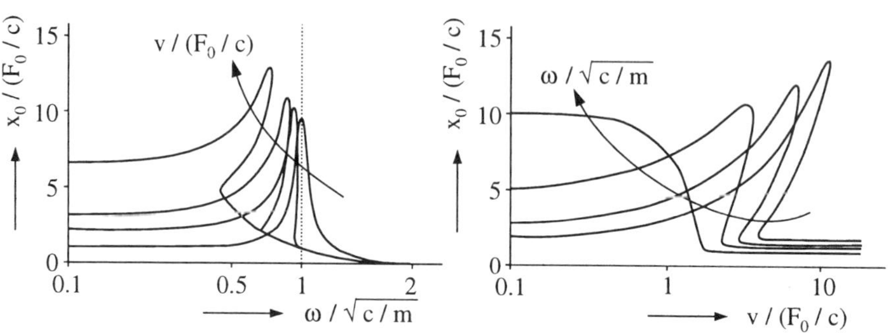

Figure 12.17: Parameter Dependencies for Backlash Rattling ($D_L = D/2\sqrt{\eta} = 0.05$)

The left diagram of Fig. 12.17 illustrates the well-known behavior of resonance curves for systems with backlash. Scanning the play characteristic (eq. 12.33 and Fig. 12.16) from zero force to nonzero force results in a resonance structure which is "more than linear" (resonance peak turns to the right); passing from the nonzero force branch to zero force we get a structure which is "less than linear" (resonance peak turns to the left). Both effects can be seen in Fig. 12.17. With increasing dimensionless play $v/(F_0/c)$ the jump phenomenon is significantly intensified when increasing or decreasing the excitation frequency $\omega/\sqrt{c/m}$.

The right diagram of Fig. 12.17 indicates a very strong jump behavior depending on the backlash itself. If the backlash v is on the order of magnitude of the spring deflection F_0/c caused by the excitation force amplitude F_0, we get a steep descent of the amplification factor $x_0/(F_0/c)$, where the character of this descent depends on the excitation frequency. With decreasing $\omega/\sqrt{c/m}$ we see a significant shift of the amplification peaks to the right. The amplification $x_0/(F_0/c)$ is proportional to the noise amplitudes. Therefore, the results of Fig. 12.17 give a physical interpretation of the rattling phenomena as measured in the ship reversal gear. In addition this could be confirmed by a simulation with a complete model, which indicates that the neglections in our simplified model are correctly estimated.

12.3 Dynamics of a Synchronizer

12.3.1 Introduction

Synchronizing units in transmission systems of vehicles are usually applied to even out the angular speed difference between the main shaft of a transmission and the gear to be engaged. A process of that kind must be very fast and precise, particularly in trucks, where the torques to synchronize the gears are very large. This is one important argument to apply double-cone synchronizers instead of a single-cone configuration. For the same axial shift force, double-cone synchronizers transmit nearly twice the torque of a single-cone synchronizer. On the other hand, double-cone synchronizers very often are more sensitive with respect to locking safety, which is an important operational aspect of synchronizers. They should not unlock before the speed difference at the friction surface equals zero, which often is addressed as safety with regard to premature unlocking.

The following analysis gives an estimate of premature unlocking on the basis of a plane model. Such a plane model cannot map all effects taking place in reality because the components of a synchronizer will exhibit out-of-plane motion due to tolerances and asymmetries of the design. Nevertheless, synchronizer dynamics represents an interesting application of the theory. Its effects are impacts, and some additional stick-slip phenomena are modeled as well [86].

12.3.2 Operation of a Synchronizer

Figure 12.18 portrays the main features of a synchronizer, and Fig. 12.19 gives some details of the entrainment configuration. The synchronizing process is initiated by an axial force, generated by hand, pneumatically or hy-

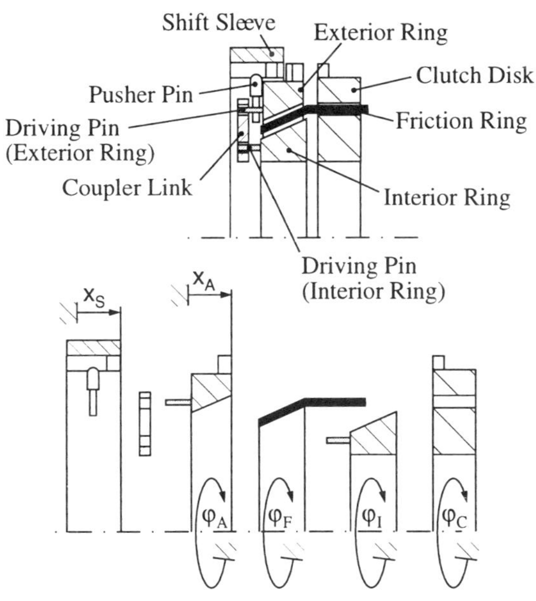

Figure 12.18: Mechanical Model of a Synchronizer

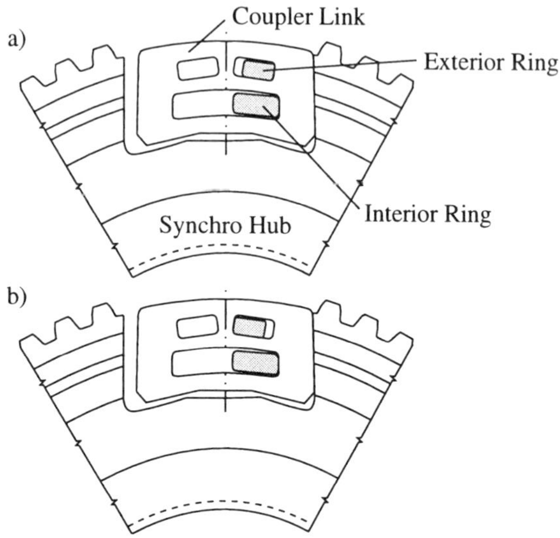

Figure 12.19: Entrainment Pin in the Clutch Component: (a) Locking Position I, (b) Locking Position II

draulically, which acts on the sleeve. If everything works without any disturbances we get the following sequence of events:

- The axial force accelerates the sleeve in the direction to the exterior ring.
- As a consequence, the pusher pins, which are moved by the sleeve, contact the exterior ring. The unit exterior ring/friction ring/interior ring is pressed against the clutch disk.
- The above process and the generating axial force produce a frictional torque moving the exterior and interior rings into locking position I (Fig. 12.19a), where the pins of both rings contact the coupler link at one side only.
- With increasing axial force contact between the sleeve teeth and exterior ring teeth is established. In the first instant locking position I will be maintained, which means that only the frictional torque at the exterior ring withstands a possible unlocking movement.
- In a second instant the frictional torque at the exterior ring will not be sufficient to prevent an unlocking motion. This motion of the exterior ring ends at the other side of the backlash of the coupler link. The then attained locking position II (Fig. 12.19b) is characterized by contact

of the coupler link and interior ring with one side of the synchronizing body and by contact of the exterior ring's driving pin with the other side of the coupler link.

- In this position the frictional torque of the interior ring counteracts an unlocking movement. The unlocking safety of the interior and exterior rings blocks the axial motion of the sleeve as long as the angular speeds are not the same.

- After synchronization of the angular speeds the torque at the friction ring changes sign, which opens the locked teeth at the exterior ring and allows switching of the stage, which means a connection of sleeve and the dogteeth on the clutch disk.

12.3.3 Mechanical and Mathematical Models

Figure 12.18 portrays the mechanical model with a total of six degrees of freedom:

$$\boldsymbol{q} = (x_S,\, x_A,\, \varphi_A,\, \varphi_F,\, \varphi_I,\, \varphi_C)^T \in \mathbb{R}^6 . \tag{12.38}$$

There are two translational degrees of freedom: x_S the axial displacement of the sleeve and x_A the axial displacement of the exterior ring. In addition, the exterior ring possesses a rotational degree of freedom φ_A. Further rotational degrees of freedom are φ_F, φ_I, φ_C, which represent angular rotations of the friction ring, the interior ring and the clutch, respectively.

The equations of motion are given by eqs. (6.1), and in the case of classical impact theory by eqs. (7.11) and (7.12). The forces relevant for our problem are depicted in Fig. 12.20.

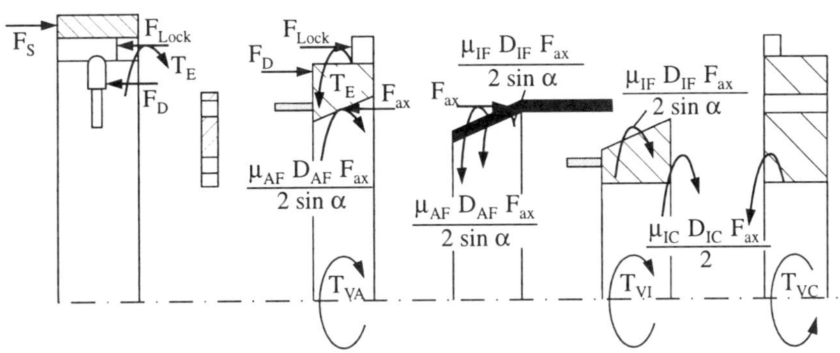

Figure 12.20: Forces on the Components of the Mechanical Model

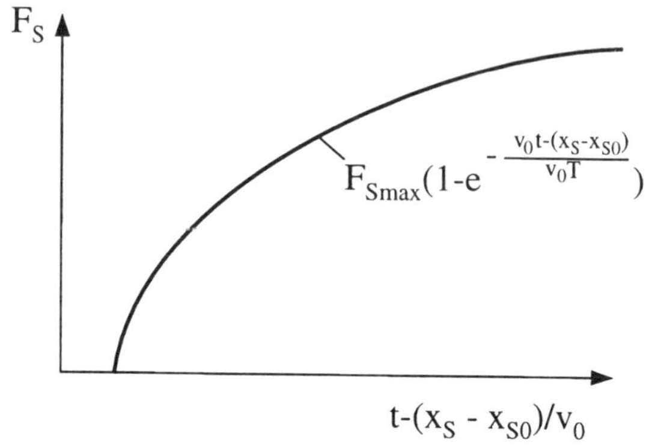

Figure 12.21: Axial Shift Force F_S (t = Time, v_0 = Pneumatic Piston Velocity)

For the shift force F_S on the sleeve we assume that it will be generated pneumatically. A typical diagram from measurements is shown in Fig. 12.21. The forces on the pusher pins emerge from the springs pressing the pin into corresponding notches of the sleeve. Pushing the sleeve in the axial direction, the pusher pins transmit axial force to the exterior ring while moving in a radial direction. Figure 12.22 gives a principal impression of the axial forces F_D acting on the pins (Fig. 12.20).

Later we must consider the forces and torques resulting from oil drag, which will be modeled as velocity-dependent damping with measured coefficients. Disturbance forces come in from other gear stages and from deviations

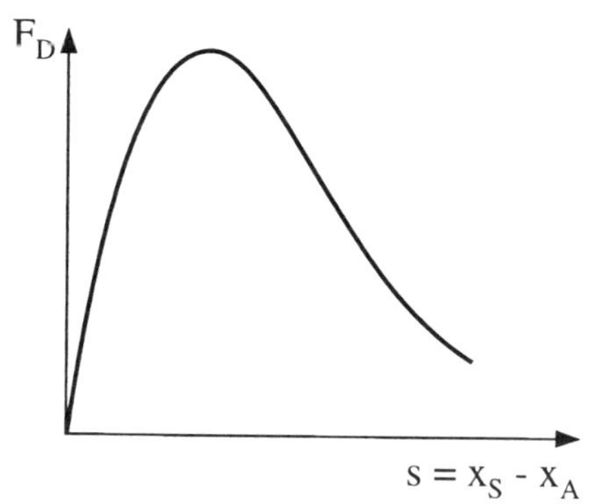

Figure 12.22: Pusher Pin Force F_D in Dependence on the Axial Displacement s

from a constant angular speed at the transmission exit. These forces and torques usually can be modeled as periodic forces.

For the evaluation of the equations of motion we need all constraint equations. The following contact possibilities exist (F = frictional contact and stick-slip, I = impulsive contact):

• sleeve–inertial	F	
• sleeve–ext. ring	F	I
• ext. ring–inertial		I
• ext. ring–int. ring		I
• int. ring–inertial		I
• friction ring–ext. ring	F	
• friction ring–int. ring	F	
• clutch disk–int. ring	F	
• clutch disk–friction ring		I

"Inertial" is related to the rotating input shaft coordinate system, in which the synchro hub is fixed. Any possible combination of these contacts may occur. For the evaluation of the $\boldsymbol{W}$-matrices in the equations of motion we need the constraint equation g_i, which acts as an indicator in a passive contact situation and as a constraint in an active one.

For the contact sleeve–exterior ring we get

$$g_1 = (x_A - x_S)n_x + \frac{1}{2}D_A\varphi_A n_z \geq 0\,, \tag{12.39}$$

which is the relative distance in the mesh of the teeth of the sleeve and exterior ring. Constraining relative motion of this contact by stiction, we have

$$g_2 = (x_S - x_A)n_z + \frac{1}{2}D_A\varphi_A n_x = 0\,. \tag{12.40}$$

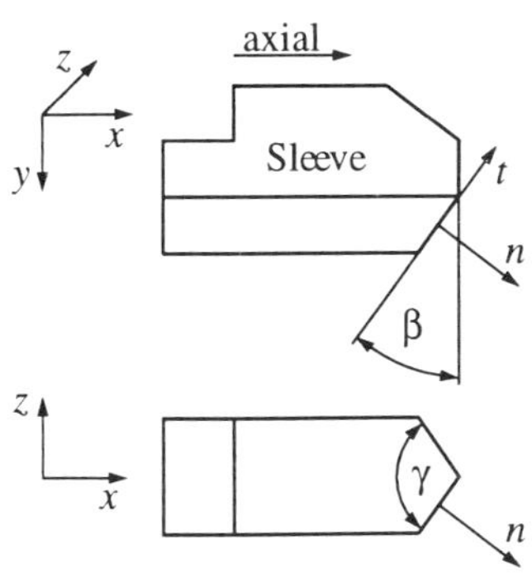

Figure 12.23: Geometry of Locked Teeth

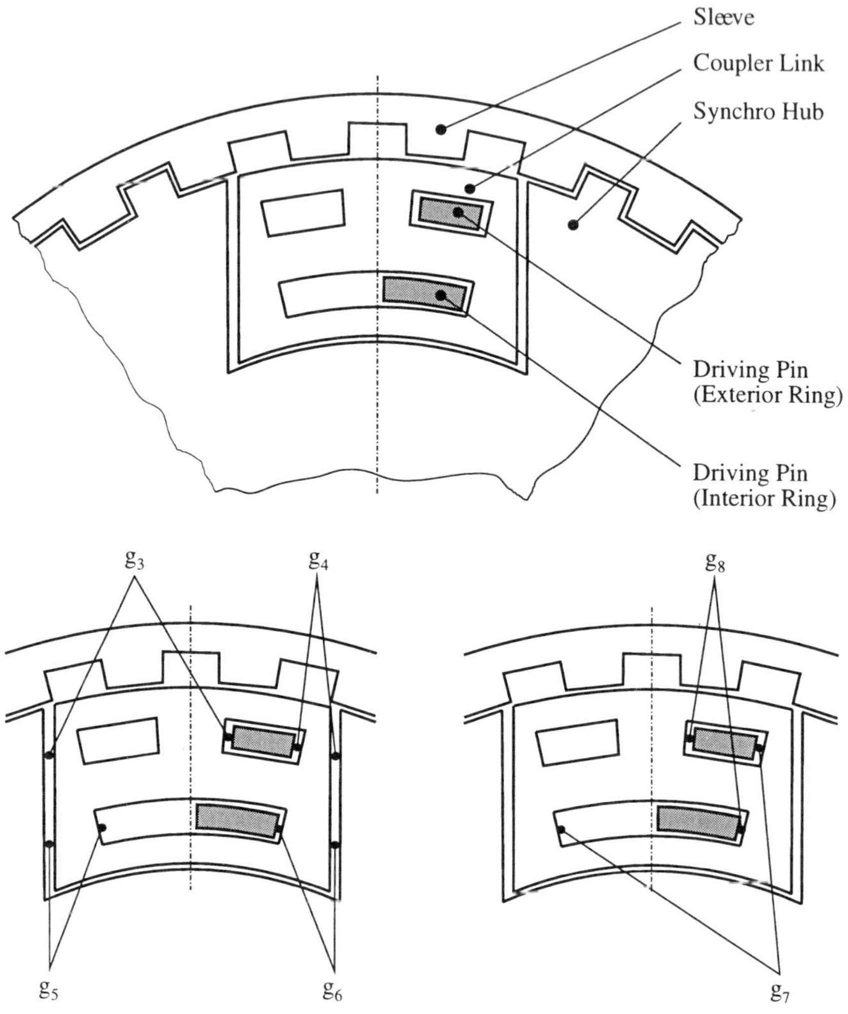

Figure 12.24: Details of Contact Constraints

The components of the normal and tangential vectors $\boldsymbol{n}$ and $\boldsymbol{t}$ follow from Fig. 12.23:

$$
\begin{aligned}
\boldsymbol{n} &= (n_x,\, n_y,\, n_z)^T & \boldsymbol{t} &= (t_x,\, t_y,\, t_z)^T \\
n_x &= +a \sin\tfrac{\gamma}{2} & t_x &= +\sin\beta \\
n_y &= +a \sin\tfrac{\gamma}{2}\cdot\tan\beta & t_y &= -\cos\beta \\
n_z &= -a \cos\tfrac{\gamma}{2} & t_z &= 0
\end{aligned}
\tag{12.41}
$$

with

$$a = \left(1 + \tan^2 \beta \cdot \sin^2 \frac{\gamma}{2}\right)^{-1/2} .$$

The contacts between the coupler link and the two rings are determined by the backlashes in the link, which results in (see Fig. 12.24)

$$\begin{aligned} g_3 &= -\varphi_A && \geq 0 \\ g_4 &= +\varphi_A - \varphi_{A\,\min} && \geq 0 \\ g_5 &= -\varphi_I && \geq 0 \\ g_6 &= +\varphi_I - \varphi_{I\,\min} && \geq 0 . \end{aligned} \tag{12.42}$$

In addition, the relative rotations of the inner and outer rings are limited by the backlashes within the coupler link. With respect to the small link mass, it does not possess own degrees of freedom. Nevertheless its geometry enters into the constraint equations (see Fig. 12.24):

$$\begin{aligned} g_7 &= +(\varphi_A - \varphi_I) - \Delta\varphi_{AI,\min} \geq 0 , \\ g_8 &= -(\varphi_A - \varphi_I) + \Delta\varphi_{AI,\max} \geq 0 , \end{aligned} \tag{12.43}$$

where $\Delta\varphi_{AI,\min}$, $\Delta_{AI,\max}$ are the smallest and largest angular displacements as allowed by the coupler link (Fig. 12.19). In a similar way we define the constraints between the friction ring and the clutch:

$$\begin{aligned} g_9 &= +(\varphi_C - \varphi_F) - \Delta\varphi_{CF,\min} \geq 0 , \\ g_{10} &= -(\varphi_C - \varphi_F) + \Delta\varphi_{CF,\max} \geq 0 . \end{aligned} \tag{12.44}$$

The vectors $\boldsymbol{w}$ needed for establishing the $\boldsymbol{W}$-matrices in eqs. (6.1), (7.11), (7.12) follow from the constraint equations (12.39) to (12.44) in a straightforward way ($\boldsymbol{w}_i^T = \partial g_i / \partial \boldsymbol{q}$):

$$\begin{aligned} \boldsymbol{w}_1^T &= \left(-n_x,\ +n_x,\ \frac{1}{2} D_A n_z,\ 0,\ 0,\ 0\right) \\ \boldsymbol{w}_2^T &= \left(+n_z,\ -n_z,\ \frac{1}{2} D_A n_x,\ 0,\ 0,\ 0\right) \\ \boldsymbol{w}_3^T &= (0,\ 0,\ -1,\ 0,\ 0,\ 0) \\ \boldsymbol{w}_4^T &= (0,\ 0,\ +1,\ 0,\ 0,\ 0) \\ \boldsymbol{w}_5^T &= (0,\ 0,\ 0,\ -1,\ 0,\ 0) \\ \boldsymbol{w}_6^T &= (0,\ 0,\ 0,\ +1,\ 0,\ 0) \\ \boldsymbol{w}_7^T &= (0,\ 0,\ +1,\ 0,\ -1,\ 0) \\ \boldsymbol{w}_8^T &= (0,\ 0,\ -1,\ 0,\ +1,\ 0) \\ \boldsymbol{w}_9^T &= (0,\ 0,\ 0,\ -1,\ 0,\ +1) \\ \boldsymbol{w}_{10}^T &= (0,\ 0,\ 0,\ +1,\ 0,\ -1) . \end{aligned} \tag{12.45}$$

Corresponding to the current contact situation these constraint vectors combine to form the constraint matrix $\boldsymbol{W}$.

Safety with regard to untimely unlocking will be considered only in a quasi-static way. A quantitative measure is given by

$$s = \frac{T_R}{T_E} , \tag{12.46}$$

where for locking position I T_R is the friction torque on the outer ring and T_E is the torque generated by the shift force from the sleeve. The torque T_R counteracts unlocking, whereas T_E favors unlocking. From Fig. 12.25 we may establish the static equilibrium equations in the x- and z-directions, which gives F_N and F_U. The torques are

$$\begin{aligned} T_E &= \tfrac{1}{2} D_A F_U , \\ T_R &= \left(\frac{F_S - \mu_{Sx} F_U}{2 \sin \alpha} \right) (\mu_{AF} D_{AF} + \mu_{IF} D_{IF} + \mu_{IC} D_{IC} \sin \alpha) , \end{aligned} \tag{12.47}$$

where μ_{ij} are the sliding friction coefficients and D_{ij} are the diameters of the corresponding components. From eqs. (12.46) and (12.47) we finally get

$$s = \left(\frac{\mu_{SA} + \tan \frac{\gamma}{2}}{1 - \mu_{SA} \tan \frac{\gamma}{2}} \right) \left(\frac{D_{AF}\mu_{AF} + D_{IF}\mu_{IF} + D_{IC}\mu_{IC} \sin \alpha}{D_A \sin \alpha} \right). \tag{12.48}$$

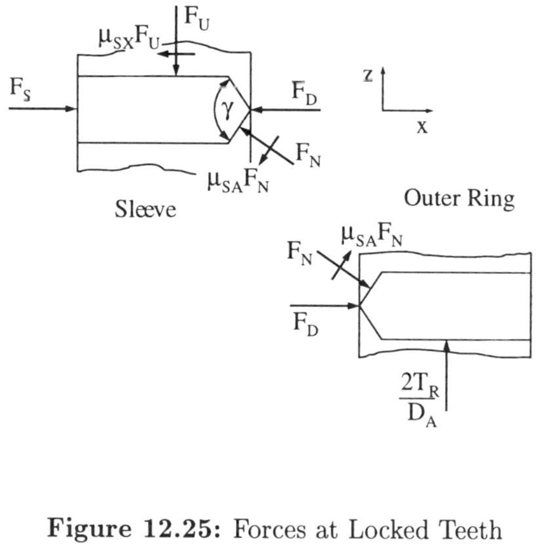

Figure 12.25: Forces at Locked Teeth

The safety s thus increases with increasing μ-values and with increasing angle γ. Increasing friction coefficients result in an increasing friction torque T_R, which prevents unlocking. An increasing angle γ results in a decreasing torque T_E, which assists unlocking. It should be mentioned that the overall safety factor s from eq. (12.46) might not give sufficient information with which to form an opinion of the safety. In this case the detailed behavior of all components during synchronization must be considered.

12.3.4 An Example

We investigated in detail a double-cone synchronizer the configuration of which corresponds exactly to the design shown in Figs. 12.18 and 12.19 [65]. The interest in this specific case was focused on the beginning of the synchronization process, so, no complete synchronization is shown. Figure 12.26 portrays some typical features of a starting synchronization.

At the beginning the sleeve's motion is retarded by the forces of the pusher

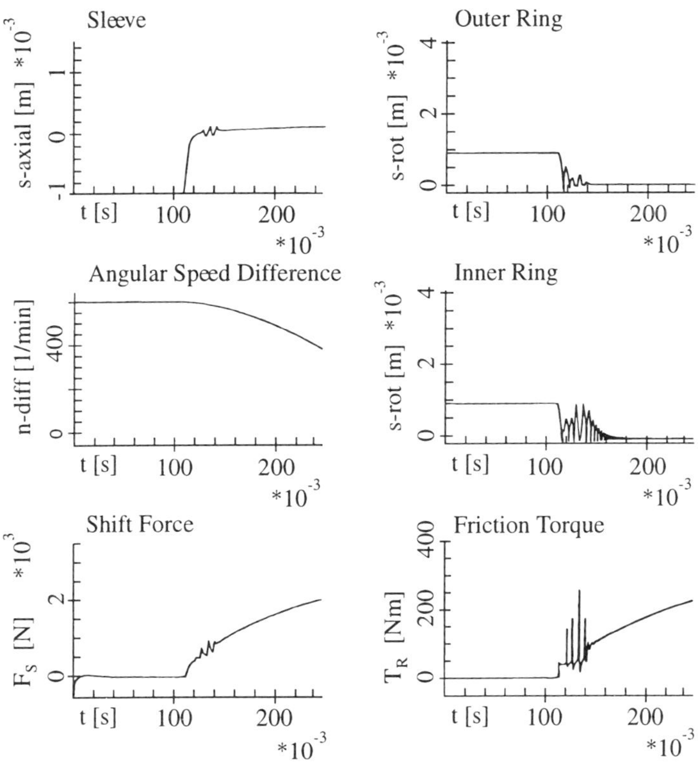

Figure 12.26: Simulation Results

pins in such a way, that the first contact with the teeth does not take place before 120 ms. This is a contact sleeve/outer ring (Fig. 12.18) in a very short instant of time and with a resulting backstroke between outer ring and synchro hub. The sleeve is not retarded anymore and moves forward not realizing the locking position I (Fig. 12.19). A series of impacts takes place from 120 to 160 ms with recoils of sleeve and outer ring. The inner ring performs an impulsive motion with respect to the synchro hub and via the coupler link with respect to the exterior ring. The limited backlash of the coupler link prevents larger backstrokes between the inner and outer rings. After about 140 ms locking postion I is nevertheless established. The shift force F_S closely follows the prescribed curve with some impact-driven deviations. The friction torque reflects the various impacts in the system by building up some significant torque peaks. The pattern of the speed difference does not show any influence of the impulsive processes. Stick-slip phenomena were not visible in the simulation results, although they are included in the model. Obviously synchronization is dominated by impulsive processes.

13

APPLICATIONS WITH COULOMB'S FRICTION LAW

13.1 Turbine Blade Damper

13.1.1 Problem and Model

For a reduction of blade vibration amplitudes in airborne gas turbines special damper devices are used which usually are arranged between the blade platforms of neigboring blades (Fig. 13.1).

These dampers possess a cross section like a parabola or a circle and are formed out of sheet steel. The dampers are arranged between the platforms of the blades in such a way that they are pressed by centrifugal forces against the oblique plane, which will vibrate together with the blades. This motion causes some relative movement between damper and oblique plane thus generating dissipation by dry friction. The design goal for dampers of that kind consists of maximizing dissipation (i.e., relative motion) or, to put it another way, to minimize stiction time [28, 29, 79, 80].

The mechanical model in Fig. 13.1 consists of two platforms (masses m_1, m_3) with turbine blades (masses m_2, m_4) connected by a friction damper (mass m_5, moment of inertia I_5). The displacements of the platforms and of the turbine blades are described by the translational coordinates (q_1, q_2, q_3, q_4). In the unconstrained case the damper possesses two translational degrees of freedom q_6, q_7 and one rotational degree of freedom q_5. The vector of the minimal coordinates is

$$\boldsymbol{q} = (q_1, q_2, q_3, q_4, q_5, q_6, q_7)^T \in \mathbb{R}^7 . \tag{13.1}$$

A constant radial force F_z, representing the centrifugal force, affects the damper and presses it against the oblique planes of the platforms. The excitation of the turbine blades is taken into account by the forces $F_i(t) =$

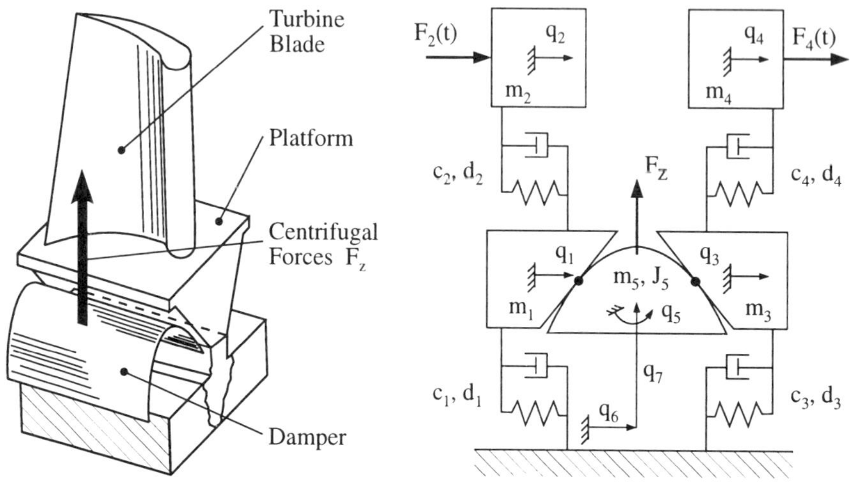

Figure 13.1: Turbine Blade Damper and Mechanical Model

$A\sin(\omega t + \varphi_i)$, $i = 2, 4$, which result from an excitation by periodically changing gas forces. From Fig. 13.1, and with the above-mentioned agreements, we are able to establish the mass matrix $\boldsymbol{M}$ and the vector $\boldsymbol{h}$ of all external forces (see eqs. 6.1):

$$\boldsymbol{M} = \begin{pmatrix} m_1 & 0 & 0 & 0 & 0 & 0 & 0 \\ 0 & m_2 & 0 & 0 & 0 & 0 & 0 \\ 0 & 0 & m_3 & 0 & 0 & 0 & 0 \\ 0 & 0 & 0 & m_4 & 0 & 0 & 0 \\ 0 & 0 & 0 & 0 & J_S & 0 & 0 \\ 0 & 0 & 0 & 0 & 0 & m_5 & 0 \\ 0 & 0 & 0 & 0 & 0 & 0 & m_5 \end{pmatrix} \tag{13.2}$$

$$\boldsymbol{h} = \begin{pmatrix} -(c_1 + c_2)q_1 + c_2 q_2 - (d_1 + d_2)\dot{q}_1 + d_2\dot{q}_2 \\ c_2 q_1 - c_2 q_2 + d_2\dot{q}_1 - d_2\dot{q}_2 + F_2(t) \\ -(c_3 + c_4)q_3 + c_4 q_4 - (d_3 + d_4)\dot{q}_3 + d_4\dot{q}_4 \\ c_4 q_3 - c_4 q_4 + d_4\dot{q}_3 - d_4\dot{q}_4 + F_4(t) \\ 0 \\ 0 \\ F_z \end{pmatrix} \tag{13.3}$$

For the derivation of the matrices and vectors of the constraints we refer to the example in Section 4.6. All results presented there can be transferred to the model in Fig. 13.1 with only minor changes in nomenclature. In particular the terms in eqs. (4.50), (4.52), (4.54), (4.55), in this context, are written

$$\begin{pmatrix} \boldsymbol{w}_{N1}^T \\ \boldsymbol{w}_{N2}^T \end{pmatrix} = \frac{1}{\sqrt{1+a^2}} \begin{pmatrix} -a & 0 & 0 & 0 & (\sin q_5 - a\cos q_5)\left(p_2 - \frac{1}{2p_1} - p_1\sigma_1^2\right) & a & -1 \\ 0 & 0 & a & 0 & (\sin q_5 + a\cos q_5)\left(p_2 - \frac{1}{2p_1} - p_1\sigma_2^2\right) & -a & -1 \end{pmatrix}$$

$$\begin{pmatrix} \boldsymbol{w}_{T1}^T \\ \boldsymbol{w}_{T2}^T \end{pmatrix} = \frac{1}{\sqrt{1+a^2}} \begin{pmatrix} -1 & 0 & 0 & 0 & (\cos q_5 + a\sin q_5)\left(-p_2 - p_1\sigma_1^2\right) & 1 & a \\ 0 & 0 & -1 & 0 & (\cos q_5 - a\sin q_5)\left(-p_2 - p_1\sigma_2^2\right) & 1 & -a \end{pmatrix}$$

$$\begin{pmatrix} \overline{w}_{N1} \\ \overline{w}_{N2} \end{pmatrix} \tag{13.4}$$

$$= \frac{\dot{q}_5^2}{\sqrt{1+a^2}} \begin{pmatrix} (\cos q_5 + a\sin q_5)\left(p_2 - \frac{1}{2p_1} - 3p_1\sigma_1^2 - 8p_1^3\sigma_1^4\right) \\ (\cos q_5 - a\sin q_5)\left(p_2 - \frac{1}{2p_1} - 3p_1\sigma_2^2 - 8p_1^3\sigma_2^4\right) \end{pmatrix}$$

$$\begin{pmatrix} \overline{w}_{T1} \\ \overline{w}_{T2} \end{pmatrix} = \frac{\dot{q}_5^2}{\sqrt{1+a^2}} \begin{pmatrix} (\cos q_5 + a\sin q_5)\left((2p_2p_1 - 1)\sigma_1 - 2p_1^2\sigma_1^3\right) \\ (\cos q_5 - a\sin q_5)\left((2p_2p_1 - 1)\sigma_2 - 2p_1^2\sigma_2^3\right) \end{pmatrix}$$

where subscripts 1 and 2 denote the left and right contact points in Fig. 13.1, respectively. Furthermore, σ_1 and σ_2 are the values of the parabola contact parameters which are given for the left contact point by eq. (4.48),

$$\sigma_1 = \frac{-1}{2p_1} \cdot \frac{a\cos q_5 - \sin q_5}{a\sin q_5 + \cos q_5},$$

and for the right contact point when substituting the angle of inclination of the plane $a = \cot\gamma$ by $-a$:

$$\sigma_2 = \frac{-1}{2p_1} \cdot \frac{-a\cos q_5 - \sin q_5}{-a\sin q_5 + \cos q_5}.$$

As a whole the configuration consists of a coupled two-point contact problem, where the normal and tangential forces of the two contact points influence each other and depend on the system's dynamics.

To get appropriate initial values for the numerical integration we determine first, with $F_2 = F_4 = 0$, the frictionless equilibria of the damper using equs. (6.1) for constrained motion. Frictionless contact ($\mu = 0$) means that the tangential constraints do not exist; thus only the equations of normal constraints have to be considered (subscript N in eqs. 13.4). Therefore we get from eqs. (6.1) that

$$\begin{aligned} \boldsymbol{M}\ddot{\boldsymbol{q}} - \boldsymbol{h} - \boldsymbol{W}_N\boldsymbol{\lambda}_N &= \boldsymbol{0} \in \mathbb{R}^7 \qquad (13.5)\\ \ddot{\boldsymbol{g}}_N = \boldsymbol{W}_N^T\ddot{\boldsymbol{q}} + \overline{\boldsymbol{w}}_N &= \boldsymbol{0} \in \mathbb{R}^2 \\ \dot{\boldsymbol{g}}_N = \boldsymbol{W}_N^T\dot{\boldsymbol{q}} &= \boldsymbol{0} \\ \boldsymbol{g}_N &= \boldsymbol{0}\,. \qquad (13.6) \end{aligned}$$

Equilibria (subscript 0) are characterized by vanishing velocities and accelerations ($\ddot{\boldsymbol{q}}_0 = 0$, $\dot{\boldsymbol{q}}_0 = 0$), whereas the displacements and normal forces ($\boldsymbol{q}_0, \boldsymbol{\lambda}_{N0}$) are constant. From eqs. (13.5, 13.6) we get nine nonlinear algebraic equations for ($\boldsymbol{q}_0 \in \mathbb{R}^7$, $\boldsymbol{\lambda}_{N0} \in \mathbb{R}^2$), which read

$$\left.\begin{aligned} \boldsymbol{h}(\boldsymbol{q}_0) + \boldsymbol{W}_N\boldsymbol{\lambda}_{N0} = \boldsymbol{0} \in \mathbb{R}^7\,, \\ \boldsymbol{g}_N(\boldsymbol{q}_0) = \boldsymbol{0} \in \mathbb{R}^2\,. \end{aligned}\right\} \qquad (13.7)$$

13.1.2 Results and Verification

An evaluation of the eqs. (13.7) in terms of the damper's parabola parameter p_1 yields as an example the angular displacements q_{50}, which are depicted in Fig. 13.2.

For small values of p_1 only the symmetric equilibrium $q_{50} = 0$ exists. When the parabola parameter p_1 is enlarged, $q_{50} = 0$ becomes unstable and two nonsymmetric equilibria arise. Before starting the numerical integration for a given value of p_1 the equilibrium $\boldsymbol{q}_0$ is computed and then used as the initial displacement for the simulation of the system under the influence of friction. It turns out that in many cases computer time is reduced by more than two orders of magnitude due to the fact that the stationary situation

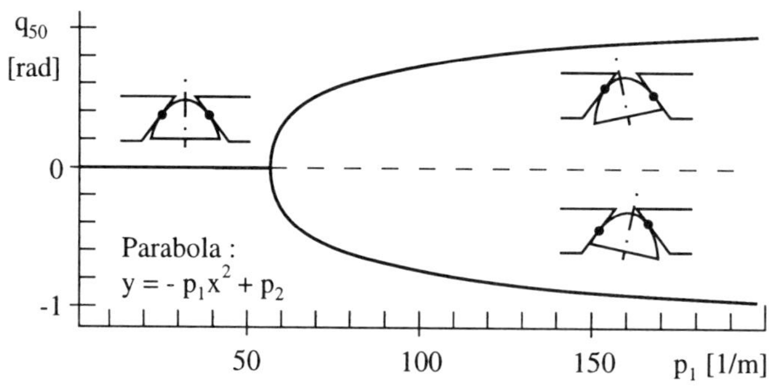

Figure 13.2: Equilibrium Positions of the Damper

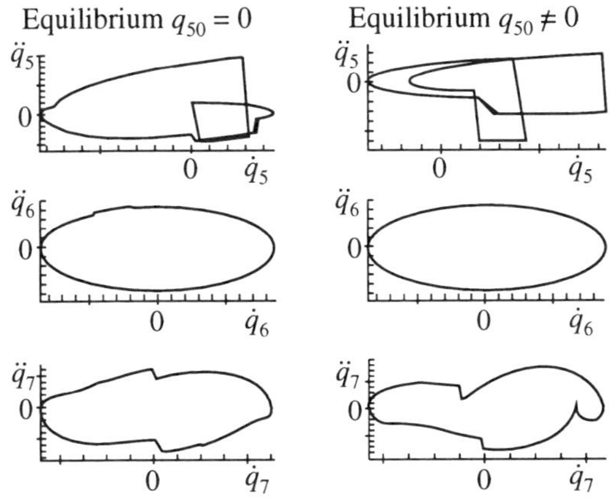

Figure 13.3: Dynamical Behavior of Damper

is known in advance and need not be realized by time-consuming numerical integration.

The aim of the design of a friction damper is to obtain the largest possible dissipation of energy by friction. Thus, the damper has to exhibit sufficiently long phases of sliding and show a low tendency to lock up. Using an original data set from industry, Fig. 13.3 shows the computed damper accelerations as a function of their velocities for a symmetric and a nonsymmetric equilibrium. The damper shows a different dynamic behavior, depending on the equilibrium, and is working in both cases, which can be seen on the unsteady

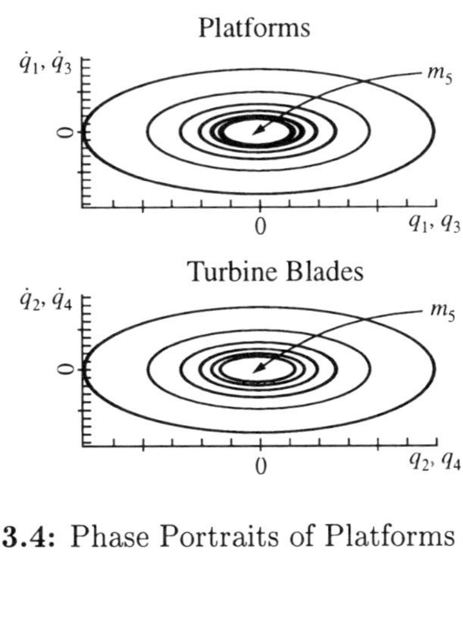

Figure 13.4: Phase Portraits of Platforms and Blades

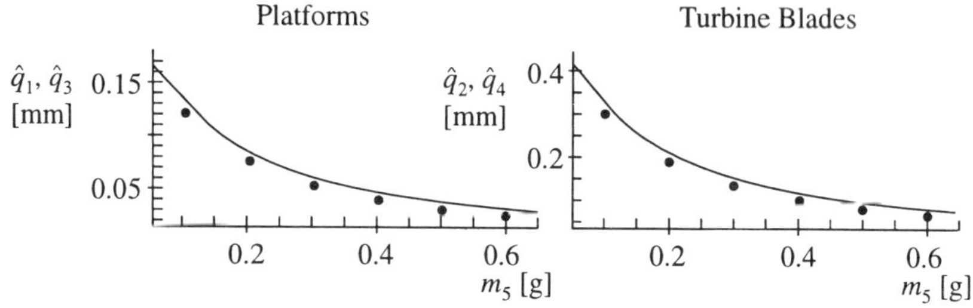

Figure 13.5: Comparison of Linear and Nonlinear Simulation

changes in accelerations due to stick-slip transitions.

Another task is to examine how the blade and platform amplitudes can be reduced by installing different dampers. In Fig. 13.4 the phase portraits of the blade and platform coordinates are plotted for different damper masses m_5.

Obviously, an increased damper mass leads to a reduction of the amplitudes, which unfortunately results mainly from the changes in the system mass and is not an effect of dry friction produced by the damper. Figure 13.5 shows the platform and blade amplitudes in dependency on the damper mass m_5. The dots correspond to the amplitudes $\hat{q}$ of the coordinates of Fig. 13.3. The solid line results from a simulation of a linear system which is achieved by the assumption that the damper is not working. Thus, the damper together with the two platforms is regarded as one rigid body and the model of Fig. 13.1 becomes linear. The difference between the dots and the solid line then shows exactly the reduction of the amplitudes by friction. For that special data set the damper shows a nice nonlinear friction-induced behavior, but

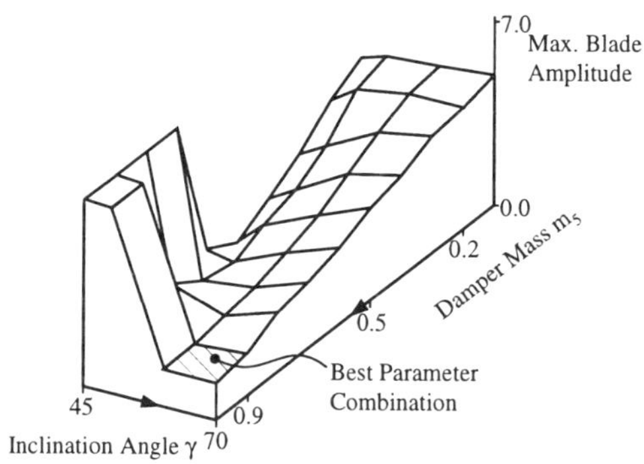

Figure 13.6: Parameter Evaluation for Optimal Damping [29]

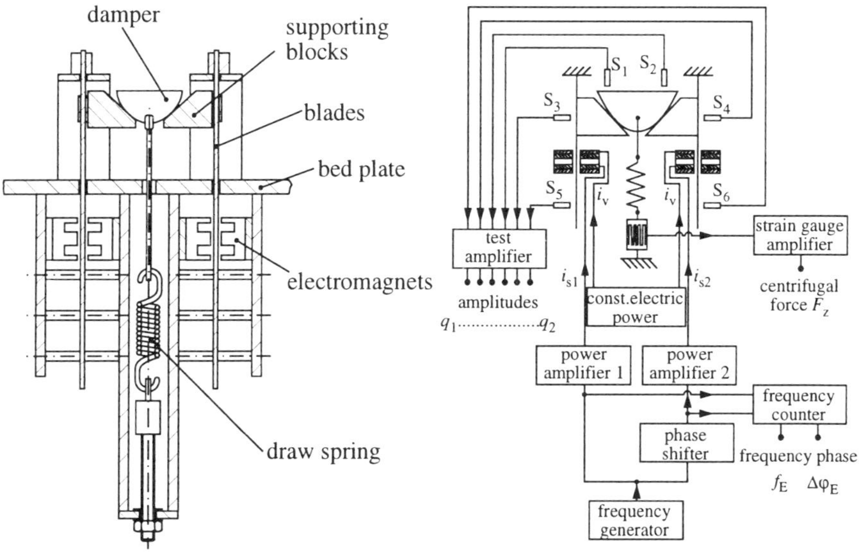

Figure 13.7: Laboratory Test Setup for a Turbine Blade Damper [29]

is not able to influence the amplitudes of the system significantly. A more rigorous variation of the most important parameters like inclination angle γ and damper mass m_5 shows, however, that significant optima can be realized by suitable adaptation of those parameters. Figure 13.6 gives an impression for a certain data set which corresponds to an industrial application [29].

The theoretical model as presented above was verified by a laboratory test setup (Fig. 13.7). For these experiments the centrifugal force is replaced by a spring force, the excitation realized electromagnetically, the blades represented by two bars. The Geometries of the damper and the blade platforms are the same as for real dampers, only larger. The amplitudes of the damper and the blades are measured by six inductive displacement transducers, the spring force by a strain gauge arrangement.

Figure 13.8 gives one typical comparison with theory. For practical considerations it is important to know at what combination of excitation amplitude and frequency the damper leaves its stiction state and starts to move at least in one contact point. Such diagrams define the field of action of the damper under consideration. Figure 13.8 shows three curves for three values of the spring force (centrifugal force) and for a contact angle of $\gamma = 50°$. The excitations force amplitude $F_{2,4}$ [N] must be augmented for larger centrifugal (spring) forces F_S. The very small $F_{2,4}$-values for an excitation frequency at 61 Hz can be explained in the following way. The values of $f_E = 61$ Hz and

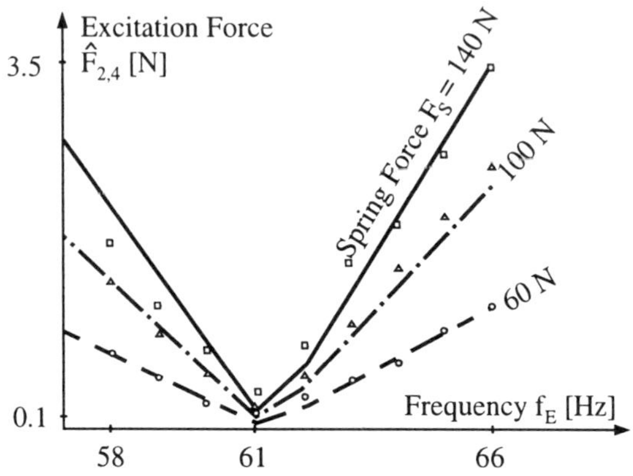

Figure 13.8: Comparison Theory (——, - - -, - · · -) and Measurements (□, △, ○) [29]

$f_E = 62$ Hz correspond to eigenfrequencies of the blocked blade-damper system with three degrees of freedom only. A transition from stiction to sliding can be achieved by especially small excitation forces when exciting by the resonance frequencies. This is the reason for the minimum in Fig. 13.8.

13.2 Friction Clutch Vibrations

13.2.1 Introduction

Clutch vibrations in car drive train systems belong to the classical self-excited vibrations. In comparison to other friction-driven vibrations they are nonstationary, and they are usually generated by shifting a stage in a changeover gear by hand and by operating the clutch for this purpose. In detail the clutch disks are separated, the changeover gear is shifted, the clutch disks are put together again. At the same time and in most cases the accelerator pedal will be pushed down to accelerate the car. The combination of closing the clutch disks and accelerating the car can generate self-excited vibrations, which of course significantly depend on the clutch disk linings and on the layout of the complete transmission system.

The vibration process is well known. The clutch disks close and transmit the driving torque. Due to torsional and/or axial vibrations of the transmission system, due to manufacturing tolerances and radial or angular displacements of the two disks and, above all, due to the frictional characteristic of the material pairings (see Fig. 1.1), the stiction situation between the two clutch disks may break down, and a sliding motion is started. More precisely, if the dynamical constraint forces acting on the disks become larger than the static friction forces, the clutch disks start a relative angular motion. This is continued as long as the relative angular velocity continues to exist. A transition sliding to static friction will occur if the relative angular velocity becomes zero and at the same time the static friction forces become larger than the corresponding tangential constraint forces as generated by the dynamics of the transmission system. A decreasing form of the friction characteristic (Stribeck curve, Fig. 1.1) is an important requirement for maintaining such a self-excited vibration which takes its energy from the rotation of the transmission. The transitions from static to sliding friction and vice versa act as a switch operating with the frequency of the self-sustained vibration, where this frequency in our case usually is in the near neighborhood of a system's eigenfrequency.

Friction clutch vibrations take place in a low-frequency range, around 5 to 15 Hz. Therefore, modeling can be achieved by considering a limited number of degrees of freedom and one contact only. From this we have one constraint equation and thus one constraint force. Complementarity considerations are not necessary.

13.2.2 Mechanical and Mathematical Models

Most investigations on clutch dynamics are performed by taking into account only torsional degrees of freedom. We shall do that also and consider models with a minimum of two and a maximum of nine masses for the complete transmission system. On the other hand, results from theory and measurements indicate that in certain cases more extended models with translational and additional rotational degrees of freedom might be helpful, especially in connection with investigations related to tolerance influences.

Considering torsional vibrations only, we start with the model portrayed in Fig. 13.9, which is mostly self-explanatory. The input torque represents the torque at the crankshaft, and the output torque the one at the wheels. Between input and output we have arranged the configuration of a classical rear-axle drive. The distribution of masses and stiffnesses follows more or less the sequence of the components. By experience it is reasonable to combine masses and stiffnesses of adjacent parts. By dynamic reduction it is also possible to reduce the model of Fig. 13.9 to a five-, four- or two-mass model without losing too much information with respect to clutch vibrations. We found, however, that a two-mass model can describe the friction clutch vibrations only qualitatively. For more quantitative information one needs in our case a model with more than two masses.

The equations of motion for the model of Fig. 13.9 are simply derived by applying the moment of momentum law to every body i, which yields

$$\begin{aligned} I_i\ddot{\varphi}_i &= -d_i(\dot{\varphi}_i - \dot{\varphi}_{i-1}) - d_{i+1}(\dot{\varphi}_i - \dot{\varphi}_{i+1}) \\ &\quad -c_i(\varphi_i - \varphi_{i-1}) - c_{i+1}(\varphi_i - \varphi_{i+1}) \\ &\quad +T_i(\varphi_i,\, \dot{\varphi}_i,\, t) \\ &\quad (i = 1, 2, \ldots, n) \end{aligned} \tag{13.8}$$

(φ_i is the torsional angular displacement of body i; I_i is the moment of inertia; c_i, d_i are the spring and damper constants; T_i is any torque characteristic acting between the bodies under consideration.) To include the friction-driven vibrations in the friction clutch we must consider the following simplified process (see Chapters 5 and 6). The friction clutch may have the subscript $i = i_c$.

The relative velocity between the two clutch disks is

$$\omega_r = \dot{\varphi}_{i_c} - \dot{\varphi}_{i_c - 1}\,, \tag{13.9}$$

which indicates three states (see, for example, eqs. 6.10)

$$\begin{aligned} &\bullet\ \text{state } 0, \quad \text{stiction with } \omega_r = 0\,, \\ &\bullet\ \text{state} + 1, \ \text{sliding with } \ \omega_r > 0\,, \\ &\bullet\ \text{state} - 1, \ \text{sliding with } \ \omega_r < 0\,. \end{aligned} \tag{13.10}$$

The transitions between these states are governed by the following inequalities:

- transition sliding/stiction

$$(\omega_r \le \varepsilon_\omega \wedge \Delta T > \varepsilon_M)$$

- transition stiction/sliding

$$(\omega_r > \varepsilon_\omega \wedge \Delta T \le \varepsilon_M)\,. \tag{13.11}$$

The torque difference ΔT corresponds to

$$\Delta T = \mu_0 F_{N,\max} r_m - |T_{CC}|\,, \tag{13.12}$$

where μ_0 is the static friction coefficient, $F_{N,\max}$ the maximum contact force in the friction clutch, r_m an averaged disk radius and T_{CC} the constraint torque acting on the clutch disks. The numbers ε_ω, ε_M are numerical lower limits for interpolation of the above inequalities, which is performed by the simple bisection method. The constraint torque T_{CC} can be evaluated by considering the constraint vectors $\boldsymbol{w}$ as introduced below ($T_{CC} = r_m\lambda$):

$$\boldsymbol{q} = (\varphi_1, \varphi_2, \ldots, \varphi_n)^T$$

$$\dot{g}_T = r_m\,(\dot{\varphi}_{i_c} - \dot{\varphi}_{i_c-1}) = r_m\,(\dot{q}_{i_c} - \dot{q}_{i_c-1})$$

$$\boldsymbol{w}^T = \frac{\partial \dot{g}_T}{\partial \dot{\boldsymbol{q}}} = (0,\ \ldots,\ 0,\ \underset{(i_{c-1})}{\underset{\uparrow}{-r_m}},\ \underset{(i_c)}{\underset{\uparrow}{+r_m}},\ 0,\ \ldots,\ 0) \tag{13.13}$$

$$\boldsymbol{M}\ddot{\boldsymbol{q}} - \boldsymbol{h} + \boldsymbol{w}\lambda = 0$$

$$\boldsymbol{w}^T\ddot{\boldsymbol{q}} = 0$$

$$\boldsymbol{M} = \operatorname{diag}\{I_i\}\ ; \quad i = 1, \ldots, n$$

$$\begin{aligned}
\boldsymbol{h} &= (h_1, \ldots, h_n)\\
h_1 &= -(d_1 + d_2)\dot{q}_1 + d_2\dot{q}_2 - (c_1 + c_2)q_1 + c_2 q_2 + T_{\text{in}}\\
h_i &= d_i\dot{q}_{i-1} - (d_i + d_{i+1})\dot{q}_i + d_{i+1}\dot{q}_{i+1}\\
&\quad + c_i q_{i-1} - (c_i + c_{i+1})q_i + c_{i+1}q_{i+1} + T_i(q_i, \dot{q}_i, t)\,;\\
&\quad i = 2, \ldots, n-1\\
h_n &= d_n\dot{q}_{n-1} - d_n\dot{q}_n + c_n q_{n-1} - c_n q_n\\
&\quad + T_n(q_n, \dot{q}_n, t) + T_{\text{out}}\,.
\end{aligned}$$

Before simulating the above model we must have further information on the input and output torques as well as on coupling torques. The input torque will be represented by

$$T_{\text{in}} = k_{\text{in}} + d_{\text{in}}(\omega_{CS} - \omega_{MH}) + \sum_j s_{\text{in},j}\sin(j\omega_{\text{in}}t + \varphi_{\text{in},j}) \tag{13.14}$$

(ω_{CS} is crankshaft speed, ω_{MH} is motor housing speed, ω_{in} is speed of the first motor order, j is number of motor order, k_{in} is constant motor torque, d_{in} is slope of the motor characteristic in the operation point, $s_{\text{in},j}$ is torque amplitude of the jth motor order, $\varphi_{\text{in},j}$ is phase shift.) The output torque is assumed constant because velocities should be small and wind influence will be neglected. Therefore

$$T_{\text{out}} = k_{\text{out}}\,. \tag{13.15}$$

The friction torque characteristic between the clutch disks is approximated by

$$T_C = T_{SL} + (T_{ST} - T_{SL})\exp\left(-\frac{v_r}{k_r}\right) \tag{13.16}$$

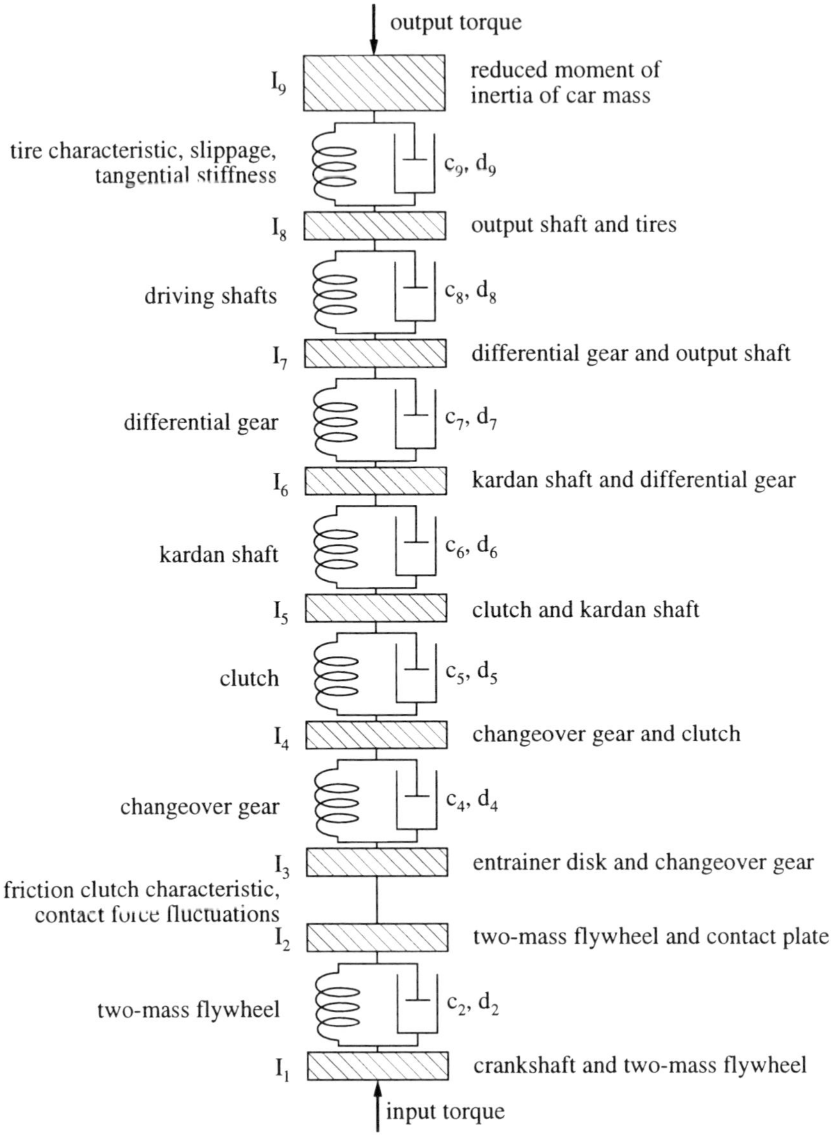

Figure 13.9: Nine-Mass Torsional Model for Friction Clutch Vibrations

(T_{ST} is maximum stiction torque, T_{SL} is sliding torque for $v_r \to \infty$, v_r is relative velocity ($v_r = r_m \omega_r$), k_r is relative nominal velocity.) The torque characteristic is shown in Fig. 13.10 (Stribeck curve).

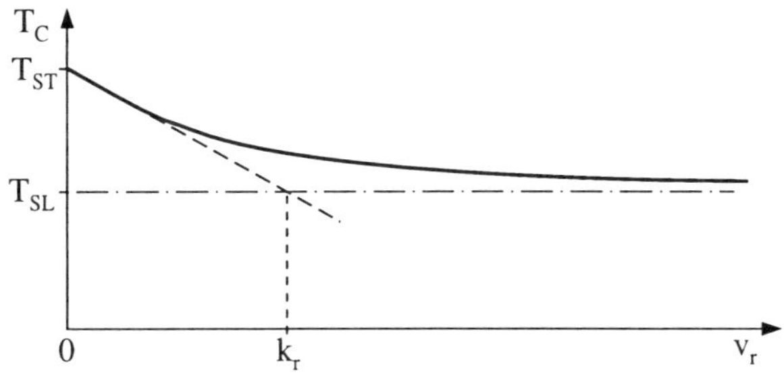

Figure 13.10: Torque Characteristic for Friction Clutch

The torque transmitted in reality by the friction clutch will be additionally influenced by fluctuations of the contact force and a series of other unbalances. They are taken into account by multiplying T_C (eq. 13.16) by some periodic terms:

$$
\begin{aligned}
T_{C,\text{total}} = T_C \Big[\; & 1 + s_{\text{Mot}} \sin \varphi_{\text{Mot}} + s_{CL} \sin(\omega_r t + \varphi_{CL}) \\
& + s_{GR} \sin(\omega_{GI} t) + \sum_{j=1}^{5} s_{CA,j} \sin(j\omega_A t + \varphi_{CA,j}) \Big]
\end{aligned}
\tag{13.17}
$$

(s_{Mot} is torque fluctuation from motor, φ_{Mot} is motor angle, s_{CL} is torque fluctuation due to misalignment, φ_{CL} is phase shift with regard to first motor order, s_{GR} is torque fluctuation at the gearbox entrance, ω_{GI} is angular speed at the gearbox entrance, $s_{CA,j}$ are torque fluctuations due to axial vibrations (jth order), ω_A is frequency of the axial vibration, $\varphi_{CA,j}$ is phase shift.)

Figure 13.11 depicts a typical tire characteristic $F = F(s)$, the force trans-

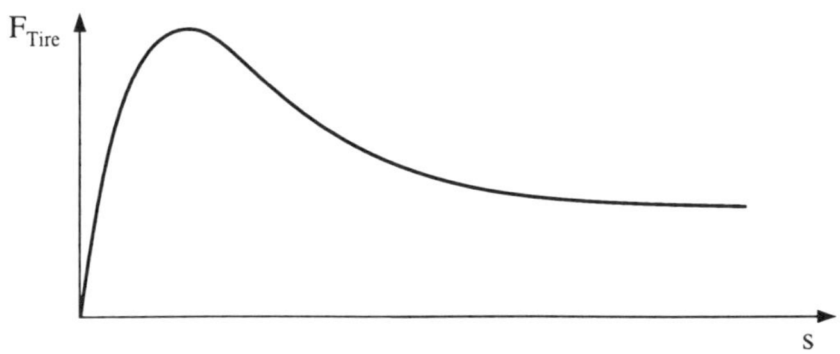

Figure 13.11: Tire Force versus Slippage

mitted dependent on the slippage s. The slippage is defined by

$$s = \frac{r\omega_{\text{Tire}} - v_{\text{Car}}}{v_{\text{Nom.}}} \tag{13.18}$$

with $v_{\text{Nom.}} = \max\left(|r\omega_{\text{Tire}}|, |v_{\text{Car}}|\right)$, ω_{Tire} the angular velocity of the tire, v_{Car} the car velocity, and r the effective tire radius.

If we consider a starting process with small slippage values, we may assume we are in the linear part of the slippage characteristic, which allows some simplifications, namely

$$F_{\text{Tire}} \approx k_s s\,. \tag{13.19}$$

The torque on the tire will be $T_{\text{Tire}} = rF_{\text{Tire}} = rk_s s$, which with definition (13.18) becomes

$$T_{\text{Tire}} = d_{\text{Tire}}\left(\omega_{\text{Tire}} - \frac{v_{\text{Car}}}{r}\right) \tag{13.20}$$

with $d_{\text{Tire}} = r^2 k_s / v_{\text{Nom.}}$.

13.2.3 Results

Many simulations and parameter variations have been carried out in cooperation with a German car manufacturer. Figure 13.12 portrays a typical situation for a six-cylinder engine and an upper-middle-class car. Torque and relative velocity of the friction clutch indicate a nice self-excited oscillation at the beginning, and after 3–4 s these vibrations end. Even three reversals of the relative angular velocity can be detected. The car's acceleration follows these vibrations, which start with a frequency of about 10 Hz and then pass over to a lower frequency of about 5 Hz.

With the above-presented model a series of parameter variations were performed to find out the most sensitive components with respect to these friction-driven vibrations. Sensitivity was considered for an area measure of the car's acceleration. If we draw the two envelope curves at the acceleration of Fig. 13.12, the enclosed area between them is taken as a measure for the vibrations.

As a result we find that only a few parameters are able to influence these vibrations. First of all, the form of the friction clutch characteristic, which means the disk linings (Fig. 13.10). Good behavior can be expected for small differences $(T_{ST} - T_{SL})$ (Fig. 13.10) and for large k_r, both requirements very much limited by possible realization. Another sensitive parameter is the differential gear ratio, which can be explained as an inertia effect, because the car mass is transmitted to the drive train with the square of that gear ratio. These two parameters possess the most influence.

A group of four further parameters influences the frictional vibrations as well, but to a smaller extent, namely the fluctuations s_{Mot}, s_{CL}, $s_{CA,j}$ of

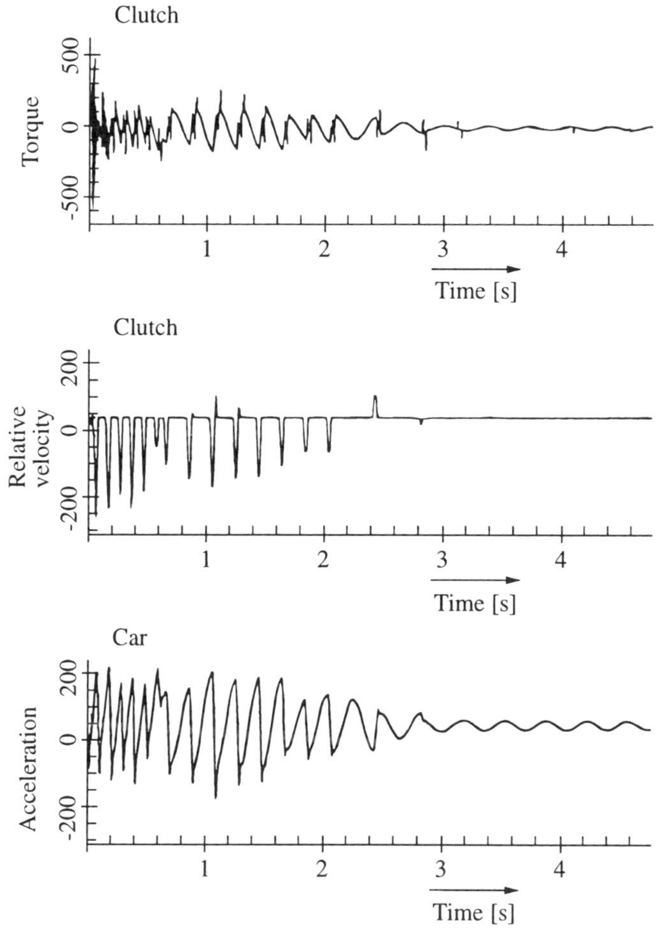

Figure 13.12: Friction Clutch Vibrations

eq. (13.17) and the stiction torque T_{ST} of eq. (13.16). All other parameters do not possess much influence. Therefore, the design of a car's transmission system with robustness against friction clutch vibrations should concentrate on the disk linings and on the possibilities of the differential gear ratio.

14

APPLICATIONS WITH IMPACTS AND FRICTION

14.1 Woodpecker Toy

14.1.1 Introduction

A woodpecker hammering down a pole is a typical system combining impacts, friction and jamming. As a matter of fact all research on unsteady processes in machines started years ago at the authors' institute with a woodpecker analysis [71]. At that time we were not yet able to deal with systems including impacts with friction. Therefore the friction losses were determined experimentally. In the meantime the contents of Chapter 8 offer a sound basis to deal with problems of that kind [25].

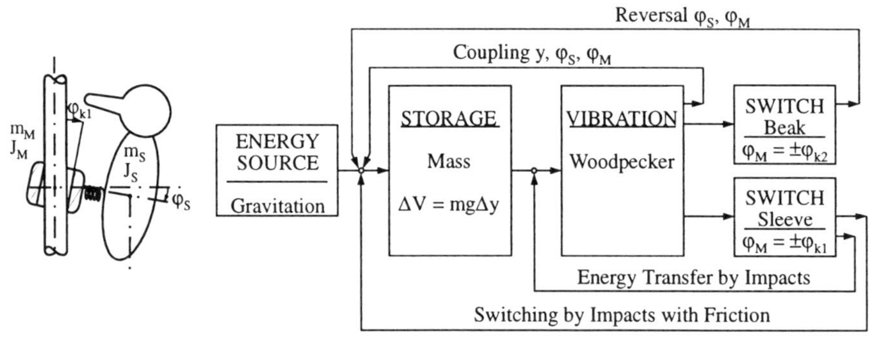

Figure 14.1: Self-Sustained Vibration Mechanism [71]

A woodpecker toy consists of a sleeve, a spring and the woodpecker. The hole of the sleeve is slightly larger than the diameter of the pole, thus allowing a kind of pitching motion interrupted by impacts with friction.

The motion of the woodpecker can be described by a limit cycle behavior as illustrated in Fig. 14.1. The gravitation represents an energy source, the energy of which is transmitted to the woodpecker mass by the y-motion. The woodpecker itself oscillates and possesses a switching function by the beak for quick φ_S reversal and by the jammed sleeve, which transmits energy to the spring by jamming impacts.

A typical sequence of events is portrayed in Fig. 14.2. We start with jamming in a downward postion, moving back again due to the deformation of the spring, and including a transition from one to three degrees of freedom between phases 1 and 2. Step 3 is jamming in an upward position (1 DOF) followed by a beak impact which supports a quick reversal of the φ-motion. Steps 5 to 7 are then equivalent to steps 3 to 1.

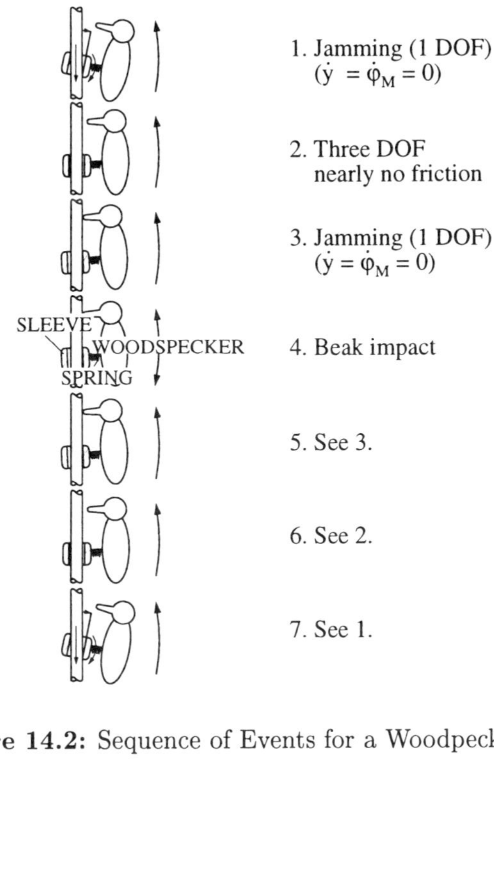

Figure 14.2: Sequence of Events for a Woodpecker Toy

14.1.2 Mechanical and Mathematical Models*

The system possesses three degrees of freedom $\boldsymbol{q} = (y, \varphi_M, \varphi_S)^T$, where φ_S and φ_M are the absolute angles of rotation of the woodpecker and the sleeve, respectively, and y describes the vertical displacement of the sleeve (Fig. 14.3): Horizontal deviations are negligible. The diameter of the hole in the sleeve is slightly larger than the diameter of the pole. Due to the resulting clearance, the lower or upper edge of the sleeve may come into contact with the pole. This is modeled by constraints 2 and 3. Further contact may occur when the beak of the woodpecker hits the pole, which is expressed by constraint 1. The special geometrical design of the toy enables us to assume only small deviations of the displacements. Thus a linearized evaluation of the system's kinematics is sufficient and leads to the dynamical terms and constraint magnitudes listed below. For the dynamics of the woodpecker we apply the theory Chapter 8 for impacts with friction [25], but we assume that no tangential impulses are stored during the impulsive processes ($\Lambda_{TSi} = 0$ in eqs. 8.31 and 8.37). The mass matrix $\boldsymbol{M}$, the force vector $\boldsymbol{h}$ and the constraint vectors $\boldsymbol{w}$ follow from Fig. 14.3 in a straightforward manner. They are

$$\boldsymbol{M} = \begin{pmatrix} (m_S + m_M) & m_S l_M & m_S l_G \\ m_S l_M & (J_M + m_S l_M^2) & m_S l_M l_G \\ m_S l_G & m_S l_M l_G & (J_S + m_S l_G^2) \end{pmatrix}$$

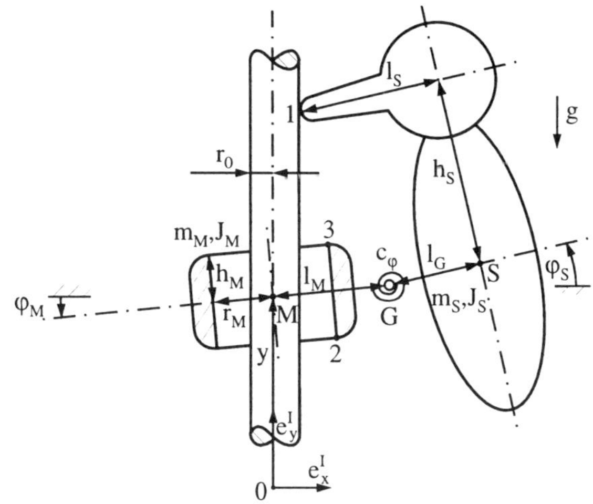

Figure 14.3: Woodpecker Model

*From [25]. Reprinted by permission of Kluwer Academic Publishers.

$$
\begin{aligned}
\boldsymbol{h} &= \begin{pmatrix} -(m_S+m_M)g \\ -c_\varphi(\varphi_M-\varphi_S)-m_S g l_M \\ -c_\varphi(\varphi_S-\varphi_M)-m_S g l_G \end{pmatrix}; \quad \boldsymbol{q} = \begin{pmatrix} y \\ \varphi_M \\ \varphi_S \end{pmatrix} \\
\boldsymbol{w}_{N1} &= \begin{pmatrix} 0 \\ 0 \\ -h_S \end{pmatrix}; \quad \boldsymbol{w}_{N2} = \begin{pmatrix} 0 \\ h_M \\ 0 \end{pmatrix}; \quad \boldsymbol{w}_{N3} = \begin{pmatrix} 0 \\ -h_M \\ 0 \end{pmatrix} \\
\boldsymbol{w}_{T1} &= \begin{pmatrix} 1 \\ l_M \\ l_G-l_S \end{pmatrix}; \quad \boldsymbol{w}_{T2} = \begin{pmatrix} 1 \\ r_M \\ 0 \end{pmatrix}; \quad \boldsymbol{w}_{T3} = \begin{pmatrix} 1 \\ r_M \\ 0 \end{pmatrix}.
\end{aligned}
\tag{14.1}
$$

14.1.3 Results

We consider theoretically and experimentally a woodpecker toy with the following data set:

Dynamics: $m_M = 0.0003$; $J_M = 5.0 \cdot 10^{-9}$; $m_S = 0.0045$; $J_S = 7.0 \cdot 10^{-7}$; $c_\varphi = 0.0056$; $g = 9.81$.
Geometry: $r_0 = 0.0025$; $r_M = 0.0031$; $h_M = 0.0058$; $l_M = 0.010$; $l_G = 0.015$; $h_S = 0.02$; $l_S = 0.0201$.
Contact: $\varepsilon_{N1} = 0.5$; $\varepsilon_{N2} = \varepsilon_{N3} = 0.0$; $\mu_1 = \mu_2 = \mu_3 = 0.3$; $\varepsilon_{T1} = \varepsilon_{T2} = \varepsilon_{T3} = \nu_1 = \nu_2 = \nu_3 = 0.0$.

Using these parameters, the contact angles of the sleeve and the woodpecker result in $|\varphi_M| = 0.1$ rad and $\varphi_S = 0.12$ rad, respectively. Before discussing the dynamical behavior obtained by a numerical simulation, some results from an analytical investigation of the system may be presented.

First, we assume that constraint 2 is sticking. The coordinates φ_M and y then are fixed to certain constant values ($\varphi_M = -0.1$ rad), and the system has only one degree of freedom (φ_S) with an equilibrium position at $\varphi_{S0} = -0.218$ rad. Sticking at that position is only possible if $\mu_2 \geq 0.285$. Such values of μ_2 simultaneously correspond to a locking effect of the system in the sense that no vertical force, acting on the woodpecker's center of mass, could lead to a transition to sliding, however strong it would be. Undamped oscillations around this equilibrium with a frequency of 9.10 Hz influence the contact forces and lead the system to change into another state if the amplitudes are large enough.

The second analytically investigated system state is the unconstrained motion, where each of the distances is positive and hence the system has three degrees of freedom. Besides the fourfold zero eigenvalue which describes the rotational and translational free-body motion, a complex pair of eigenvalues with a frequency of $f = 72.91$ Hz exists. The corresponding part of the eigenvector related to the coordinates $\boldsymbol{q} = (y, \varphi_M, \varphi_S)^T$ is given by $\boldsymbol{u} = (-0.086, 10.7, -1.0)^T$ and shows the ratio of the amplitudes.

The limit cycle of the system, computed by a numerical simulation, is

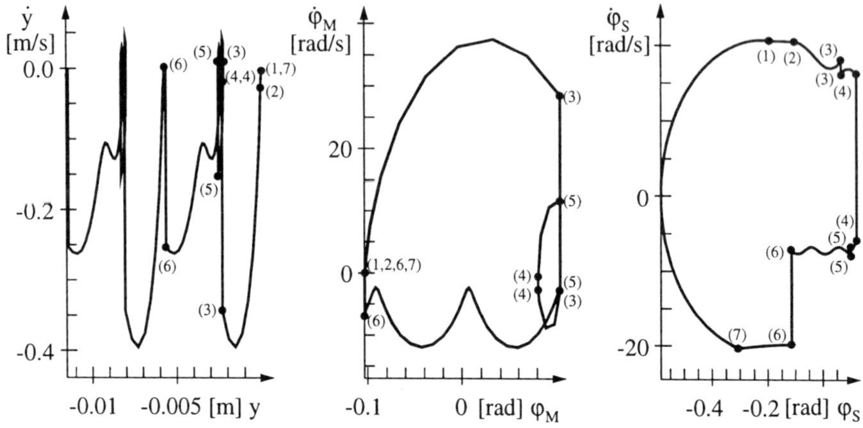

Figure 14.4: Phase Space Portraits

depicted in Fig. 14.4. We start our discussion at point (6) where the lower edge of the sleeve hits the pole. This completely inelastic frictional impact leads to continual contact of the sleeve with the pole. After a short episode of sliding (6)–(7) we observe a transition of the sleeve to sticking (7). The angle of the woodpecker is now large enough to ensure continual sticking of the sleeve by the self-locking mechanism. In that state the system has only one degree of freedom, and the large 9.10-Hz oscillation can be observed where the woodpecker swings down and up until it reaches point (1). At (1) the tangential constraint becomes passive and the sleeve slides up to point (2) where contact is lost. Note that the spring is not free of stresses in this situation; thus during the free-flight phase (2)–(3) the high-frequency oscillation ($f = 72.91$ Hz) of the unbound system occurs in the phase space plots. In this state the sleeve moves downward (y decreases), and the first part of the falling height Δy at one cycle is achieved. At (3) the upper edge of the sleeve hits the pole with a frictional, completely inelastic impact. Contact, however, is not maintained due to the loaded spring. Point (4) corresponds to a partly elastic impact of the beak against the pole. After that collision the velocity $\dot{\varphi}_S$ is negative and the woodpecker starts to swing downward. At (5) the upper edge of the sleeve hits the pole a second time with immediate separation. Then the system is unbound and moving downward (5)–(6), where the second part of the falling height is achieved and the 72.91-Hz frequency can be observed once more.

Note that due to the completely filled mass matrix $\boldsymbol{M}$, an impact in one of the constraints affects each of the coordinates, which can be seen by the velocity jumps in the phase portraits and time histories of Figs. 14.4 and 14.5.

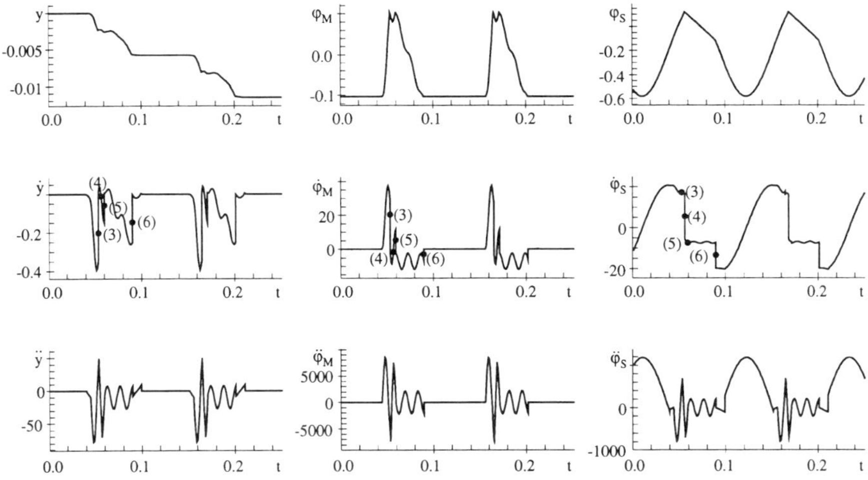

Figure 14.5: Time History of the Coordinates

Table 14.1 summarizes all of the state transitions during one cycle, and Table 14.2 compares the amounts of dissipated energy. The main dissipation results from the beak impact and lower sleeve impact, which contribute 88% of the dissipation. The remaining 12% are shared by the upper sleeve impacts and phases of sliding which are nearly negligible. The frequency of the computed

Phase Plot Point	State Transitions			
(1) Constraint 2	Sticking	→	Sliding	
(2) Constraint 2	Sliding	→	Separation	
(3) Constraint 3	Separation	→	Separation	First upper sleeve impact
(4) Constraint 1	Separation	→	Separation	Beak impact
(5) Constraint 3	Separation	→	Separation	Second upper sleeve impact
(6) Constraint 2	Separation	→	Sliding	Lower sleeve impact
(7) Constraint 2	Sliding	→	Sticking	

Table 14.1: Transitions

Change in Potential Energy	$\Delta V = 2.716 \cdot 10^{-4}$	100.00%
First upper sleeve impact	$\Delta T_{(3)} = -0.223 \cdot 10^{-4}$	8.21%
Second upper sleeve impact	$\Delta T_{(5)} = -0.046 \cdot 10^{-4}$	1.69%
Beak impact	$\Delta T_{(4)} = -1.370 \cdot 10^{-4}$	50.44%
Lower sleeve impact	$\Delta T_{(3)} = -1.032 \cdot 10^{-4}$	38.00%
Phases of sliding	$\Delta T_{(G)} = -0.045 \cdot 10^{-4}$	1.66%

Table 14.2: Amounts of Dissipated Energy

limit cycle in Fig. 14.4 amounts to $f = 8.98$ Hz and is slightly different from the measured value of $f = 9.2$ Hz.

The total falling height during one cycle can be seen in the left diagram of Fig. 14.4. The computed and measured values are $\Delta y = 5.7$ mm and $\Delta y = 5.3$ mm, respectively.

14.2 Drilling Machine*

14.2.1 Introduction

Commercial percussion drilling machines possess for their operation in the percussion drill mode two cam disks with some teeth. By moving the upper disk at the motor side with constant angular speed, for example, and by pressing the drill into the material, a hammering process is started that consists of a combined drill/impact motion. Machines of that type must be designed in such a way that the removal rate in hard materials is optimized while the acceleration load at the handle is minimized. These requirements can be achieved by proper design of the cams especially with respect to the number of cams and their geometric form. The goal of the subsequent investigation is therefore to find the optimum number and shape of the cams [22].

For this purpose we assume that in the contact processes between the two cam disks free-flight phases, impacts and sliding friction might occur. No stick-slip phenomena and no impacts with friction, in the sense of Chapter 8, will take place. On the other hand, we shall include a model of the operator's arm [58], which is important to achieve correct information about the acceleration load on the handle.

*From [22]. Reprinted by permission of Kluwer Academic Publishers.

14.2.2 Mechanical and Mathematical Models

The mechanical model shown in Fig. 14.6 basically is composed of two parts, the percussion drilling machine and a hand-arm model. The drilling machine consists of three rigid bodies: the mass m_0 contains the drill and the first percussion disk (disk 0), body 1 shows the second disk (disk 1) together with the gearbox and the motor, and body 2 corresponds to the flexible mounted handle of the machine.

The displacements of these three bodies are described by the absolute translational coordinates x_0, x_1 and x_2. The motor of the drilling machine rotates with constant angular velocity Ω and activates the drill. The elasticity of the gear unit is taken into account by the spring-damper element (c_φ, d_φ). So the drilling machine possesses two rotational degrees of freedom, where φ_1 describes the inertial angular displacement, and φ_0 denotes the relative angular displacement between motor and drill that results from the elasticity of the gear unit. The two percussion disks with radius r sit on the bodies 0 and 1. Both disks hold cams with similar geometrical forms, which is described by a function $e(\sigma)$ on its outer surface. During operation the cams slip on each

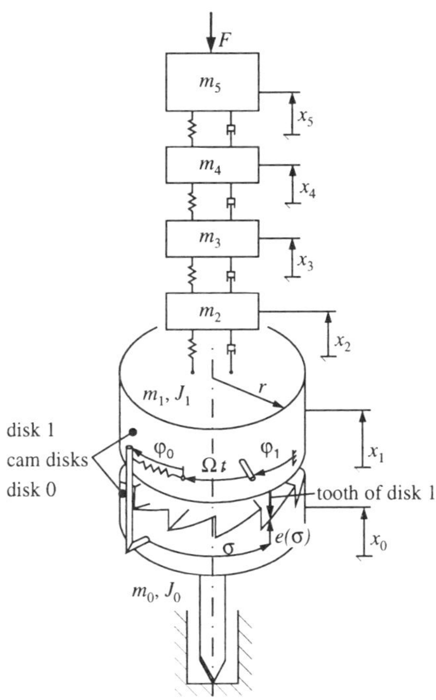

Figure 14.6: Mechanical Model

other until they lose contact on their tops. The following free-flight state of the drilling machine continues until the two disks hit each other again because of the pressing force F. The resulting impact causes the drilling feed.

During the free-flight state the two cam disks are separated and the drill does not touch the environment. In that state the motion of the system is described by a linear differential equation system:

$$\boldsymbol{M}\ddot{\boldsymbol{q}} + \boldsymbol{D}\dot{\boldsymbol{q}} + \boldsymbol{C}\boldsymbol{q} - \boldsymbol{h} = 0 \quad \in \mathbb{R}^8 , \tag{14.2}$$

with $\boldsymbol{q}$ being the set of generalized coordinates specified in Fig. 14.6:

$$\boldsymbol{q} = [x_0\ x_1\ x_2\ x_3\ x_4\ x_5\ \varphi_0\ \varphi_1]^T . \tag{14.3}$$

The matrices $\boldsymbol{M}, \boldsymbol{D}$ and $\boldsymbol{C}$ represent the masses within the system, the dampings, and the stiffness coefficients, respectively. They are obtained in a straightforward manner directly from Fig. 14.6. The vector $\boldsymbol{h}$ contains the pressing force and, together with the remaining terms, is

$$\boldsymbol{M} = \begin{bmatrix} m_0 & 0 & & & & & & \\ 0 & m_1 & 0 & & & & & \\ & 0 & m_2 & 0 & & & & \\ & & 0 & m_3 & 0 & & & \\ & & & 0 & m_4 & 0 & & \\ & & & & 0 & m_5 & 0 & \\ & & & & & 0 & J_0 & J_0 \\ & & & & & & J_0 & J_0 + J_1 \end{bmatrix} , \tag{14.4}$$

$$\boldsymbol{D} = \begin{bmatrix} 0 & 0 & & & & & & \\ 0 & d_1 & -d_1 & & & & & \\ & -d_1 & d_1 + d_2 & -d_2 & & & & \\ & & -d_2 & d_2 + d_3 & -d_3 & & & \\ & & & -d_3 & d_3 + d_4 & -d_4 & & \\ & & & & -d_4 & d_4 & 0 & \\ & & & & & 0 & d_\varphi & 0 \\ & & & & & & 0 & 0 \end{bmatrix} , \tag{14.5}$$

$$\boldsymbol{C} = \begin{bmatrix} 0 & 0 & & & & & & \\ 0 & c_1 & -c_1 & & & & & \\ & -c_1 & c_1 + c_2 & -c_2 & & & & \\ & & -c_2 & c_2 + c_3 & -c_3 & & & \\ & & & -c_3 & c_3 + c_4 & -c_4 & & \\ & & & & -c_4 & c_4 & 0 & \\ & & & & & 0 & c_\varphi & 0 \\ & & & & & & 0 & 0 \end{bmatrix} , \tag{14.6}$$

$$\boldsymbol{h} = [0\ 0\ 0\ 0\ 0\ \ -F\ 0\ 0]^T . \tag{14.7}$$

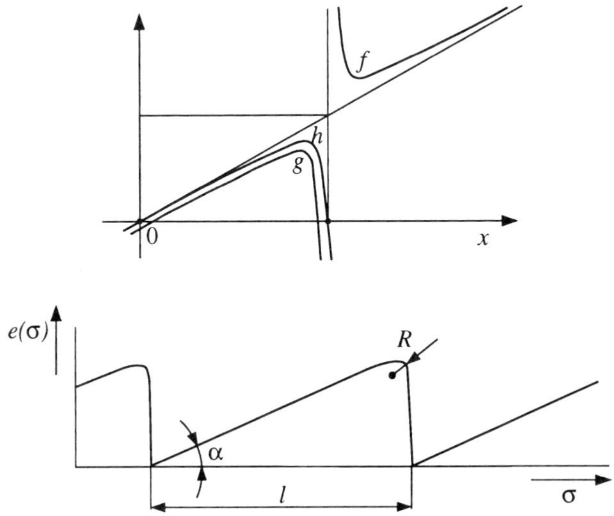

Figure 14.7: Contour Function

Two possible contact points are modeled in order to investigate the impulse transfer from the upper cam disk via the lower cam disk and the tool to the environment. The first contact point is between the cams of the disks; the second one is between the tool and the bottom of the drill hole.

To be able to state the constraint magnitudes of the contact between the cams, we have to describe their contours by a function $e(\sigma)$ which is continuous in its second derivative. This continuity is necessary in order to avoid jumps in the contact forces during the sliding process. In Fig. 14.7 the function

$$f(x) = cx \left(\frac{x}{x-b} \right)^a \tag{14.8}$$

is plotted, which fulfills this requirement. Using the linear transformations

$$f(x) \to g(x) \to h(x) \to e(\sigma) \tag{14.9}$$

we fit $f(x)$ to the searched outline $e(\sigma)$. The parameters a, b and c in (14.8) can be used to determine the length of the cams l, the angle of inclination α and the radius of curvature R at the top of the cams.

The variable σ depends on the generalized coordinate φ_0 and explicitly on the time t due to the rotation of the motor. This can be seen from Fig. 14.6 or, more clearly, from the winding-off of the contact kinematics in Fig. 14.8. With it, the contact parameter σ is easily determined to be

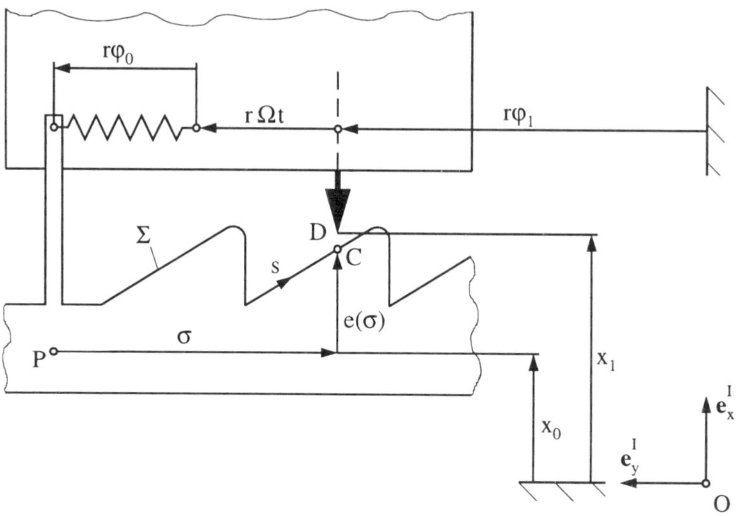

Figure 14.8: Contact Kinematics, Cams

$$\sigma = r(\Omega t + \varphi_0 - k) \tag{14.10}$$

where k takes into account the numbers of cams which have already passed point D during operation. Following Section 4.5 the contour vector, the tangent and the normal vectors are

$$\begin{aligned} {}_I\boldsymbol{r}_{P\Sigma} &= \begin{pmatrix} e(\sigma) \\ -\sigma \\ 0 \end{pmatrix} \quad ; \quad {}_I\boldsymbol{t} = \frac{1}{\sqrt{1+e'^2}} \begin{pmatrix} e'(\sigma) \\ -1 \\ 0 \end{pmatrix} \quad ; \\ {}_I\boldsymbol{n} &= \frac{1}{\sqrt{1+e'^2}} \begin{pmatrix} -1 \\ -e'(\sigma) \\ 0 \end{pmatrix} , \end{aligned} \tag{14.11}$$

thus the angle of inclination $\overline{\alpha}$ at the contact point C is given by the relation

$$\tan\overline{\alpha} = e'(\sigma) \Rightarrow \frac{1}{\sqrt{1+e'^2}} = \cos\overline{\alpha} \,. \tag{14.12}$$

Finally the absolute velocities of the contact points C and D have to be stated. They are straightforwardly derived from Fig. 14.8 and are

$${}_I\boldsymbol{v}_C = \begin{pmatrix} \dot{x}_0 \\ r(\dot{\varphi}_0 + \dot{\varphi}_1 + \Omega) \\ 0 \end{pmatrix} \quad ; \quad {}_I\boldsymbol{v}_D = \begin{pmatrix} \dot{x}_1 \\ r\dot{\varphi}_1 \\ 0 \end{pmatrix} . \tag{14.13}$$

Now we have to state the distance g_{N1} between the contact points. This can be done according to eq. (4.23) or, much easier by the relation

$$g_{N1}^* = x_1 - x_0 - e(\sigma) \tag{14.14}$$

(cp. Fig. 14.8). It should be mentioned that g_{N1}^* does not correspond to the natural distance g_{N1} between the contact points but shows the same properties. If the cams are separated then $g_{N1}^* > 0$ holds, while $g_{N1}^* = 0$ indicates the closed contact. The relative velocities in the normal and tangential directions are derived by means of eq. (4.34):

$$\dot{g}_{N1} = \boldsymbol{n}^T(\boldsymbol{v}_C - \boldsymbol{v}_D) \quad ; \quad \dot{g}_{T1} = \boldsymbol{t}^T(\boldsymbol{v}_C - \boldsymbol{v}_D) \,, \tag{14.15}$$

which leads, after insertion of (14.11), (14.12), (14.13), to the expressions

$$\begin{aligned} \dot{g}_{N1} &= (-\dot{x}_0 + \dot{x}_1 - e' r\dot{\varphi}_0 - e' r\Omega)\cos\overline{\alpha} \\ \dot{g}_{T1} &= (e'\dot{x}_0 - e'\dot{x}_1 - r\dot{\varphi}_0 - r\Omega)\cos\overline{\alpha} \,. \end{aligned} \tag{14.16}$$

The changes of the relative velocities, $\ddot{g}_{N1}$ and $\ddot{g}_{T1}$, follow from a time differentiation of (14.16). For this purpose we denote

$$\begin{aligned} L_N &:= (-\dot{x}_0 + \dot{x}_1 - e' r\dot{\varphi}_0 - e' r\Omega) \\ L_T &:= (e'\dot{x}_0 - e'\dot{x}_1 - r\dot{\varphi}_0 - r\Omega) \end{aligned} \tag{14.17}$$

in order to get

$$\begin{aligned} \ddot{g}_{N1} &= \dot{L}_N \cos\overline{\alpha} + L_N(\cos\overline{\alpha})^{\bullet} \\ \ddot{g}_{T1} &= \dot{L}_T \cos\overline{\alpha} + L_T(\cos\overline{\alpha})^{\bullet} \,. \end{aligned} \tag{14.18}$$

Since the relative accelerations are needed only for detecting the transitions from sticking to sliding and for contact to separation, i.e., for contact states with vanishing relative velocities, eq. (14.18) simplifies under this restriction to

$$\begin{aligned} \ddot{g}_{N1}(\dot{g}_{N1} = 0) &= \dot{L}_N \cos\overline{\alpha} \\ \ddot{g}_{T1}(\dot{g}_{T1} = 0) &= \dot{L}_T \cos\overline{\alpha} \,, \end{aligned} \tag{14.19}$$

where the relations ($\dot{g}_{N1} = 0 \Leftrightarrow L_N = 0$), ($\dot{g}_{T1} = 0 \Leftrightarrow L_T = 0$) have been used. After resubstitution and differentiation of L_N and L_T from (14.17) the accelerations (14.19) become

$$\begin{aligned} \ddot{g}_{N1} &= (-\ddot{x}_0 + \ddot{x}_1 - e' r\ddot{\varphi}_0 - e''\dot{\sigma} r\dot{\varphi}_0 - e''\dot{\sigma} r\Omega)\cos\overline{\alpha} \\ \ddot{g}_{T1} &= (e'\ddot{x}_0 - e'\ddot{x}_1 - r\ddot{\varphi}_0 + e''\dot{\sigma}\dot{x}_0 - e''\dot{\sigma}\dot{x}_1)\cos\overline{\alpha} \end{aligned} \tag{14.20}$$

with $\dot{\sigma}$ being the time derivative of the contact parameter (14.10):

$$\dot{\sigma} = r(\Omega + \dot{\varphi}_0) \,. \tag{14.21}$$

On the other hand, the relative velocities and accelerations are given by eqs. (4.38) and (4.45):

$$\dot{g}_{N1} = \boldsymbol{w}_{N1}^T \dot{\boldsymbol{q}} + \tilde{w}_{N1} \qquad \dot{g}_{T1} = \boldsymbol{w}_{T1}^T \dot{\boldsymbol{q}} + \tilde{w}_{T1} \, ; \tag{14.22}$$

$$\ddot{g}_{N1} = \boldsymbol{w}_{N1}^T \ddot{\boldsymbol{q}} + \overline{w}_{N1} \qquad \ddot{g}_{T1} = \boldsymbol{w}_{T1}^T \ddot{\boldsymbol{q}} + \overline{w}_{T1} \, . \tag{14.23}$$

Comparing (14.16) with (14.22) and (14.20) with (14.23) and concerning the structure of the coordinate vector $\boldsymbol{q}$ (14.3), all the constraint terms are identified to be

$$\begin{aligned} \boldsymbol{w}_{N1}^T &= (-1\;1\;0\;0\;0\;0\;\; -e'r\;\; 0)\cos\overline{\alpha} \\ \boldsymbol{w}_{T1}^T &= (e'\;\; -e'\;0\;0\;0\;0\;\; -r\;\; 0)\cos\overline{\alpha} \\ \tilde{w}_{N1} &= -e'r\Omega\cos\overline{\alpha} \\ \tilde{w}_{T1} &= -r\Omega\cos\overline{\alpha} \\ \overline{w}_{N1} &= -e''r^2(\Omega+\dot{\varphi}_0)^2\cos\overline{\alpha} \\ \overline{w}_{T1} &= e''r(\Omega+\dot{\varphi}_0)(\dot{x}_0-\dot{x}_1)\cos\overline{\alpha} \, . \end{aligned} \tag{14.24}$$

The frictionless contact constraint between the drill and the environment is trivial. From Fig. 14.6 we get immediately

$$\begin{aligned} g_{N0} &= x_0 \geq 0 \; ; \quad \tilde{w}_{N0} = \overline{w}_{N0} = 0 \, ; \\ \boldsymbol{w}_{N0}^T &= (1\;0\;0\;0\;0\;0\;0\;0) \, . \end{aligned} \tag{14.25}$$

Equations (14.2)–(14.7) completely describe the free-flight state of the system where the cam disks are separated and the drill is not in contact with the environment. Due to the pressing force the cam disks move closer and closer during operation until they hit each other. This event is modeled by a cascade of two completely inelastic impacts in succession. During the first collision cam disk 1 hits disk 0, which stands idle in the axial direction (see Fig. 14.9). Only constraint 1 is taken into account in the impact equations, which are solved by using Newton's impact law with $\varepsilon_{N1} = 0$ (see Chapter 7). After this impact both disks are in contact and are moving in the axial direction toward the bottom of the drill hole. There the second impact takes place. The velocities shortly after the first impact and just before the second are the same. Again the second impact is modeled inelastically, but now both constraints (0 and 1) are considered in the evaluation process with $\varepsilon_{N1} = \varepsilon_{N0} = 0$. As a result we get the relative velocities after that impact which are equal to zero and the transferred impulses $\boldsymbol{\Lambda}_N$ of eq. (7.11), Section 7.2. The removal rate V_S per impact which is achieved by percussion drilling is obtained proportionally to the second power of the impulse transferred to the drill hole:

$$\boldsymbol{\Lambda}_N = \begin{pmatrix} \Lambda_{N1} \\ \Lambda_{N0} \end{pmatrix} \; ; \quad V_S = k\Lambda_{N0}^2 \; ; \quad k = \text{const.} \tag{14.26}$$

With that we attain the removal rate per time unit by averaging the impacts which take place during the simulation time.

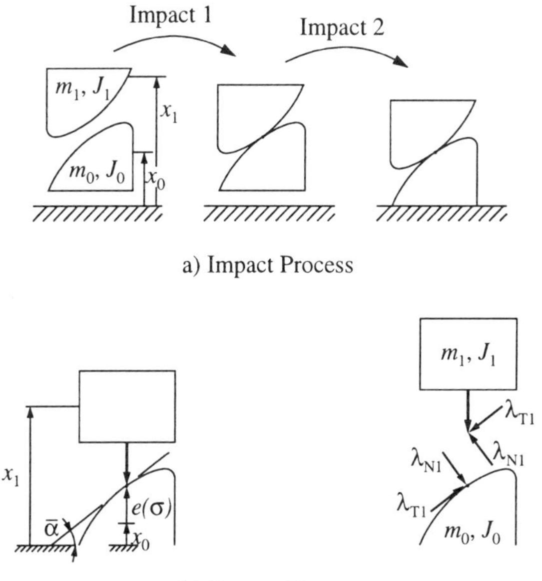

Figure 14.9: Contact Dynamics

When the impact process has ended, the frictional contact between the cams may open due to the force regime or may remain closed. The second case turned out to show only sliding without sticking. Thus the friction force λ_{T1} can be expressed by their normal force λ_{N1} with Coulomb's law for sliding, which results together with the frictionless unilateral contact constraint of the drill (index 0) in the expression $(\boldsymbol{W}_N + \boldsymbol{W}_G\overline{\overline{\boldsymbol{\mu}}}_G)\boldsymbol{\lambda}_N$ of eq. (5.11). Additionally we have $n_N = 2$, $n_H = 0$, $\boldsymbol{W}_H = \boldsymbol{\lambda}_H = 0$. Solving the underlying linear equation system (5.17), (5.18) enables us to determine the unknown accelerations $\ddot{\boldsymbol{q}}$ as well as the contact forces λ_{N1}, λ_{T1}, λ_{N0} which are depicted for the cam's contact in Fig. 14.9. The transitions between the contact states of the cams are evaluated by supervising certain kinematic and kinetic indicator magnitudes which are the distance indicator g^*_{N1} form (14.14), the normal contact force λ_{N1} in Fig. 14.9, and the relative normal acceleration $\ddot{g}_{N1}$ from (14.23).

Figure 14.10 shows all possible transitions of the cam's contact. The free-flight state Z_1 is valid as long as the distance between the disks, indicated by g^*_{N1}, is greater than zero. When this distance disappears, $g^*_{N1} = 0$, the two inelastic impacts Z_2 take place. Impacts lead to jumps in the velocities; therefore the normal force λ_{N1} may contain a discontinuity at the moment of impact, too. Thus, two transitions after the impact are possible: For a

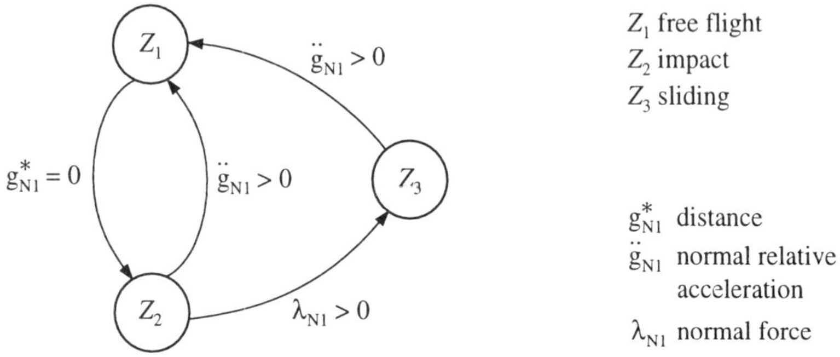

Figure 14.10: Transitions in the Cam's Contact

positive normal relative acceleration ($\ddot{g}_{N1} > 0$) the system returns to the free-flight state Z_1, and for a positive normal force ($\lambda_{N1} > 0$) sliding Z_3 takes place. During sliding the normal force decreases continuously to zero. Then the two disks lose contact and again the state free-flight Z_1 is valid.

14.2.3 Results

Figure 14.11 shows the behavior of the machine for percussion drilling in concrete. The removal rate is plotted as a function of the rotational speed

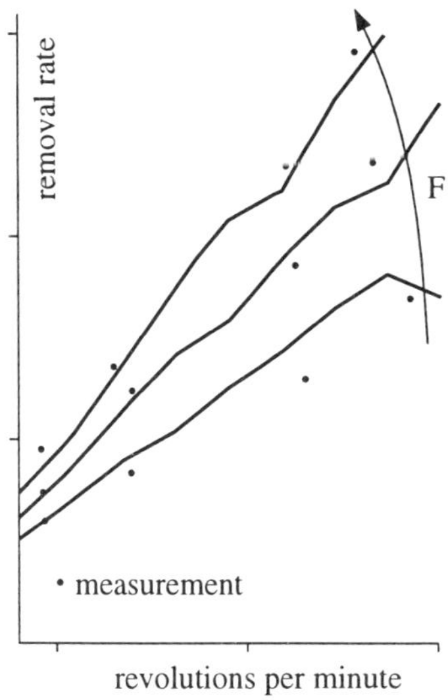

Figure 14.11: Drilling in Concrete

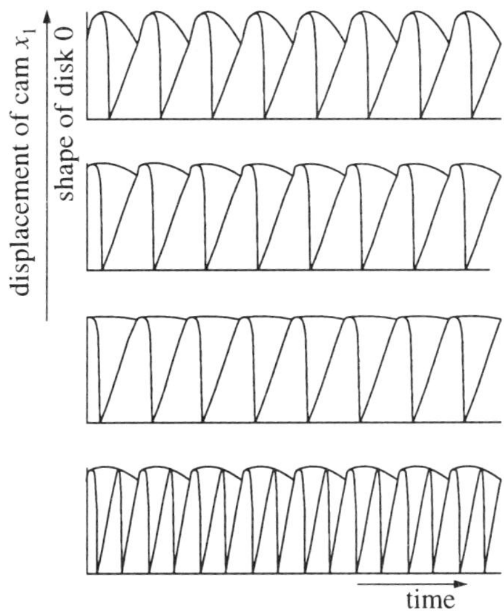

Figure 14.12: Time Behavior of Displacement x_1

of the motor for different pressing forces. We can see that the removal rate grows with increasing speed of rotation and increasing pressing forces. This can be explained by the fact that, on the one hand, the percussion frequency and, on the other hand, the impulse transferred by the impact grow. The lines result from calculation and are confirmed by measurements, represented by the dots.

In Fig. 14.12 the shape of disk 0 and the time behavior of the coordinate x_1 which corresponds to the axial displacement of disk 1 is plotted for different rotational speeds. The kinks and the different smoothing out on the top of the cams result from the arising impacts and effects of dry friction. With an increasing number of revolutions the contact point of the cams moves closer to the top, until a state is reached where the cams jump from one top to the next. With a further increase of the rotational speed only every second cam is hit. This transition is shown in the phase space portraits, Fig. 14.13. It should be kept in mind that mechanical systems with impulsive and stick-slip processes are highly nonlinear due to unsteady transitions and changing of degrees of freedom. Therefore, such systems may be called structure variant. In Fig. 14.13 the first two pictures correspond to the case that every cam of the disks is hit. It is easy to make out the sliding and the free-flight state, followed by an impact. By increasing the rotational speed, picture (b), the hitting point of the cams moves upward and the limit cycle gets smaller. The maximal altitude of x_1 in the free-flight state, however, remains nearly constant. Further increasing of the rotational speed leads to a bifurcation of

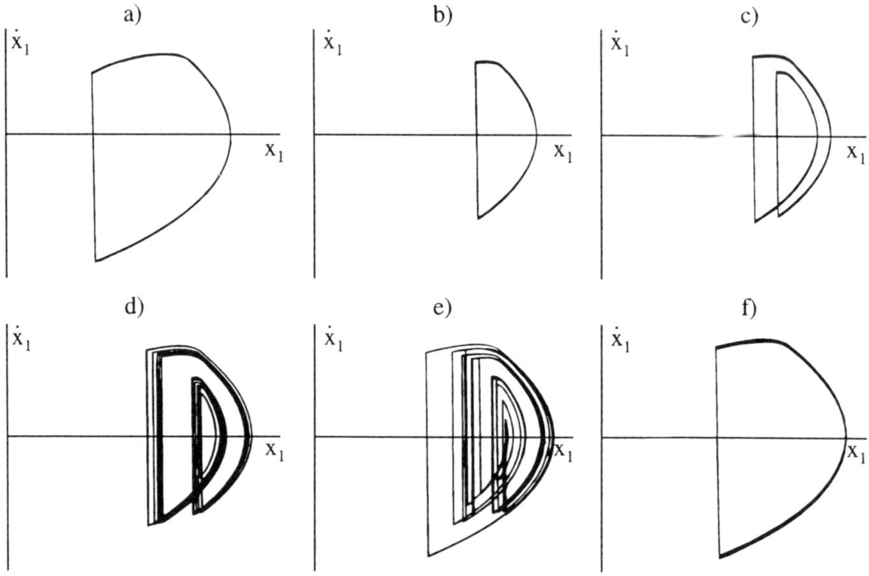

Figure 14.13: Phase Space Portrait

the solution. Picture (c) shows a two-periodic solution. The cams are hit at two different points which are alternately closer to the top and closer to the ground of the cams. If the rotational speed rises even more, the solution bifurcates again by period doubling. Picture (d) shows a periodic solution of higher order which becomes irregular in picture (e) and finally changes in picture (f) to the solution where only every second cam is hit. The same behavior is observed at transitions from the second cam hit to the third one.

The aim of this examination is to analyze the influence of the geometric design of the disks on the removal rate and handle vibrations, where the removal rate was calculated as a function of the impulses (eq. 14.26), and the handle vibrations were determined proportional to the root-mean-square value of the Fourier coefficients of the handle accelerations. The angle of inclination α and the number of cams of one disk were chosen as design parameters. Figure 14.14 shows the result of this variation for parameter intervals where only that solution exists where every cam is hit. With an increasing number of cams the hitting point of the cam disks moves toward the top of the cams. Thereby the frequency of impact increases and the handle amplitude decreases, which is shown by an increasing removal rate and decreasing handle vibrations. If the angle of inclination is enlarged, a more powerful impulse in axial direction takes place. The result is a better removal rate, but also slightly higher handle vibrations.

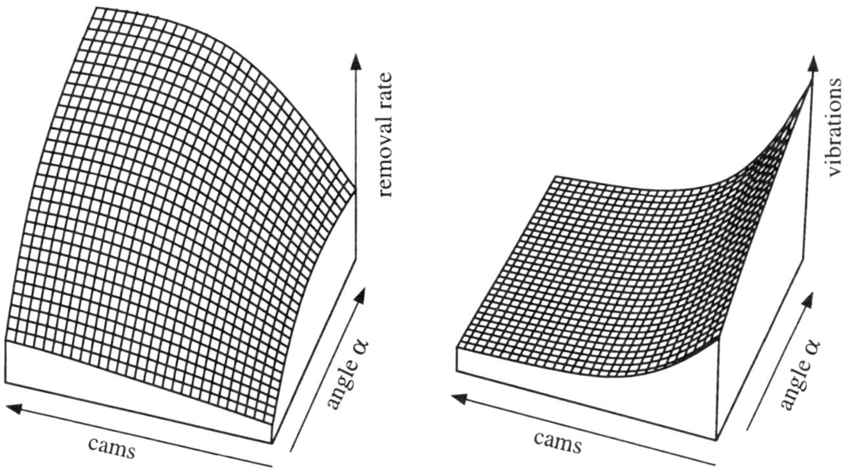

Figure 14.14: Removal Rate and Handle Vibrations

14.3 Electropneumatic Drilling Machine

14.3.1 Introduction

The design of a professional drilling machine must follow two guidelines. First, the removal rate in hard material like concrete should be optimal; second, the vibration loads at the handle should be kept below the acceleration loads as defined by the usual work regulations. Considerations of that kind are the starting point for analyzing the vibrations of drilling machines more perfectly while providing some means to study parameter influences in a quicker and broader way [92].

We shall consider electropneumatic drilling machines, and, similar to the percussion machines in Section 14.2, we shall focus our attention on an optimization of both the removal rate and the vibration load at the handle. Therefore we include in our investigations a man's arm model. Electropneumatic drilling machines operate in a completely different manner than percussion drilling machines. In addition they are characterized by numerous impact and friction contacts which realistically can be modeled only by the methods presented in Part 1.

The operating principle of an electropneumatic drilling machine may be seen from Fig. 14.15. The drilling machine consists of five bodies: gearbox (1) including handle, motor and driving piston; primary piston (2); secondary piston (3); tool (4); and pressure cylinder (5), including tool guide. The motion of each body is described by one translational degree of freedom. The

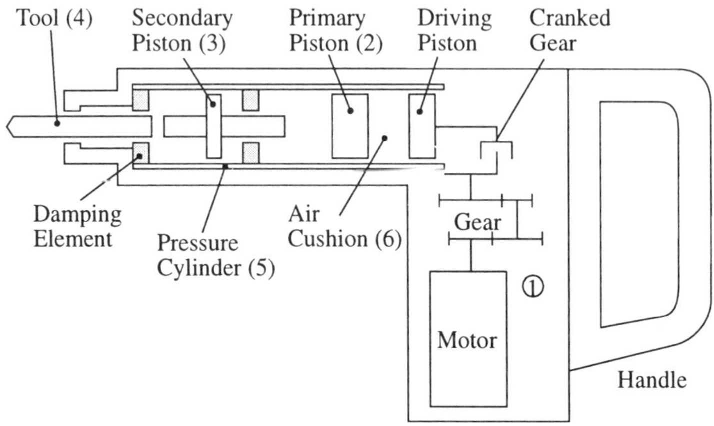

Figure 14.15: Basic Components of an Electropneumatic Drilling Machine

constant rotary motion of the motor is transformed via a crank mechanism to a given sinusoidal translational motion of the driving piston that excites the primary piston through compression and decompression of the enclosed air cushion (6). Thus, the primary piston is supplied with energy with the cycle of crank rotation and partially transfers its kinetic energy through a sequence of impacts to the secondary piston and the tool. The motion of the secondary piston is restrained in one direction by a backstop, and for assembly reasons there is backlash in the coupling connecting bodies (5) and (1). Finally, in all axial guides there is dry friction resulting from sealing rings. In addition to the hysteresis characteristic for the tool-environment model, the unsteady force elements are characterized by two rigid contacts, three elastic contacts and five frictional guides.

14.3.2 Mechanical and Mathematical Models

The first step for modeling machines or mechanisms as multibody systems consists of the definition of their degrees of freedom and existing force characteristics. To this extent we classify the system to consist of ideal rigid joints, rigid joints with backlash, rigid joints with friction, and force couplings with given force-time, force-distance or force-velocity characteristics which may be discontinuous. The basic idea for the general description of such systems then starts from a set of minimum generalized coordinates $\boldsymbol{q} \in \mathbb{R}^f$ that describe the kinematics of the system when it is only subjected to the subset of ideal rigid joints. In this stage the mechanical model consists in general of several individual subsystems.

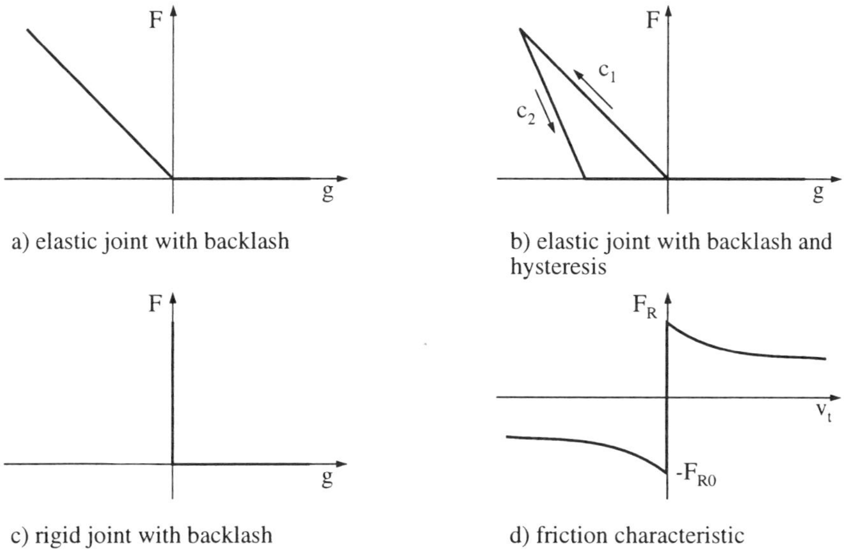

Figure 14.16: Nonlinear Force Characteristics Related to Contact Problems

In the next step we need to specify the forces acting between the bodies. The nonlinear force characteristics related to contact problems in multibody systems can be organized into three classes:

- Elastic joints with backlash (Fig. 14.16a)
  For elastic joints with backlash the transferred forces are determined by the local deformations $g(\boldsymbol{q})$. As long as there is no contact $g > 0$ the transferred force must vanish, whereas there is a linear or nonlinear dependency on the deformation $g < 0$ for two gearing bodies. A modified nonlinear force-distance graph with hysteresis is shown in Fig. 14.16(b) where the force also depends on the phases of compression and decompression, respectively.

- Rigid joints with backlash (Fig. 14.16c)
  If contact deformations can be neglected, there are unilaterally acting constraints $g(\boldsymbol{q}) \geq 0$. The joint forces for contacting bodies $g = 0$ can be calculated by the additional requirement that the relative acceleration $\ddot{g} = 0$ must vanish, which yields algebraic constraints for the solution space of the equations of motion. Thus, the mechanical model becomes structure variant since the number of constraints and, equivalently, the number of degrees of freedom may change with time.

- Frictional guides (Fig. 14.16d)
  According to the character of the reaction forces we need to distinguish

sliding friction and static friction. If sliding friction is present, $v_t \neq 0$, there is an additional active force $F_R(v_t)$ which is directed opposite to the relative motion. For static friction a passive force acts on the system dynamics that correlates to the kinematical constraint $v_t = 0$. The passive force is further limited by the maximum static friction force $|F_R| \leq F_{R0}$. For simplicity we consider frictional guides where F_{R0} can be assumed to be independent of the normal contact force F_N.

For the case under consideration the equations of Chapters 6 and 7 apply, particularly eqs. (6.1), (6.8), (6.36) and eqs. (7.11), (7.12). The relevant vectors and matrices follow from Fig. 14.17 in a straightforward way:

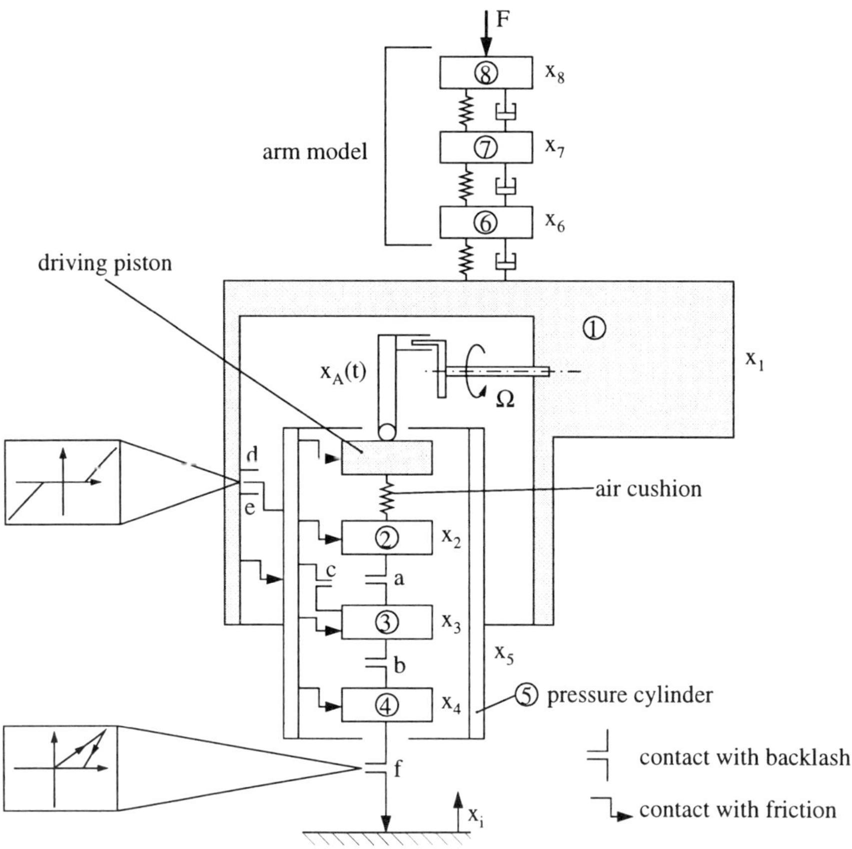

Figure 14.17: Mech. Model of an Electropneumatic Drilling Machine [92]

- Generalized coordinates:

$$\boldsymbol{q} = (x_1\ x_2\ x_3\ x_4\ x_5\ x_6\ x_7\ x_8)^T \tag{14.27}$$

- Mass matrix:

$$\boldsymbol{M} = \operatorname{diag}\{m_i\}\ ;\ \ i = 1, \ldots, 8 \tag{14.28}$$

- Smooth part $\boldsymbol{h}_1$ of applied forces $\boldsymbol{h} = \boldsymbol{h}_1 + \boldsymbol{h}_2$:

$$\mathbf{h}_1 = \begin{bmatrix} c_1(x_6 - x_1) + d_1(\dot{x}_6 - \dot{x}_1) - d_5(\dot{x}_1 + \dot{x}_A - \dot{x}_5) \\ -m_A\ddot{x}_A + F_A(x_A,\, x_2) \\ -d_5(\dot{x}_2 - \dot{x}_5) - F_A(x_A,\, x_2) \\ 0 \\ 0 \\ d_5(\dot{x}_1 + \dot{x}_A - \dot{x}_5) + d_5(\dot{x}_2 - \dot{x}_5) \\ c_2(x_7 - x_6) + d_2(\dot{x}_7 - \dot{x}_6) - c_1(x_6 - x_1) - d_1(\dot{x}_6 - \dot{x}_1) \\ c_3(x_8 - x_7) + d_3(\dot{x}_8 - \dot{x}_7) - c_2(x_7 - x_6) - d_2(\dot{x}_7 - \dot{x}_6) \\ -F - c_3(x_8 - x_7) - d_3(\dot{x}_8 - \dot{x}_7) \end{bmatrix} \tag{14.29}$$

c_i : Stiffness coefficients

d_i : Damping coefficients

$x_A(t)$: Kinematic excitation from driving unit

m_A : Condensed mass of driving unit

F_A : Force from adiabatic air cushion

F : Pressing force

- Nonsmooth part $\boldsymbol{h}_2$ of applied forces $\boldsymbol{h} = \boldsymbol{h}_1 + \boldsymbol{h}_2$:

$$\mathbf{h}_2 = \begin{bmatrix} 0 & 1 & -1 & 0 \\ 0 & 0 & 0 & 0 \\ -1 & 0 & 0 & 0 \\ 0 & 0 & 0 & 1 \\ 1 & -1 & 1 & 0 \\ 0 & 0 & 0 & 0 \\ 0 & 0 & 0 & 0 \\ 0 & 0 & 0 & 0 \end{bmatrix} \cdot \begin{bmatrix} F_c \\ F_d \\ F_e \\ F_f \end{bmatrix}\ ; \qquad \begin{aligned} g_c &= x_5 - x_3 \\ g_d &= x_1 - x_5 \\ g_e &= x_5 - x_1 + \varepsilon \\ g_f &= x_4 \end{aligned} \tag{14.30}$$

$$F_i(g_i, \dot{g}_i) = \begin{cases} -c_i g_i - d_i \dot{g}_i & : g_i < 0, \ -c_i g_i - d_i \dot{g}_i > 0 \\ 0 & : \text{else} \end{cases} \quad ; \quad i = c, d, e$$

$$F_f(g_f, \dot{g}_f) = \begin{cases} -c_{f1} g_f - d_{f1} \dot{g}_f & : g_f < 0, \ \dot{g}_f < 0 \\ -c_{f2}(g_f + a) - d_{f2} \dot{g}_f & : g_f + a < 0, \ \dot{g}_f > 0 \,, \\ & \quad -c_{f2}(g_f + a) - d_{f2} \dot{g}_f > 0 \\ 0 & : \text{else} \end{cases}$$

c_i : Stiffness coefficients

d_i : Damping coefficients

Subscript c : Damping element between pressure cylinder (5) and secondary piston (3)

Subscript d, e : Force element with backlash ε between casing (1) and pressure cylinder (5)

Subscript f : Hysteresis characteristic between tool (4) and environment

- Unilateral frictionless contact-impact constraints:

$$\begin{aligned} g_{Na} &= x_2 - x_3 \geq 0 \,; \quad \tilde{w}_{Na} = \overline{w}_{Na} = 0 \\ g_{Nb} &= x_3 - x_4 \geq 0 \,; \quad \tilde{w}_{Nb} = \overline{w}_{Nb} = 0 \\ \boldsymbol{w}_{Na} &= (0 \ 1 \ -1 \ 0 \ 0 \ 0 \ 0 \ 0)^T \\ \boldsymbol{w}_{Nb} &= (0 \ 0 \ 1 \ -1 \ 0 \ 0 \ 0 \ 0)^T \end{aligned} \tag{14.31}$$

Subscript a : Contact constraint between primary piston (2) and secondary piston (3)

Subscript b : Contact constraint between secondary piston(3) and tool (4)

- Coulomb friction elements with given normal forces:

$$\begin{aligned} \dot{g}_{T1} &= \dot{x}_4 - \dot{x}_5 & &; \ \tilde{w}_{T1} = \overline{w}_{T1} = 0 \\ \dot{g}_{T2} &= \dot{x}_3 - \dot{x}_5 & &; \ \tilde{w}_{T2} = \overline{w}_{T2} = 0 \\ \dot{g}_{T3} &= \dot{x}_2 - \dot{x}_5 & &; \ \tilde{w}_{T3} = \overline{w}_{T3} = 0 \\ \dot{g}_{T4} &= \dot{x}_1 + \dot{x}_A(t) - \dot{x}_5 & &; \ \tilde{w}_{T4} = \dot{x}_A(t); \ \overline{w}_{T4} = \ddot{x}_A(t) \\ \dot{g}_{T5} &= \dot{x}_1 - \dot{x}_5 & &; \ \tilde{w}_{T5} = \overline{w}_{T5} = 0 \end{aligned} \tag{14.32}$$

$$\boldsymbol{w}_{T1} = (0\ 0\ 0\ 1\ \ -1\ 0\ 0\ 0)$$

$$\boldsymbol{w}_{T2} = (0\ 0\ 1\ 0\ \ -1\ 0\ 0\ 0)$$

$$\boldsymbol{w}_{T3} = (0\ 1\ 0\ 0\ \ -1\ 0\ 0\ 0)$$

$$\boldsymbol{w}_{T4} = (1\ 0\ 0\ 0\ \ -1\ 0\ 0\ 0)$$

$$\boldsymbol{w}_{T5} = (1\ 0\ 0\ 0\ \ -1\ 0\ 0\ 0)$$

Subscript 1 : Coulomb friction between tool (4) and pressure cylinder (5)
Subscript 2 : Coulomb friction between secondary piston (3) and pressure cylinder (5)
Subscript 3 : Coulomb friction between primary piston (2) and pressure cylinder (5)
Subscript 4 : Coulomb friction between driving piston and pressure cylinder (5)
Subscript 5 : Coulomb friction between pressure cylinder (5) and casing (1)

We have a total of five masses of the machine and three masses of the arm-hand model, each mass with one degree of freedom. In addition the machine possesses five contacts with backlash and thus with impulsive motion, and another five contacts with friction and thus with stick-slip phenomena. The numerical solution of the equations of motion follows the rules presented in Chapters 6 and 7, which include a rather costly formulation as a linear complementarity problem.

14.3.3 Simulations

The simulations were performed for an electropneumatic drilling machine of a southern European manufacturer. Two goals could be realized: a reduction of the vibration load on the handle and an improvement of the machine's operation, particularly with respect to the removal rate. In the following we present some typical results.

Figure 14.18 portrays a characteristic motion of the most important machine components. The curves describe the absolute changes of position, and the penetration of secondary piston and damping element indicate compression of the springs. The sequence of impacts is as follows. The primary piston pushes the secondary piston, which hits the tool and bounces back. It then pushes the primary piston into the reverse direction and transmits an upward velocity to it. On the other side the tool pushes into the material and is thrown back with about 50% of its energy. It again hits the secondary mass, which transmits its energy partially to the damping element.

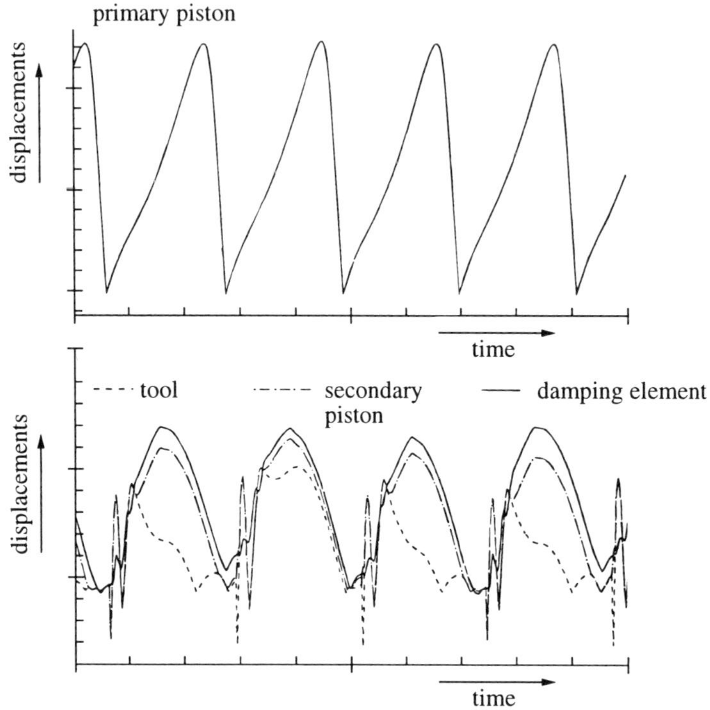

Figure 14.18: Motion of the Machine Components

Within the free-motion range of the damping element several short impacts between tool and secondary piston will take place which cannot be seen in Fig. 14.18 due to a rough time resolution. By the jarring blow of tool and secondary piston the machine housing performs an upward motion before it will be pressed down again by the arm force. This sequence of events is confirmed by experience.

As a second result we compare simulations and measurements for the machine housing. The upper diagram of Fig. 14.19 portrays the acceleration time history of the handle, which is proportional to the sum of forces applied to the gearbox. Five cycles of crank rotations are shown. The thin graph corresponds to the model shown in Fig. 14.17, and positive values represent an acceleration toward the hand-arm system. In the time history the amplitudes are dominated by the occurrence of maximum air pressure and the kickback reaction of the tool and secondary piston in the backstop (c). This causes body (5) to bounce between the stops (d) and (e), which gives rise to acceleration peaks of the gearbox in both directions.

Figure 14.19 also depicts results derived from a model without backlash in

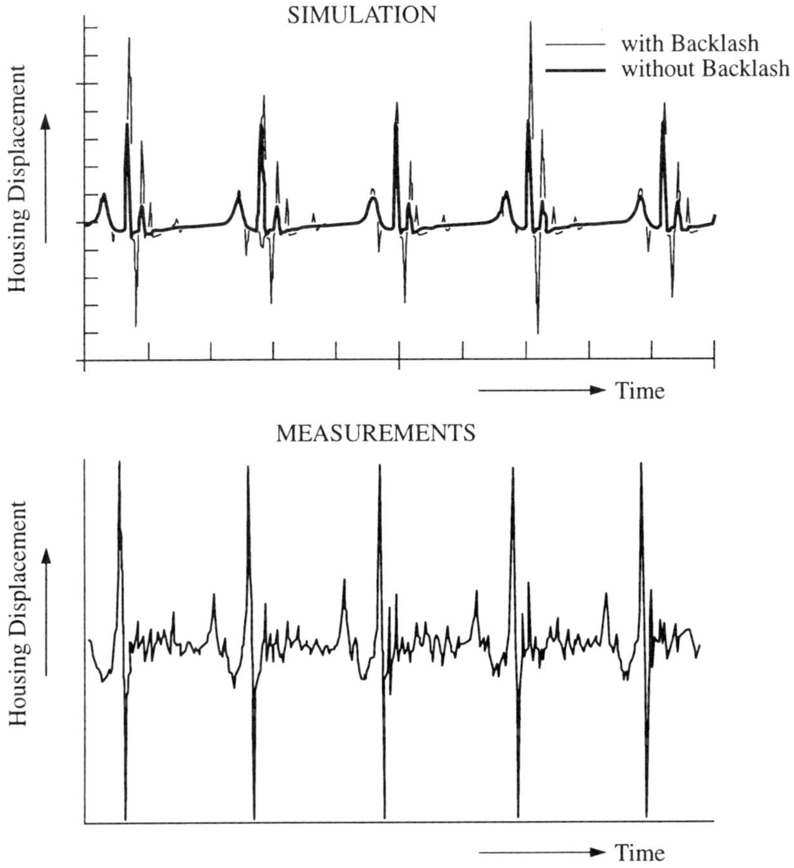

Figure 14.19: Comparison of Simulations and Measurements

the coupling between bodies (1) and (5). The comparison shows that dynamic loads in the presence of backlash produce excessive joint forces which, on the one hand, result in high-frequency vibrations of the handle and, on the other hand, cause accelerated wear of the stops. In the lower diagram of Fig. 14.19 we have contrasted experimental results with our numerical data which show good correspondence to our model. This correlation proves that it is essential to accurately model unsteady force characteristics inherent to machine dynamics in order to get a better insight to real system behavior.

14.4 Landing Gear Dynamics

14.4.1 Introduction

The dynamics of a landing airplane includes some unsteady aspects not only with regard to the external landing impact but also with respect to some internal carriage properties. Large forces are generated by wheel spin-up and, at least as a possibility, by impacts and stick-slip phenomena within the shock absorbers [9, 85, 98].

As the transfer element between the runway and the airplane, the landing gear greatly influences aircraft ground behavior. While the dynamical response of an airplane on a rough and uneven runway can be examined by a linear model, determining the forces in the landing gear itself and the process of wheel spin-up require a nonlinear system description. Consideration of hard impact landings especially requires a more accurate nonlinear model. In addition, the airplane and the carriage must be considered as elastic components because otherwise the forces transmitted to a rigid structure would be too large. Therefore we obtain a stucture-variant model including impacts and stick-slip processes and assuming an elastic airplane, at least approximately. On the other hand, we shall restrict ourselves to a two-dimensional model.

14.4.2 Models

MECHANICAL MODEL

The model consists of the following rigid and elastic bodies:

$i = 1 \rightarrow$ elastic fuselage,
$i = 2 \rightarrow$ elastic main carriage,
$i = 3 \rightarrow$ elastic nose carriage,
$i = 4 \rightarrow$ rigid main carriage wheels,
$i = 5 \rightarrow$ rigid nose carriage wheels.

The principal configuration is depicted in Fig. 14.20. For the shock absorber it will be assumed that their bending deformation is homogeneous over the variable length, which means that the two components of the absorber do not bend in a different way. It turns out that this will be a sufficient approximation. The details of the shock absorber will be presented later.

During operation we have to consider various continuous and discontinuous nonlinearities. The tires possess a stiffness characteristic which increases progressively with the tire deformation. Moreover, the spin-up process of the wheels is governed by a highly nonlinear friction-slippage relationship. The shock absorber itself reveals some nonlinear features. The two gas chambers

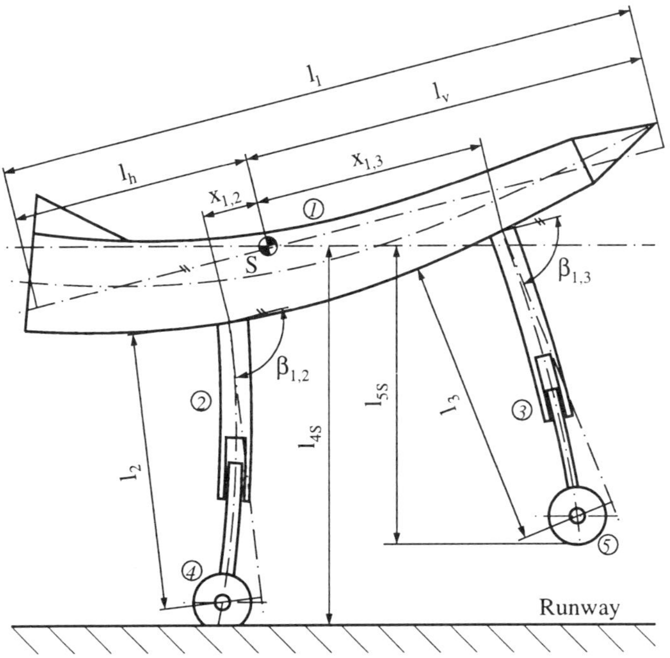

Figure 14.20: Planar Mechanical Model of a Landing Elastic Airplane [9]

follow a polytropic compression and expansion law, the fluid flow losses induce a quadratic damping behavior, and the friction forces from ground reaction forces are nonlinear also.

Discontinuous nonlinearities enter through the upper and lower stops within the shock absorber and through stick-slip processes between the two shock absorber cylinders (Fig. 14.25). Mechanically and mathematically they follow the theory as presented in Part 1.

Equations of Motion

For deriving the equations of motion we introduce the coordinate systems as defined in Fig. 14.21. The degrees of freedom resulting from this model include rigid degrees of freedom (number f_r) and elastic degrees of freedom (number f_e). With $\boldsymbol{q}_r \in \mathbb{R}^{f_r}$ and $\boldsymbol{q}_e \in \mathbb{R}^{f_e}$ we obtain

$$\begin{aligned} \boldsymbol{q}^T &= [\boldsymbol{q}_r^T, \boldsymbol{q}_e^T] \\ &= [x, y, \alpha, y_{2,2}, y_{2,3}, \varphi_2, \varphi_3, \boldsymbol{q}_1^T, \boldsymbol{q}_2^T, \boldsymbol{q}_3^T] \in \mathbb{R}^{f=f_r+f_e} \end{aligned} \tag{14.33}$$

$$\begin{aligned} \boldsymbol{q}_i^T &= [q_{i1}, q_{i2}, \ldots, q_{in_{ai}}] \; , \quad i = 1,2,3 \\ n_{ai} &: \text{ number of shape functions for body } i \\ f &: \text{ number of minimal coordinates.} \end{aligned}$$

For establishing the equations of motion (Section 3.1) we first have to evaluate the absolute velocities of each mass element dm_i in each body i, preferably in a body-fixed frame (Fig. 14.21). In addition to Section 2.3 we must supplement the absolute velocities by elastic deformation terms [13]. We get

$$\begin{aligned} {}_R\boldsymbol{v}_{Ei} &= {}_R\boldsymbol{v}_{0i} + {}_R\tilde{\boldsymbol{\omega}}_{0i}({}_R\boldsymbol{x}_i + {}_R\boldsymbol{s}_i) + {}_R\dot{\boldsymbol{s}}_i \; , \\ {}_R\boldsymbol{\omega}_{Ei} &= {}_R\boldsymbol{\omega}_{0i} + {}_R\dot{\boldsymbol{\varphi}}_i \; , \end{aligned} \tag{14.34}$$

written in the reference coordinate system of the fuselage (Fig. 14.21). The magnitudes ${}_R\boldsymbol{v}_{0i}$, ${}_R\boldsymbol{\omega}_{0i}$ are applied velocities of the mass element dm_i under consideration, ${}_R\boldsymbol{s}_i$ is the elastic deformation at point ${}_R\boldsymbol{x}_i$, and ${}_R\boldsymbol{\varphi}_i$ is the angular displacement of the mass element due to rotation. All deformation vectors are assumed to be small. Any transformation from the R-system in any body-fixed K-system can be achieved by the matrix [13]

$$\boldsymbol{A}_{KR} = (\boldsymbol{E} + \tilde{\boldsymbol{\varphi}}_i)^T \, , \tag{14.35}$$

$${}_K\boldsymbol{\omega}_{Ei} = \boldsymbol{A}_{KR} \; {}_R\boldsymbol{\omega}_{Ei} = (\boldsymbol{E} + \tilde{\boldsymbol{\varphi}}_i)^T ({}_R\boldsymbol{\omega}_{0i} + {}_R\dot{\boldsymbol{\varphi}}_i) \, , \tag{14.36}$$

with the transformation of $\boldsymbol{\omega}$ as an example.

The elastic deformations are approximated by a Ritz approach with cubic spline functions [13]:

$$\begin{aligned} \boldsymbol{s}_i(x_i, t) &= \begin{pmatrix} 0 \\ w_i(x_i, t) \\ 0 \end{pmatrix} , \\ \boldsymbol{\varphi}_i(x_i, t) &= \left(\frac{\partial \boldsymbol{s}_i(x_i, t)}{\partial x_i} \right) = \begin{pmatrix} 0 \\ \overline{\boldsymbol{w}}_i'^T(x_i)\boldsymbol{q}_i(t) \\ 0 \end{pmatrix} , \end{aligned} \tag{14.37}$$

$$\text{with} \qquad (\;)' = \left(\frac{\partial}{\partial x_i} \right) , \qquad w_i(x_i, t) = \overline{\boldsymbol{w}}_i^T(x_i)\boldsymbol{q}_i(t) \; .$$

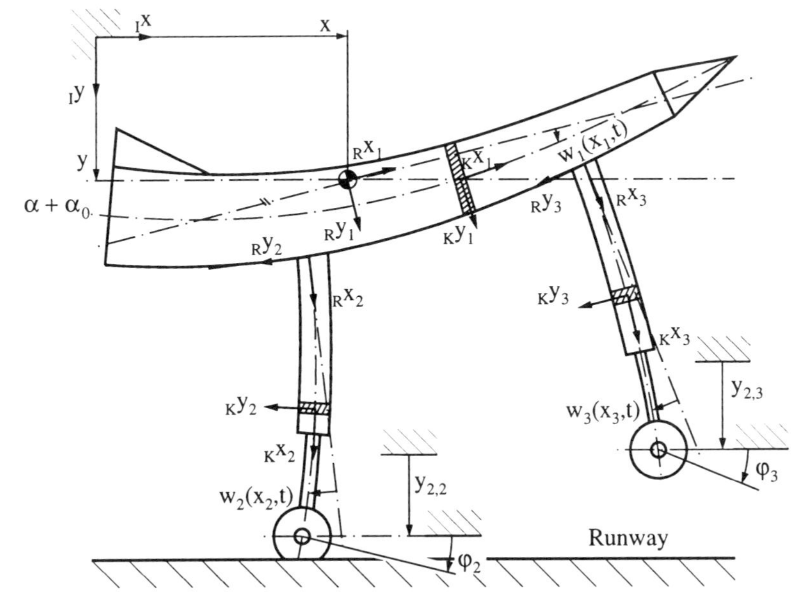

Figure 14.21: Coordinates and Degrees of Freedom

Since $\boldsymbol{v}_{0i} = \boldsymbol{v}_{0i}(\dot{\boldsymbol{q}}_r, \dot{\boldsymbol{q}}_e)$, $\boldsymbol{\omega}_{0i} = \boldsymbol{\omega}_{0i}(\dot{\boldsymbol{q}}_r, \dot{\boldsymbol{q}}_e)$ but $\boldsymbol{s} = \boldsymbol{s}(\boldsymbol{q}_e)$, $\boldsymbol{\varphi} = \boldsymbol{\varphi}(\boldsymbol{q}_e)$, we may write eq. (14.34) in the following form:

$$
\begin{aligned}
{}_R\boldsymbol{v}_{Ei} &= {}_R\left[\left(\frac{\partial \boldsymbol{v}_{0i}}{\partial \dot{\boldsymbol{q}}_r}\right) + (\tilde{\mathbf{x}}_i + \tilde{\mathbf{s}}_i)^T \left(\frac{\partial \boldsymbol{\omega}_{0i}}{\partial \dot{\boldsymbol{q}}_r}\right)\right] \dot{\boldsymbol{q}}_r \\
&\quad + {}_R\left[\left(\frac{\partial \boldsymbol{v}_{0i}}{\partial \dot{\boldsymbol{q}}_e}\right) + (\tilde{\mathbf{x}}_i + \tilde{\mathbf{s}}_i)^T \left(\frac{\partial \boldsymbol{\omega}_{0i}}{\partial \dot{\boldsymbol{q}}_e}\right) + \left(\frac{\partial \dot{\boldsymbol{s}}_i}{\partial \dot{\boldsymbol{q}}_e}\right)\right] \dot{\boldsymbol{q}}_e\,, \\
{}_K\boldsymbol{\omega}_i &= {}_K\left(\frac{\partial \boldsymbol{\omega}_{0i}}{\partial \dot{\boldsymbol{q}}_r}\right)\dot{\boldsymbol{q}}_r + {}_K\left[\left(\frac{\partial \boldsymbol{\omega}_{0i}}{\partial \dot{\boldsymbol{q}}_e}\right) + (\boldsymbol{E} + \tilde{\boldsymbol{\varphi}}_i)^T \left(\frac{\partial \dot{\boldsymbol{\varphi}}_i}{\partial \dot{\boldsymbol{q}}_e}\right)\right] \dot{\boldsymbol{q}}_e\,, \\
{}_R\boldsymbol{v}_i &= {}_R\boldsymbol{J}_{Ti}\dot{\boldsymbol{q}}\,, \\
{}_K\boldsymbol{\omega}_i &= {}_K\boldsymbol{J}_{Ri}\dot{\boldsymbol{q}}\,,
\end{aligned}
\tag{14.38}
$$

with $\boldsymbol{q}$ from eq. (14.33). The Jacobians follow from direct comparison.

The equations of motion follow from the d'Alembert-Jourdain principle in the well-known form

$$
\sum_{i=1}^{n} \int_{K_i} \left\{ \left(\frac{\partial \boldsymbol{v}}{\partial \dot{\boldsymbol{q}}}\right)^T [d\dot{\boldsymbol{p}} - d\boldsymbol{f}^e] + \left(\frac{\partial \boldsymbol{\omega}}{\partial \dot{\boldsymbol{q}}}\right)^T \left[d\dot{\boldsymbol{L}} - d\boldsymbol{l}^e\right] \right\}_i = 0\,, \tag{14.39}
$$

where the momentum $d\boldsymbol{p}_i$ and the moment of momentum $d\boldsymbol{L}_i$ of a mass element dm_i with moment of inertia $d\boldsymbol{I}_i$ are given by

$$\begin{aligned} d\boldsymbol{p}_i &= dm_i \boldsymbol{v}_i = dm_i\, \boldsymbol{J}_{Ti}\dot{\boldsymbol{q}} \\ d\boldsymbol{L}_i &= d\boldsymbol{I}_i \boldsymbol{\omega}_i = d\boldsymbol{I}_i\, \boldsymbol{J}_{Ri}\dot{\boldsymbol{q}}\,. \end{aligned} \tag{14.40}$$

The external forces and torques are $d\boldsymbol{f}^e$, $d\boldsymbol{l}^e$. The sum is taken over all five bodies of the model, Fig. 14.21 ($n = 5$). Note that the terms in eqs. (14.39) may be evaluated in different coordinate frames due to the fact that the scalar products of the columns of the Jacobians and the theorems of momentum or moment of momentum are independent from the chosen coordinate system. Therefore, we appoint for $d\dot{\boldsymbol{p}}_i$ the reference system R and for $d\dot{\boldsymbol{L}}$ the body-fixed system K. This is especially convenient with respect to the moments of inertia. Combining (14.38) and (14.40) yields

$$\begin{aligned} {}_R d\dot{\boldsymbol{p}}_i &= \boldsymbol{A}_{RI}\frac{d}{dt}\left(\boldsymbol{A}_{IR} dm_i\, {}_R\boldsymbol{J}_{Ti}\dot{\boldsymbol{q}}\right) \\ &= \left(\boldsymbol{A}_{RI}\boldsymbol{A}_{IR}\, {}_R\boldsymbol{J}_{Ti}\ddot{\boldsymbol{q}} + \boldsymbol{A}_{RI}\left(\boldsymbol{A}_{IR}\, {}_R\dot{\boldsymbol{J}}_{Ti} + \dot{\boldsymbol{A}}_{IR}\, {}_R\boldsymbol{J}_{Ti}\right)\dot{\boldsymbol{q}}\right) dm_i \\ &= \left({}_R\boldsymbol{J}_{Ti}\ddot{\boldsymbol{q}} + \left({}_R\dot{\boldsymbol{J}}_{Ti} + {}_R\tilde{\boldsymbol{\omega}}_{IR}\, {}_R\boldsymbol{J}_{Ti}\right)\dot{\boldsymbol{q}}\right) dm_i \end{aligned} \tag{14.41}$$

$$\begin{aligned} {}_K d\dot{\boldsymbol{L}}_i &= \boldsymbol{A}_{KI}\frac{d}{dt}\left(\boldsymbol{A}_{IK}\, {}_K d\boldsymbol{I}_i\, {}_K\boldsymbol{J}_{Ri}\dot{\boldsymbol{q}}\right) \\ &= {}_K d\boldsymbol{I}_i\, {}_K\boldsymbol{J}_{Ri}\ddot{\boldsymbol{q}} + \left({}_K d\boldsymbol{I}_i\, {}_K\dot{\boldsymbol{J}}_{Ri} + {}_K\tilde{\boldsymbol{\omega}}_{IK}\, {}_K d\boldsymbol{I}_i\, {}_K\boldsymbol{J}_{Ri}\right)\dot{\boldsymbol{q}}\,. \end{aligned} \tag{14.42}$$

The equations of motion can now be derived by rather lengthy evaluations in considering eqs. (14.38) to (14.42) and the theories of Chapters 2 and 3. As a final result we get

$$\boldsymbol{M}\ddot{\boldsymbol{q}} + \boldsymbol{g}(\boldsymbol{q},\dot{\boldsymbol{q}}) - \boldsymbol{h}(\boldsymbol{q},\dot{\boldsymbol{q}},t) = 0 \tag{14.43}$$

with the following matrices and vectors:

- symmetric and positive-definite mass matrix ($dm_i = \varrho_i A_i(x_i)dx_i$)

$$\begin{aligned} \boldsymbol{M} = \sum_i \int_{K_i} \Bigg\{ & R_i \begin{bmatrix} \boldsymbol{J}_{T_r}^T\boldsymbol{J}_{T_r} & \boldsymbol{J}_{T_r}^T\boldsymbol{J}_{T_e} \\ \boldsymbol{J}_{T_e}^T\boldsymbol{J}_{T_r} & \boldsymbol{J}_{T_e}^T\boldsymbol{J}_{T_e} \end{bmatrix}_i A_i(x_i) \\ & + K_i \begin{bmatrix} \boldsymbol{J}_{R_r}^T\bar{\boldsymbol{I}}\boldsymbol{J}_{R_r} & \boldsymbol{J}_{R_r}^T\bar{\boldsymbol{I}}\boldsymbol{J}_{R_e} \\ \boldsymbol{J}_{R_e}^T\bar{\boldsymbol{I}}\boldsymbol{J}_{R_r} & \boldsymbol{J}_{R_e}^T\bar{\boldsymbol{I}}\boldsymbol{J}_{R_e} \end{bmatrix}_i \Bigg\} \varrho_i\, dx_i\,. \end{aligned} \tag{14.44}$$

- vector of the velocity- and position-dependent forces and torques

$$
\begin{aligned}
\boldsymbol{g}(\boldsymbol{q},\dot{\boldsymbol{q}}) = \sum_i \int_{K_i} \Bigg\{ {}_{R_i} & \left(\begin{bmatrix} \boldsymbol{J}_{T_r}^T \dot{\boldsymbol{J}}_{T_r} & \boldsymbol{J}_{T_r}^T \dot{\boldsymbol{J}}_{T_e} \\ \boldsymbol{J}_{T_e}^T \dot{\boldsymbol{J}}_{T_r} & \boldsymbol{J}_{T_e}^T \dot{\boldsymbol{J}}_{T_e} \end{bmatrix}_i \right. \\
& \left. + \boldsymbol{\omega}_i^T \otimes \begin{bmatrix} \tilde{\boldsymbol{J}}_{T_r}^T \boldsymbol{J}_{T_r} & \tilde{\boldsymbol{J}}_{T_r}^T \boldsymbol{J}_{T_e} \\ \tilde{\boldsymbol{J}}_{T_e}^T \boldsymbol{J}_{T_r} & \tilde{\boldsymbol{J}}_{T_e}^T \boldsymbol{J}_{T_e} \end{bmatrix}_i \right) A_i(x_i) \\
& + {}_{K_i} \left(\begin{bmatrix} \boldsymbol{J}_{R_r}^T \bar{\boldsymbol{I}} \dot{\boldsymbol{J}}_{R_r} & \boldsymbol{J}_{R_r}^T \bar{\boldsymbol{I}} \dot{\boldsymbol{J}}_{R_e} \\ \boldsymbol{J}_{R_e}^T \bar{\boldsymbol{I}} \dot{\boldsymbol{J}}_{R_r} & \boldsymbol{J}_{R_e}^T \bar{\boldsymbol{I}} \dot{\boldsymbol{J}}_{R_e} \end{bmatrix}_i \right. \\
& \left. \left. + \boldsymbol{\omega}_i^T \otimes \begin{bmatrix} \tilde{\boldsymbol{J}}_{R_r}^T \bar{\boldsymbol{I}} \boldsymbol{J}_{R_r} & \tilde{\boldsymbol{J}}_{R_r}^T \bar{\boldsymbol{I}} \boldsymbol{J}_{R_e} \\ \tilde{\boldsymbol{J}}_{R_e}^T \bar{\boldsymbol{I}} \boldsymbol{J}_{R_r} & \tilde{\boldsymbol{J}}_{R_e}^T \bar{\boldsymbol{I}} \boldsymbol{J}_{R_e} \end{bmatrix}_i \right) \right\} \varrho_i \, dx_i .
\end{aligned}
\tag{14.45}
$$

- vector of all external forces and torques

$$
\begin{aligned}
& \boldsymbol{h}(\boldsymbol{q},\dot{\boldsymbol{q}},t) \\
& = \sum_i \int_{K_i} \left\{ {}_{R_i} \begin{pmatrix} \boldsymbol{J}_{T_r}^T \\ \boldsymbol{J}_{T_e}^T \end{pmatrix}_i d\left({}_{R_i}\boldsymbol{f}_i^e \right) + {}_{K_i} \begin{pmatrix} \boldsymbol{J}_{R_r}^T \\ \boldsymbol{J}_{R_e}^T \end{pmatrix}_i d\left({}_{K_i}\boldsymbol{l}_i^e \right) \right\} .
\end{aligned}
\tag{14.46}
$$

The vector $\boldsymbol{g}$ of the velocity- and position-dependent forces and torques must be supplemented by the elastic restoring forces to be applied at the corresponding location. We have

$$
\begin{aligned}
\boldsymbol{K}_i \boldsymbol{q}_i & = \left[\int_{K_i} E_i \bar{I}_i \left(\frac{\partial w_i''(x_i,t)}{\partial \boldsymbol{q}_i} \right) \left(\frac{\partial w_i''(x_i,t)}{\partial \boldsymbol{q}_i} \right)^T dx_i \right] \boldsymbol{q}_i \\
& = \left[\int_{K_i} E_i \bar{I}_i \overline{\boldsymbol{w}}_i''(x_i) \overline{\boldsymbol{w}}_i''^T(x_i) \, dx_i \right] \boldsymbol{q}_i .
\end{aligned}
\tag{14.47}
$$

In evaluating these matrices and vectors all magnitudes of higher order were neglected, especially the deformation terms. In addition, all integrals over the absorber length were calculated for a medium length as an approximation, not taking into account the time dependency of the absorber lengths.

Evaluation of the Equations of Motion

The evaluation of the matrices and vectors of the above equations follows the geometry of Figs. 14.20 and 14.21 but is a tedious task. We shall repeat only the final results; for details see [9].

- Vector of the external forces

The external forces and torques will be explained in the following sections. The vector is

$$
\boldsymbol{h}(\boldsymbol{q},\dot{\boldsymbol{q}},t)=\begin{bmatrix}
-W+F_S-2F_{Sp,2}-F_{Sp,3} \\
-A-F_S\alpha_0-2F_{1,2}-F_{1,3} \\
+\left(m_{ges}-2(m_4+m_{2,u})-(m_5+m_{3,u})\right)g \\
\\
2(x_{1,2}F_{1,2}-l_{4,S}F_{Sp,2}-M_B(t)) \\
+x_{1,3}F_{1,3}-l_{5,S}F_{Sp,3}+M_A \\
\\
2(F_{1,2}-F_{R,2}+(m_{2,u}+m_4)) \\
\\
F_{1,3}-F_{R,3}+(m_{3,u}+m_5) \\
\\
2(r_4(t)F_{Sp,2}-M_B) \\
\\
r_5(t)F_{Sp,3} \\
\\
-(A+W\alpha_0)\overline{\mathbf{w}}_1(x_1=0)-M_A\overline{\mathbf{w}}_1{}'(x_1=0) \\
-2(F_{1,2}+gm_{2,o})\overline{\mathbf{w}}_1(x_{1,2}) \\
-(F_{1,3}+gm_{3,o})\overline{\mathbf{w}}_1(x_{1,3}) \\
+2(l_{02}F_{Sp,2}+M_B(t))\overline{\mathbf{w}}_1{}'(x_{1,2}) \\
+l_{03}F_{Sp,3}\overline{\mathbf{w}}_1{}'(x_{1,3}) \\
\\
2(F_{Sp,2}\overline{\mathbf{w}}_2(l_{02})+M_B(t)\overline{\mathbf{w}}_2{}'(l_{02})) \\
\\
F_{Sp,3}\overline{\mathbf{w}}_3(l_{03})
\end{bmatrix} \tag{14.48}
$$

- Mass matrix

The mass matrix is composed of four submatrices

$$
\boldsymbol{M}=\begin{bmatrix}\boldsymbol{M}_r & \boldsymbol{M}_{re} \\ \boldsymbol{M}_{re}^T & \boldsymbol{M}_e\end{bmatrix}, \tag{14.49}
$$

with the following components:

$$
\mathbf{M}_r = \begin{bmatrix}
m_{ges} & 0 & \begin{array}{l} m_2 l_{02} + 2m_4 l_{4,S_0} \\ +\frac{1}{2} m_5 l_{05} + m_5 l_{5,S_0} \end{array} & 0 & 0 & 0 & 0 \\
 & \begin{array}{l} m_{ges} - 2(m_4 + m_{2,u}) \\ -(m_5 + m_{3,u}) \end{array} & -2m_{2,o} x_{1,2} - m_{3,o} x_{1,3} & 0 & 0 & 0 & 0 \\
 & & I_{ges} & 0 & 0 & 0 & 0 \\
 & & & 2(m_4 + m_{2,u}) & 0 & 0 & 0 \\
 & \text{symmetric} & & & m_5 + m_{3,u} & 0 & 0 \\
 & & & & & 2I_4 & 0 \\
 & & & & & & I_5
\end{bmatrix} \tag{14.50}
$$

$$
\mathbf{M}_e = \begin{bmatrix}
\begin{array}{l} 2\left[(m_4 + \frac{1}{3} m_2) l_{02}^2 + \int_0^{l_{02}} I_{2,z}^* dx_2\right] \overline{\mathbf{w}}_1'(x_{1,2}) \overline{\mathbf{w}}_1'^T(x_{1,2}) \\ + \left[(m_5 + \frac{1}{3} m_3) l_{03}^2 + \int_0^{l_{03}} I_{3,z}^* dx_3\right] \overline{\mathbf{w}}_1'(x_{1,3}) \overline{\mathbf{w}}_1'^T(x_{1,3}) \\ +2m_{2,o} \overline{\mathbf{w}}_1(x_{1,2}) \overline{\mathbf{w}}_1^T(x_{1,2}) + m_{3,o} \overline{\mathbf{w}}_1(x_{1,3}) \overline{\mathbf{w}}_1^T(x_{1,3}) \\ + \left[\int_{-l_{1,h}}^{l_{1,v}} (m_1^* \overline{\mathbf{w}}_1 \overline{\mathbf{w}}_1^T + I_{1,z}^* \overline{\mathbf{w}}_1' \overline{\mathbf{w}}_1'^T) dx_1\right] \end{array} & \begin{array}{l} 2\overline{\mathbf{w}}_1'(x_{1,2}) \left[m_4 l_{02} \overline{\mathbf{w}}_2^T(l_{02}) \right. \\ + \int_0^{l_{02}} (m_2^* x_2 \overline{\mathbf{w}}_2^T \\ \left. + I_{2,z}^* \overline{\mathbf{w}}_2'^T) dx_2 \right] \end{array} & \begin{array}{l} \overline{\mathbf{w}}_1'(x_{1,3}) \left[m_5 l_{03} \overline{\mathbf{w}}_3^T(l_{03}) \right. \\ + \int_0^{l_{03}} (m_3^* x_3 \overline{\mathbf{w}}_3^T \\ \left. + I_{3,z}^* \overline{\mathbf{w}}_3'^T) dx_3 \right] \end{array} \\
 & \begin{array}{l} 2\left[\int_0^{l_{02}} (m_2^* \overline{\mathbf{w}}_2 \overline{\mathbf{w}}_2^T \right. \\ + I_{2,z}^* \overline{\mathbf{w}}_2' \overline{\mathbf{w}}_2'^T) dx_2 \\ \left. + m_4 \overline{\mathbf{w}}_2(l_{02}) \overline{\mathbf{w}}_2^T(l_{02}) \right] \end{array} & 0 \\
\text{symmetric} & & \begin{array}{l} \left[\int_0^{l_{03}} (m_3^* \overline{\mathbf{w}}_3 \overline{\mathbf{w}}_3^T \right. \\ + I_{3,z}^* \overline{\mathbf{w}}_3' \overline{\mathbf{w}}_3'^T) dx_3 \\ \left. + m_5 \overline{\mathbf{w}}_3(l_{03}) \overline{\mathbf{w}}_3^T(l_{03}) \right] \end{array}
\end{bmatrix} \tag{14.51}
$$

$$\mathbf{M}_{re} = \begin{bmatrix} \begin{array}{l} -(2m_4 + m_2) l_{02} \overline{\mathbf{w}}_1'^T(x_{1,2}) \\ -(m_5 + \frac{1}{2} m_3) l_{03} \overline{\mathbf{w}}_1'^T(x_{1,3}) \end{array} & -2m_4 \overline{\mathbf{w}}_2^T(l_{02}) - 2\int_0^{l_{02}} m_2^* \overline{\mathbf{w}}_2^T dx_2 & -m_5 \overline{\mathbf{w}}_3^T(l_{03}) - \int_0^{l_{03}} m_3^* \overline{\mathbf{w}}_3^T dx_3 \\ 2m_{2,o} \overline{\mathbf{w}}_1^T(x_{1,2}) + m_{3,o} \overline{\mathbf{w}}_1^T(x_{1,3}) & 0 & 0 \\ \begin{array}{l} -2\left[m_4 l_{02} l_{4,S_0} + \frac{1}{3} m_2 l_{02}^2 \right. \\ \qquad \left. + \int_0^{l_{02}} I_{2,z}^* dx_2 \right] \overline{\mathbf{w}}_1'^T(x_{1,2}) \\ -\left[m_5 l_{03} l_{5,S_0} + \frac{1}{3} m_3 l_{03}^2 \right. \\ \qquad \left. + \int_0^{l_{03}} I_{3,z}^* dx_3 \right] \overline{\mathbf{w}}_1'^T(x_{1,3}) \\ -2m_{2,o} x_{1,2} \overline{\mathbf{w}}_1^T(x_{1,2}) \\ -m_{3,o} x_{1,3} \overline{\mathbf{w}}_1^T(x_{1,3}) \\ -\int_{-l_{1,h}}^{l_{1,v}} (m_1^* x_1 \overline{\mathbf{w}}_1^T + I_{1,z}^* \overline{\mathbf{w}}_1'^T) dx_1 \end{array} & \begin{array}{l} -2m_4 l_{4,S_0} \overline{\mathbf{w}}_2^T(l_{02}) \\ -2\left[\int_0^{l_{02}} (m_2^* x_2 \overline{\mathbf{w}}_2^T + I_{2,z}^* \overline{\mathbf{w}}_2'^T) dx_2 \right] \end{array} & \begin{array}{l} -m_5 l_{5,S_0} \overline{\mathbf{w}}_3^T(l_{03}) \\ -\left[\int_0^{l_{03}} (m_3^* x_3 \overline{\mathbf{w}}_3^T + I_{3,z}^* \overline{\mathbf{w}}_3'^T) dx_3 \right] \end{array} \\ 0 & 0 & 0 \\ 0 & 0 & 0 \\ 0 & 0 & 0 \\ 0 & 0 & 0 \end{bmatrix} \tag{14.52}$$

- Vector of all velocity- and position-dependent forces and torques, which after linearization contains only the restoring forces of the elastic deformations:

$$
g(q,\dot{q}) = \begin{bmatrix} 0 \\ 0 \\ 0 \\ 0 \\ 0 \\ 0 \\ 0 \\ \left[\int_{-l_{1,h}}^{l_{1,v}} E_1 \overline{I}_{1,z}(x_1) \overline{\mathbf{w}}_1'' \overline{\mathbf{w}}_1''^T dx_1\right] q_1 \\ \left[\int_0^{l_{02}} E_2 \overline{I}_{2,z}(x_2) \overline{\mathbf{w}}_2'' \overline{\mathbf{w}}_2''^T dx_2\right] q_2 \\ \left[\int_0^{l_{03}} E_3 \overline{I}_{3,z}(x_3) \overline{\mathbf{w}}_3'' \overline{\mathbf{w}}_3''^T dx_3\right] q_3 \end{bmatrix} \tag{14.53}
$$

- Jacobians of the fuselage

$$
\begin{aligned}
{}_R J_{rT1} &= \begin{bmatrix} \cos(\alpha+\alpha_0) & -\sin(\alpha+\alpha_0) & \overline{\mathbf{w}}_1(x_1) & 0\;0\;0\;0 \\ \sin(\alpha+\alpha_0) & \cos(\alpha+\alpha_0) & -x_1 & 0\;0\;0\;0 \\ 0 & 0 & 0 & 0\;0\;0\;0 \end{bmatrix} \\
{}_K J_{rR1} &= \begin{bmatrix} 0 & 0 & 0 & 0\;0\;0\;0 \\ 0 & 0 & 0 & 0\;0\;0\;0 \\ 0 & 0 & -1 & 0\;0\;0\;0 \end{bmatrix} \\
{}_R J_{eT1} &= \begin{bmatrix} 0 & 0\;0 \\ \overline{\mathbf{w}}_1^T(x_1) & 0\;0 \\ 0 & 0\;0 \end{bmatrix} \\
{}_K J_{eR1}' &= \begin{bmatrix} 0 & 0\;0 \\ 0 & 0\;0 \\ \overline{\mathbf{w}}_1'^T(x_1) & 0\;0 \end{bmatrix}
\end{aligned} \tag{14.54}
$$

- Jacobians of the main carriage

$$
\begin{aligned}
{}_R\boldsymbol{J}_{rT2} &= \begin{bmatrix} \cos\gamma_2 & \sin\gamma_2 & \begin{array}{c} w_1(x_{1,2},t)\cos\delta_2 \\ -x_{1,2}\sin\delta_2 \\ +w_2(x_2,t) \end{array} & 0\;0\;0\;0 \\ -\sin\gamma_2 & \cos\gamma_2 & \begin{array}{c} -w_1(x_{1,2},t)\sin\delta_2 \\ -x_{1,2}\cos\delta_2 - x_2 \end{array} & 0\;0\;0\;0 \\ 0 & 0 & 0 & 0\;0\;0\;0 \end{bmatrix} \\
{}_K\boldsymbol{J}_{rR2} &= \begin{bmatrix} 0 & 0 & 0 & 0 & 0 & 0 & 0 \\ 0 & 0 & 0 & 0 & 0 & 0 & 0 \\ 0 & 0 & -1 & 0 & 0 & 0 & 0 \end{bmatrix} \\
{}_R\boldsymbol{J}_{eT2} &= \begin{bmatrix} \overline{\mathbf{w}}_1^T(x_{1,2})\sin\delta_2 - \overline{\mathbf{w}}_1'^T(x_{1,2})w_2(x_2,t) & 0 & 0 \\ \overline{\mathbf{w}}_1^T(x_{1,2})\cos\delta_2 + \overline{\mathbf{w}}_1'^T(x_{1,2})x_2 & \overline{\mathbf{w}}_2^T(x_2) & 0 \\ 0 & 0 & 0 \end{bmatrix} \qquad (14.55) \\
{}_K\boldsymbol{J}_{eR2} &= \begin{bmatrix} 0 & 0 & 0 \\ 0 & 0 & 0 \\ \overline{\mathbf{w}}_1'^T(x_{1,2}) & \overline{\mathbf{w}}_2'^T(x_2) & 0 \end{bmatrix}
\end{aligned}
$$

- Jabobians of the nose carriage

$$
\begin{aligned}
{}_R\boldsymbol{J}_{rT3} &= \begin{bmatrix} \cos\gamma_3 & \sin\gamma_3 & \begin{array}{c} w_1(x_{1,3},t)\cos\delta_3 \\ -x_{1,3}\sin\delta_3 \\ +w_3(x_3,t) \end{array} & 0\;0\;0\;0 \\ -\sin\gamma_3 & \cos\gamma_3 & \begin{array}{c} -w_1(x_{1,3},t)\sin\delta_3 \\ -x_{1,3}\cos\delta_3 - x_3 \end{array} & 0\;0\;0\;0 \\ 0 & 0 & 0 & 0\;0\;0\;0 \end{bmatrix} \\
{}_K\boldsymbol{J}_{rR3} &= \begin{bmatrix} 0 & 0 & 0 & 0 & 0 & 0 & 0 \\ 0 & 0 & 0 & 0 & 0 & 0 & 0 \\ 0 & 0 & -1 & 0 & 0 & 0 & 0 \end{bmatrix} \\
{}_R\boldsymbol{J}_{eT3} &= \begin{bmatrix} \overline{\mathbf{w}}_1^T(x_{1,3})\sin\delta_3 - \overline{\mathbf{w}}_1'^T(x_{1,3})w_3(x_3,t) & 0 & 0 \\ \overline{\mathbf{w}}_1^T(x_{1,3})\cos\delta_3 + \overline{\mathbf{w}}_1'^T(x_{1,3})x_3 & 0 & \mathbf{w}_3^T(x_3) \\ 0 & 0 & 0 \end{bmatrix} \qquad (14.56) \\
{}_K\boldsymbol{J}_{eR3} &= \begin{bmatrix} 0 & 0 & 0 \\ 0 & 0 & 0 \\ \overline{\mathbf{w}}_1'^T(x_{1,2}) & 0 & \overline{\mathbf{w}}_3'^T(x_3) \end{bmatrix}
\end{aligned}
$$

- Jacobians of the main carriage wheels

$$
\begin{aligned}
{}_I\boldsymbol{J}_{rT4} &= \begin{bmatrix} 1 & 0 & \begin{array}{l} w_1(x_{1,2},t)\cos(\alpha+\alpha_0) \\ -x_{1,2}\sin(\alpha+\alpha_0) \\ +w_2(l_2,t)\cos\gamma_2 \\ +l_2\sin\gamma_2 \end{array} & 0 & 0 & 0 & 0 \\ 0 & 0 & 0 & 1 & 0 & 0 & 0 \\ 0 & 0 & 0 & 0 & 0 & 0 & 0 \end{bmatrix} \\
{}_K\boldsymbol{J}_{rR4} &= \begin{bmatrix} 0 & 0 & 0 & 0 & 0 & 0 & 0 \\ 0 & 0 & 0 & 0 & 0 & 0 & 0 \\ 0 & 0 & 0 & 0 & 0 & 1 & 0 \end{bmatrix} \\
{}_I\boldsymbol{J}_{eT4} &= \begin{bmatrix} \begin{array}{l} \overline{\mathbf{w}}_1^T(x_{1,2})\sin(\alpha+\alpha_0) \\ -\overline{\mathbf{w}}_1'^T(x_{1,2})(w_2(l_2,t)\cos\gamma_2 \\ +l_2\sin\gamma_2) \end{array} & -\overline{\mathbf{w}}_2^T(l_2) & 0 \\ 0 & 0 & 0 \\ 0 & 0 & 0 \end{bmatrix} \\
{}_K\boldsymbol{J}_{eR4} &= \mathbf{0}
\end{aligned}
\tag{14.57}
$$

- Jacobians of the nose carriage wheel

$$
\begin{aligned}
{}_I\boldsymbol{J}_{rT5} &= \begin{bmatrix} 1 & 0 & \begin{array}{l} w_1(x_{1,3},t)\cos(\alpha+\alpha_0) \\ -x_{1,3}\sin(\alpha+\alpha_0) \\ +w_3(l_3,t)\cos\gamma_3 \\ +l_3\sin\gamma_3 \end{array} & 0 & 0 & 0 & 0 \\ 0 & 0 & 0 & 0 & 1 & 0 & 0 \\ 0 & 0 & 0 & 0 & 0 & 0 & 0 \end{bmatrix} \\
{}_K\boldsymbol{J}_{rR5} &= \begin{bmatrix} 0 & 0 & 0 & 0 & 0 & 0 & 0 \\ 0 & 0 & 0 & 0 & 0 & 0 & 0 \\ 0 & 0 & 0 & 0 & 0 & 0 & 1 \end{bmatrix} \\
{}_I\boldsymbol{J}_{eT5} &= \begin{bmatrix} \begin{array}{l} \overline{\mathbf{w}}_1^T(x_{1,3})\sin(\alpha+\alpha_0) \\ -\overline{\mathbf{w}}_1'^T(x_{1,3})(w_3(l_3,t)\cos\gamma_3 \\ +l_3\sin\gamma_3) \end{array} & 0 & -\overline{\mathbf{w}}_3^T(l_3) \\ 0 & 0 & 0 \\ 0 & 0 & 0 \end{bmatrix} \\
{}_K\boldsymbol{J}_{eR5} &= \mathbf{0}
\end{aligned}
\tag{14.58}
$$

The Jacobians result from the derivations of the velocities and angular velocities with respect to the minimal velocities $\dot{\boldsymbol{q}}$. The notation in the above formulas is (see Fig. 14.20)

$$\begin{aligned} \gamma_i &= \beta_{1,i} + w_1{}'(x_{1,i}, t) - (\alpha + \alpha_0) \\ \delta_i &= \beta_{1,i} + w_1{}'(x_{1,i}, t) \\ i &= 2, 3\,. \end{aligned} \tag{14.59}$$

EXTERNAL FORCES AND TORQUES

The vector $\boldsymbol{h}(\boldsymbol{q}, \dot{\boldsymbol{q}}, t)$ contains the external forces on the carriages and on the airplane. It can be developed from the expressions

$$\begin{aligned} d\boldsymbol{f}_i^e &= (\boldsymbol{f}_{Di}^e \delta(x_i - \xi_j) + \boldsymbol{f}_{Vi}^e)\, dx_i \\ d\boldsymbol{l}_i^e &= \left(\boldsymbol{l}_{Di}^e \delta(x_i - \xi_j) + \boldsymbol{l}_{V_i}^e\right) dx_i\,, \end{aligned} \tag{14.60}$$

where the Dirac functions $\delta(x_i - \xi_j)$ take into account the correct location ξ_j of the applied discrete forces and torques during the integration process. The subscript V assigns continuous loads. Figure 14.22 portrays the forces and torques acting on the components of the chosen model.

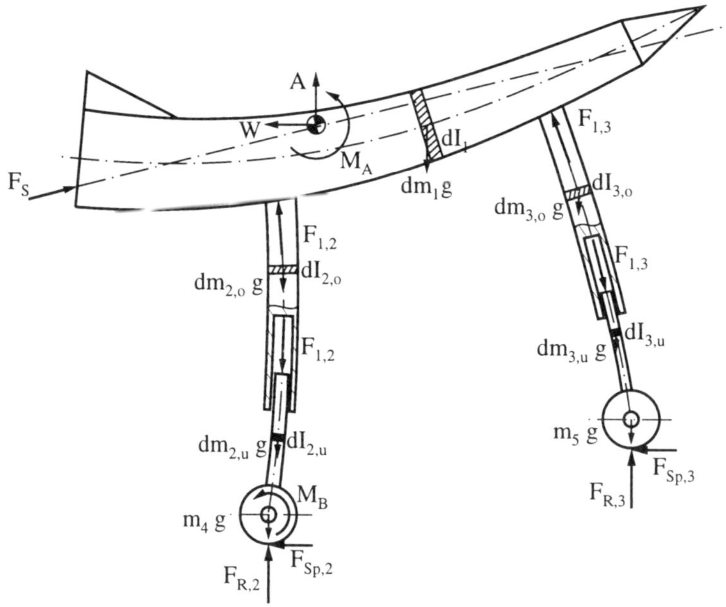

Figure 14.22: External Forces and Torques

- Aerodynamic and propulsive forces

  The aerodynamic forces and torques relate to the center of mass. They are calculated by using the velocity pressure q, a reference area F_B and the aerodynamic coefficients. For lift A, drag W and torque M_A,

$$\begin{aligned} A &= q\,F_B\,c_A \\ W &= q\,F_B\,c_W \\ M_A &= q\,F_B\,c_M\,l_\mu \end{aligned} \tag{14.61}$$

  with coefficients

$$\begin{aligned} c_A &= \frac{\partial c_A}{\partial \alpha}(\alpha - \alpha_0)\,, \\ c_W &= a_0 + a_1\alpha + a_2\alpha^2\,, \\ c_M &= c_{M0} + \left(\frac{\partial c_M}{\partial c_A}\right) c_A + \left(\frac{\partial c_M}{\partial \alpha}\right) \frac{l_\mu \dot{\alpha}}{v_\infty}\,. \end{aligned} \tag{14.62}$$

  (l_μ is reference length of a wing, v_∞ is velocity, $q = \frac{1}{2}\varrho v_\infty^2$ is velocity pressure, ϱ is density, α is angle of attack.) The propulsion force F_S results from

$$F_S = F_{S0} - k\,v\,, \tag{14.63}$$

  where F_{S0} is the idling thrust and k is a constant.

- Wheel forces

  The vertical static force on each wheel can be calculated from the nonlinear tire characteristic as given by the tire manufacturers. Figure 14.23 depicts a principal example.

  The function $F_R(s)$ is formally given by

$$F_R = \begin{cases} 0 & s(t) < 0 \\ c_1 s(t) & 0 \le s(t) < s_1 \\ c_1 s_1 + c_2(s(t) - s_1) & s_1 \le s(t) < s_2 \\ c_1 s_1 + c_2(s_2 - s_1) \\ \quad + c_3(s(t) - s_2) & s_2 \le s(t)\,. \end{cases} \tag{14.64}$$

  The tire displacements for nose and main carriages $s_i(t)$, $i = 2, 3$, follow from Fig. 14.21:

$$s(t) = y_{20} + y_2(t) + \Delta h(t)\,, \tag{14.65}$$

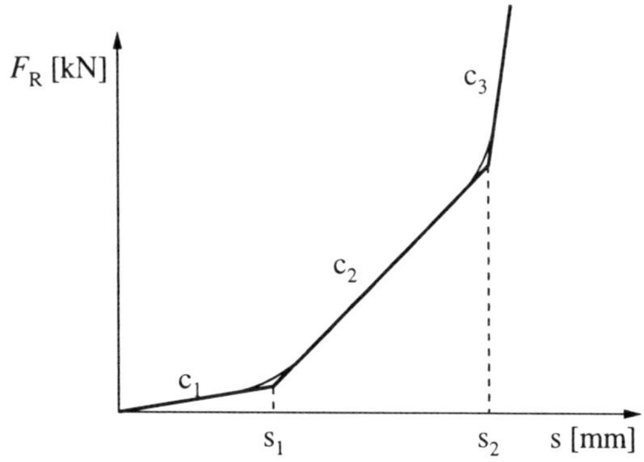

Figure 14.23: Typical Tire Characteristic

where $\Delta h(t)$ takes into account some existing roughness of the runway, and y_{20} is the displacement for static load.
The horizontal wheel force F_{S_p} resulting from spin-up can be calculated from F_R and from a friction coefficient μ_{S_p} which depends on the slippage SR (see Fig. 14.24).
The change of sign of the force will be generated automatically by the curve 14.24. It is

$$F_{Sp} = \mu_{Sp}(SR)F_R \tag{14.66}$$

with

$$\begin{aligned} SR &= \frac{\dot{x}(t) - \dot{w}(l,t) - \dot{\varphi}(t)r(t)}{\dot{x}(t) - \dot{w}(l,t)} = 1 - \frac{\dot{\varphi}(t)r(t)}{\dot{x}(t) - \dot{w}(l,t)}, \\ r(t) &= r_{00} - y_{20} - y_2(t) - \Delta h(t) \\ &= r_{00} - s(t) \end{aligned} \tag{14.67}$$

(r_{00} is radius of unloaded tire, $\dot{x} - \dot{w}$ is velocity of the wheel shaft, $\dot{\varphi} r$ is velocity due to wheel rotation.) The values of F_{S_p} and F_R are evaluated separately for main carriage and nose wheels.

- Gravity forces

  Modeling the fuselage and the carriages as elastic components, we have to consider the gravitational line loads of both parts. The fuselage is approximated by a constant line load, the carriages by piecewise constant loads. The wheel weights act upon the wheel shafts.

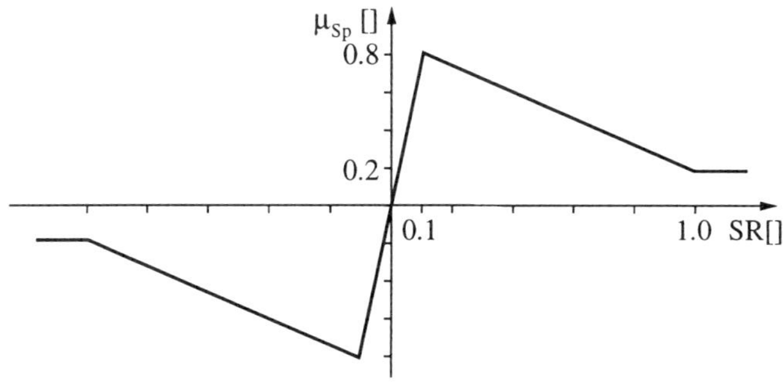

Figure 14.24: Friction Coefficient versus Slippage

- Braking torques

  For applying the brakes we regard a torque which might depend on time in any prescribed way; for example,

$$M_B(t) = M_{\text{Brake}}(t)\, H(t - t_S) \tag{14.68}$$

with

$$H(t - t_S) = \begin{cases} 0 \text{ for } t \leq t_S \\ 1 \text{ for } t > t_S \,. \end{cases}$$

TELESCOPIC SHOCK ABSORBER MODEL

The most important elements are the shock absorbers between airplane and ground. The model and force laws derived from it describe the transmission from ground to airplane and dominate the system dynamics. We consider an absorber with two gas chambers where the behavior of the gas follows a polytropic law and the fluid flow produces a quadratic damping behavior, both of course in an approximative way.

In the following the subscripts indicating the main and nose carriages are omitted. Figure 14.25 portrays the mechanical model. The two gas chambers with volumes V_{L1}, V_{L2} contain nitrogen. The other volumes are filled with oil.

If the two housing parts move with $y_1 - y_2$, the oil can flow between volumes V_1, V_2, V_4. The gas chamber with V_{L2} is a high-pressure chamber, the one with V_{L1} a low-pressure chamber. The piston in the high-pressure chamber will move only at high loads. Between the two cylinders and between

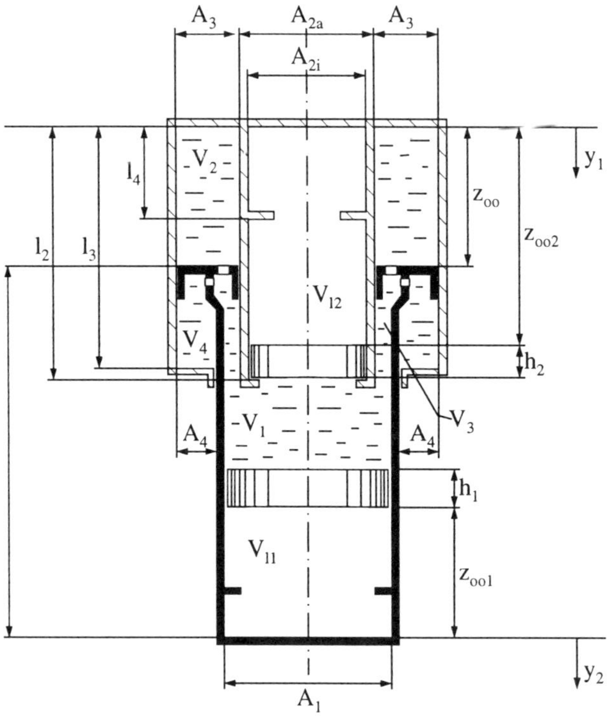

Figure 14.25: Two-Chamber Shock Absorber Model [85]

cylinders and pistons we have stops where impacts may occur. Furthermore, for large horizontal loads stick-slip phenomena or even jamming between the two cylinders may arise.

The completely unloaded situation of the absorber will be indexed by 00. The pressures then are

$$p_{L200} > p_{L100}\,, \tag{14.69}$$

and the volumes are

$$\begin{aligned} V_{L100} &= A_1 z_{001} \\ V_{L200} &= A_{2i} z_{002} \\ V_{100} &= A_1((l_1 + z_{00}) - (z_{002} + h_2) - (z_{001} + h_1)) \\ V_{200} &= A_3 z_{00} \\ V_{300} &= (A_1 - A_{2a})((z_{002} + h_2) - z_{00}) \\ V_{400} &= A_4(l_3 - z_{00})\,. \end{aligned} \tag{14.70}$$

Using y_1 and y_2 for the motion of the housing components (Fig. 14.25) the stroke is

$$\Delta y = y_1 - y_2 \,. \tag{14.71}$$

For $\Delta \dot{y} > 0$ oil flows from V_2 via V_3 to V_1 and V_4. Only the lower piston moves (Fig. 14.25). The telescopic shock absorber operates in mode I with

$$p_1 = p_{L1} \,, \quad p_{L2} = p_{L200} \,, \quad p_1 < p_{L200} \,. \tag{14.72}$$

With increasing stroke the pressure p_1 will become larger than the pressure in V_{L2}. Both gas chambers become active, and we proceed to mode II behavior with

$$p_1 = p_{L1} = p_{L2} \,, \quad p_1 > p_{L200} \,. \tag{14.73}$$

The relationship between pressures and volumes is assumed to be polytropic:

$$\frac{p_{Li}}{p_{Lj}} = \left(\frac{V_{Lj}}{V_{Li}} \right)^{n_p} \,, \quad i,j = 1,2 \tag{14.74}$$

(n_p polytropic coefficient). For the oil flow we assume that it will produce losses between V_1 and V_2 but not between V_1 and V_4. Therefore,

$$\begin{aligned} p_2 - p_1 &= \left(\frac{\lambda}{A_3} \right) \Delta(\dot{y}^2) \,\mathrm{sign}(\Delta \dot{y}) \,, \\ p_4 - p_1 &= 0 \,, \end{aligned} \tag{14.75}$$

where the loss factor λ depends also on the sign of $\Delta \dot{y}$:

$$\lambda = \lambda[\mathrm{sign}(\Delta \dot{y})] \,. \tag{14.76}$$

Considering the operational mode I (eqs. 14.72), we can establish the continuity equation

$$V_1(\Delta y) = V_{100} + [(A_3 - A_4) - (A_1 - A_{2a})] \,\Delta y \,. \tag{14.77}$$

From this and eq. (14.74) the pressure in the low-pressure chamber is given by

$$p_{L1}(\Delta y) = \frac{p_{L100}}{\left(1 - \frac{A_3 - A_4 + A_{2a}}{A_1} \left(\frac{\Delta y}{z_{001}} \right)\right)^{n_P}} \,. \tag{14.78}$$

Operational mode II is governed by eq. (14.73) and by $\Delta y \geq \Delta y_A$ with stroke Δy_A:

$$\Delta y_A = \frac{A_1 z_{001}}{A_3 - A_4 + A_{2a}} \left(1 - \left(\frac{p_{L100}}{p_{L200}} \right)^{1/n_P} \right) \tag{14.79}$$

at the transition point between modes I and II, the low- and high-pressure modes, respectively. The continuity equation for mode II is (both pistons move)

$$(A_3 - A_4 + A_{2a})(\Delta y - \Delta y_A) = A_1\,\Delta y_1 + A_{2i}\,\Delta y_2\,, \tag{14.80}$$

with constraints

$$\begin{aligned} V_{L1} &= V_{L1}|_{\Delta y = \Delta y_A} - A_1\,\Delta y_1 \\ V_{L2} &= V_{L2}|_{\Delta y = \Delta y_A} - A_{2i}\,\Delta y_2\,. \end{aligned} \tag{14.81}$$

The unknown piston movements Δy_1 and Δy_2 are evaluated in the following. Introducing

$$\gamma = \frac{\Delta y_2}{\Delta y_1} = \frac{A_1}{A_{2i}}\left(\frac{V_{L2}}{V_{L1}}\right)_{\Delta y = \Delta y_A}, \tag{14.82}$$

we can calculate the movement of the two pistons:

$$\begin{aligned} \Delta y_1 &= \frac{A_3 - A_4 + A_{2a}}{A_1 + \gamma A_{2i}}(\Delta y - \Delta y_A) \\ \Delta y_2 &= \frac{\gamma(A_3 - A_4 + A - 2a)}{A_1 + \gamma A_{2i}}(\Delta y - \Delta y_A)\,. \end{aligned} \tag{14.83}$$

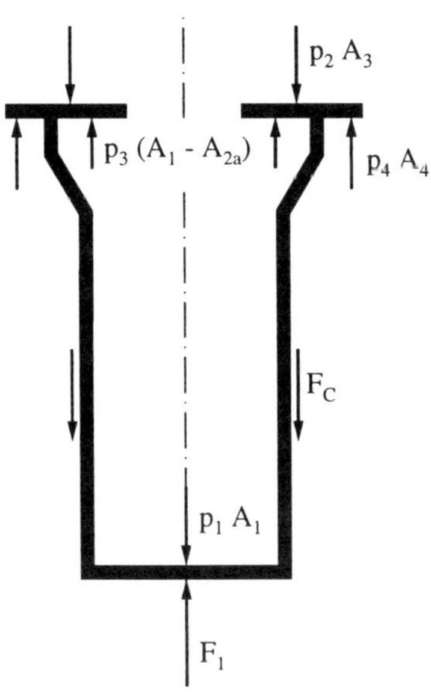

Figure 14.26: Force Equilibrium at the Shock Absorber

From this and eq. (14.74) the pressure is

$$p_{L1}(\Delta y) = \frac{p_{L200}}{\left(1 - \frac{A_3 - A_4 + A_{2a}}{V_{LA}} (\Delta y - \Delta y_A)\right)^{n_P}} , \tag{14.84}$$

with

$$V_{LA} = V_{L1}|_{\Delta y = \Delta y_A} + V_{L2}|_{\Delta y = \Delta y_A} .$$

The force law of the telescopic shock absorber will be derived by neglecting the weight of the absorber and nonsteady flow effects. Regarding these assumptions an equilibrium of forces yields (Fig. 14.26)

$$F_1 = p_1 A_1 + p_2 A_3 - p_4 A_4 - p_3 (A_1 - A_{2a}) + F_C , \tag{14.85}$$

where F_C is a Coulomb friction force. Combining the above equations we obtain

$$F_1 = \begin{cases} \dfrac{(A_3 - A_4 + A_{2a}) P_{L100}}{\left(1 - \frac{(A_3 - A_4 + A_{2a})}{A_1} \left(\frac{\Delta y}{z_{001}}\right)\right)^{n_P}} + \lambda(\Delta \dot{y})^2 \operatorname{sign}(\Delta \dot{y}) + F_C \\ \text{(mode I)} \\ \dfrac{(A_3 - A_4 + A_{2a}) P_{L200}}{\left(1 - \frac{(A_3 - A_4 + A_{2a})}{V_{LA}} (\Delta y - \Delta y_A)\right)^{n_P}} + \lambda(\Delta \dot{y})^2 \operatorname{sign}(\Delta \dot{y}) + F_C \\ \text{(mode II).} \end{cases} \tag{14.86}$$

The first term represents the gas force, the second one the damping force from oil flow losses, and the third one the Coulomb friction force. With respect to the main and nose carriages the gas and damping forces follow the same formulas, but the dry friction force is different (Fig. 14.27). On the other hand, and due to the assumed small elastic deflections, there is no significant influence of the absorber elasticity.

The Coulomb friction forces for the main (subscript 2) and the nose (subscript 3, Fig. 14.21) can be derived as

$$\begin{aligned} F_{C,2} &= \mu_G(\mid Q_1 \mid + \mid Q_2 \mid)_2 \operatorname{sign}(\Delta \dot{y}_2) \\ &= \mu_G \tfrac{l+a}{l-a} \mid F_{N,2} \mid \operatorname{sign}(\Delta \dot{y}_2) \\ &= \mu_G \tfrac{l+a}{l-a} \left(\sqrt{\tfrac{k_k^2}{1+k_k^2}} + \mu_{Sp}^2\right) F_{R,2} \operatorname{sign}(\Delta \dot{y}_2) , \end{aligned} \tag{14.87}$$

$$\begin{aligned} F_{C,3} &= \mu_G(\mid Q_1 \mid + \mid Q_2 \mid)_3 \operatorname{sign}(\Delta \dot{y}_3) \\ &= \mu_G \tfrac{l+a}{l-a} \mid F_{N,3} \mid \operatorname{sign}(\Delta \dot{y}_3) \\ &= \mu_G \tfrac{l+a}{l-a} \mid \left(\sqrt{\tfrac{k_k^2}{1+k_k^2}} - \mu_{Sp}\right) F_{R,3} \mid \operatorname{sign}(\Delta \dot{y}_3) . \end{aligned} \tag{14.88}$$

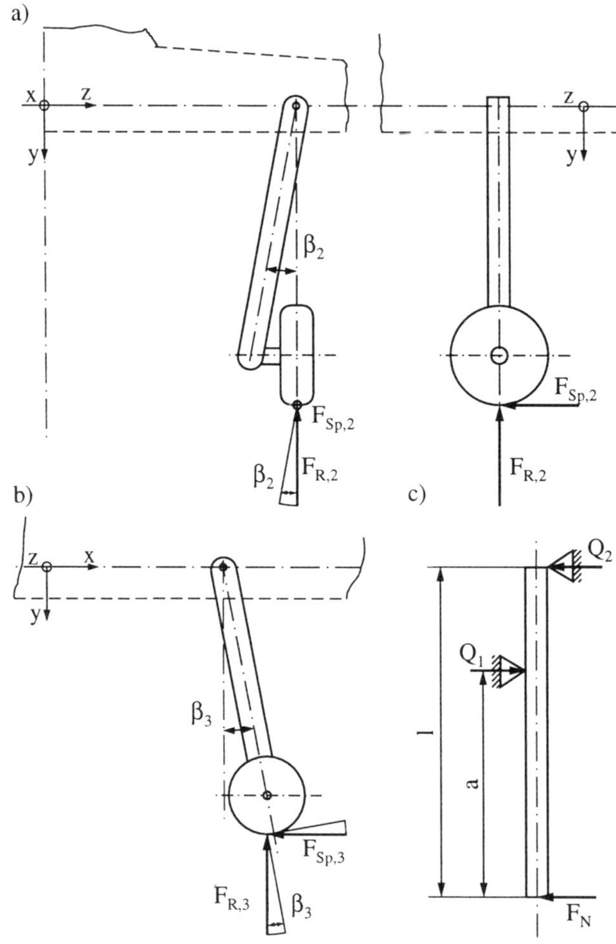

Figure 14.27: Main Carriage (a), Nose Carriage (b), System of Horizontal Forces (c)

The stroke Δy_i can be evaluated from the static stroke Δy_{0i}, from a linear combination of generalized coordinates and elastic deformation, which yields

$$\Delta y_i = \Delta y_{0i} + y(t) - x_{1,i}\alpha(t) - y_{2,i}(t) + \overline{\boldsymbol{w}}_1^T(x_{1,i})\boldsymbol{q}_1(t)\,;$$

$$\Delta \dot{y}_i = \tfrac{d}{dt}\Delta y_i = \dot{y}(t) - x_{1,i}\dot{\alpha}(t) - \dot{y}_{2,i}(t) + \overline{\boldsymbol{w}}_1^T(x_{1,i})\dot{\boldsymbol{q}}_1(t)\,; \tag{14.89}$$

$$i = 2, 3\ ,$$

with $k_k = \tan\beta$ and, therefore,

$$\sin\beta = \sqrt{\frac{\tan^2\beta}{1+\tan^2\beta}} = \sqrt{\frac{k_k^2}{1+k_k^2}}$$

($x_{1,2}$ and $x_{1,3}$ are hinge coordinates, μ_G is dry friction coefficient).

Unilateral Contacts

We consider the following unilateral contacts:

- stick-slip in the telescopic shock absorbers,
- stop at the ground side of the absorbers,
- stop at the airplane side of the absorbers.

Thus we expect impulsive and stick-slip phenomena, and the equations of Chapters 6 and 7 apply. The combination of the main and nose carriages can generate four modes of motion:

- main and nose carriages move freely,
- main carriage free, nose carriage blocked,
- main carriage blocked, nose carriage free,
- main and nose carriages blocked.

The constraint vectors $\boldsymbol{w}$ for the $\boldsymbol{W}$-matrices in eqs. (6.1), (6.5), (6.11), (6.36) must cover all possibilities. From the stroke velocities (eq. 14.89) we get (see eqs. 14.33) for stick-slip

$$\begin{aligned} \boldsymbol{w}_{k,2}^T &= (0, 1, -x_{1,2}, -1, 0, 0, 0, \overline{\boldsymbol{w}}_1^T(x_{1,2}), 0, 0) \\ \boldsymbol{w}_{k,3}^T &= (0, 1, -x_{1,3}, 0, -1, 0, 0, \overline{\boldsymbol{w}}_1^T(x_{1,3}), 0, 0) \\ \overline{\boldsymbol{w}} &= \boldsymbol{0} \\ k &= 1 \text{ or } 2\,. \end{aligned} \tag{14.90}$$

The frictionless contact-impact constraints and transitions are handled as follows:

- near ground:

$$\begin{aligned} \Delta y &= \Delta y_{\min} && \text{lower contact closed} \\ \Delta \dot{y} &< 0 && \text{relative velocity before impact} \\ \Delta \dot{y} &= 0 && \text{relative velocity after inelastic impact} \\ F_Z &< 0 && \text{contact force when contact is maintained} \end{aligned} \tag{14.91}$$

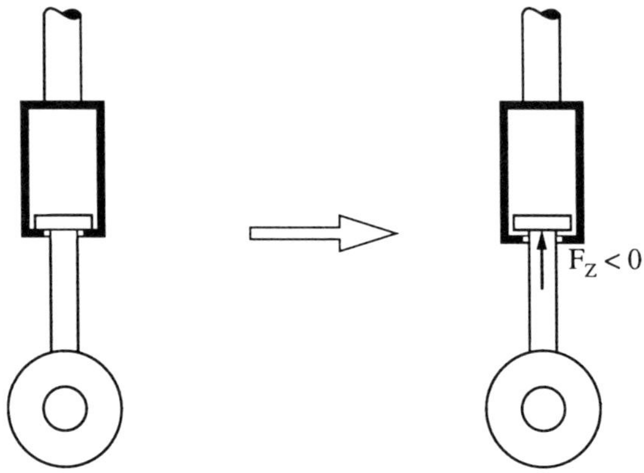

Figure 14.28: Constraint Force for a Closed Contact

- near airplane (see Fig. 14.28):

$$\begin{aligned}
\Delta y &= \Delta y_{\max} && \text{upper contact closed} \\
\Delta \dot{y} &> 0 && \text{relative velocity before impact} \\
\Delta \dot{y} &= 0 && \text{relative velocity after inelastic impact} \\
F_Z &> 0 && \text{contact force when contact is maintained}
\end{aligned} \tag{14.92}$$

Unsteadiness	Start	End
Stick-slip	$\mid F_{FS} \mid - \mid F_{C0} \mid > 0$ $\Delta \dot{y} = 0$	$\mid F_{FS} \mid - \mid F_{C0} \mid = 0$ $\Delta \ddot{y} \neq 0$
Stop top	$\left.\begin{array}{l}\Delta y = \Delta y_{\max} \\ \Delta \dot{y} > 0\end{array}\right\} F_Z > 0$	$F_Z = 0$
Stop bottom	$\left.\begin{array}{l}\Delta y = \Delta y_{\min} \\ \Delta \dot{y} < 0\end{array}\right\} F_Z < 0$	$F_Z = 0$

Table 14.3: Conditions for Contact Events

The contact constraints open when the forces F_Z are going to change sign. The procedure with regard to these equations closely follows Section 3.2 and the hammering example (Chapter 11). The conditions for the contact events (see Part 1) are shown in Table 14.3. (F_{FS} is static friction force, F_{C0} is tangential constraint force, for F_Z see Fig. 14.28.)

14.4.3 Simulations

LANDING ON A PLANE RUNWAY—RIGID AIRPLANE

The first simulations consider a landing impact on a plane surface with a medium general aviation propeller-driven airplane consisting of a rigid fuselage and rigid carriages. The touchdown velocity is 4.3 m/s, and the horizontal velocity is 60 m/s. The landing angle of attack is $\alpha = 13°$, and the angular velocity is $\dot{\alpha} = 0$. The initial values of the geometric magnitudes y, $y_{2,2}$, $y_{2,3}$ are chosen in such a way that touchdown of the main carriage wheels starts with zero forces, $F_{R,2} = 0$, $F_{S_P,2} = 0$. The nose wheel has no ground contact for these initial values ($\alpha = 13°$).

Figure 14.29 illustrates the situation of the airplane for the above landing impact. The center of mass moves about 300 mm (curve 1), and its velocity has a maximum at the beginning (curve 2). The forces acting on the fuselage from the main and nose carriages reach a maximum of about 170 kN for the main absorbers (curve 3) and a peak value of about 450 kN for the nose, where the nose load builds up later (curve 4).

The motion of the carriages is of course similar (Fig. 14.30). The velocity diagrams (curves 6, 8) nicely indicate the impulsive and stick-slip processes within the shock absorbers. Typically the main absorbers behave much more

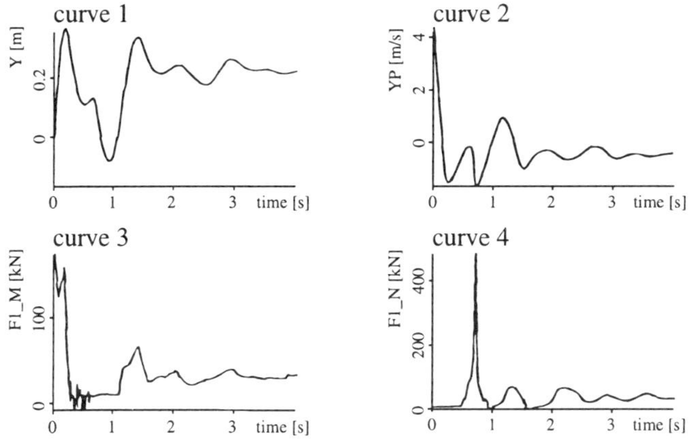

Figure 14.29: Rigid Airplane Landing Impact—Airplane Results

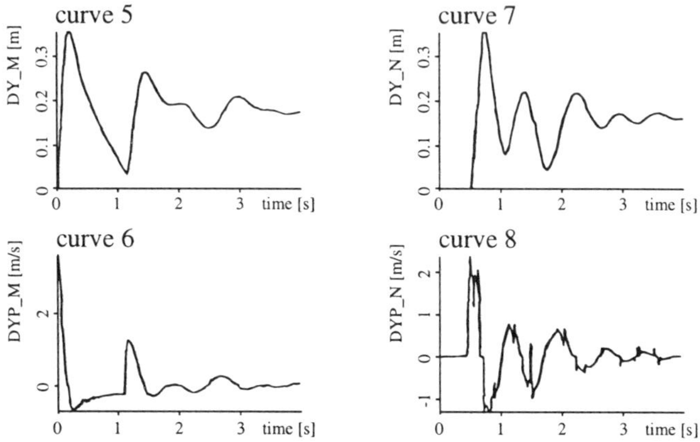

Figure 14.30: Rigid Airplane Landing Impact—Carriage Results

smoothly than the nose absorber, which shows a series of impacts and friction phenomena.

Landing on a Plane Runway—Elastic Airplane

Elasticity has been considered by a Ritz approach (eqs. 14.37) and by taking into account the first five eigenfrequencies for the fuselage and the first two eigenfrequencies for the main and nose carriages. All other data were the same as for the rigid airplane. Results are partially different from those in

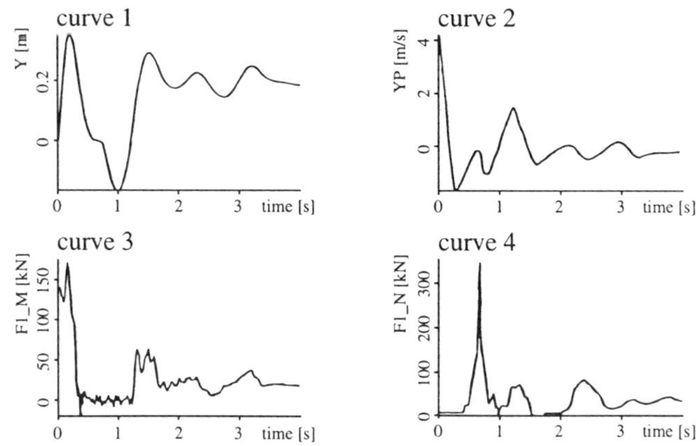

Figure 14.31: Elastic Airplane Landing Impact—Airplane Results

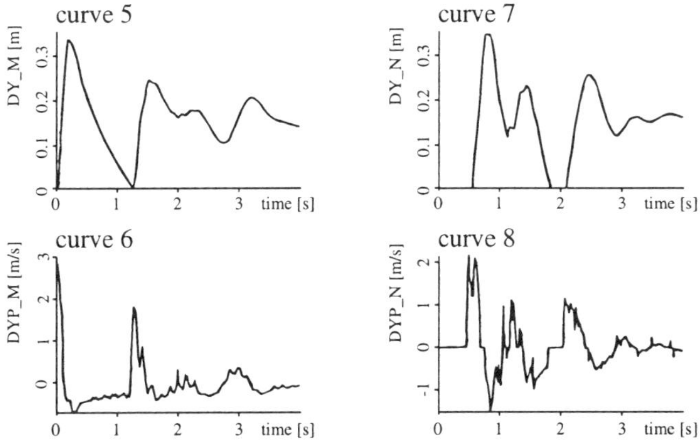

Figure 14.32: Elastic Airplane Landing Impact—Carriage Results

Figs. 14.29, 14.30 (see Figs. 14.31, 14.32).

The displacements are more or less the same, the nose wheel performs an internal impact at the upper stop (curve 7), and in the force and stroke velocity diagrams the lowest elastic fuselage eigenfrequency (~7 Hz) can be detected. The elasticity neither increases nor decreases the tendencies for impacts and stick-slip (curves 8 in Figs. 14.30, 14.32), but it reduces significantly the maximum load on the nose carriage.

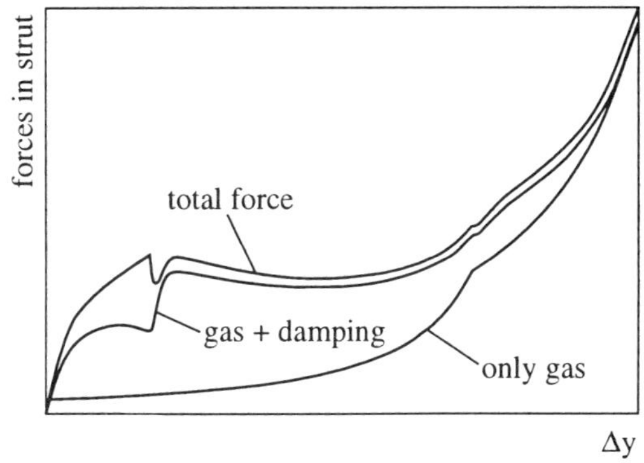

Figure 14.33: Shock Absorber Force-Stroke Diagram for Main Landing Gear

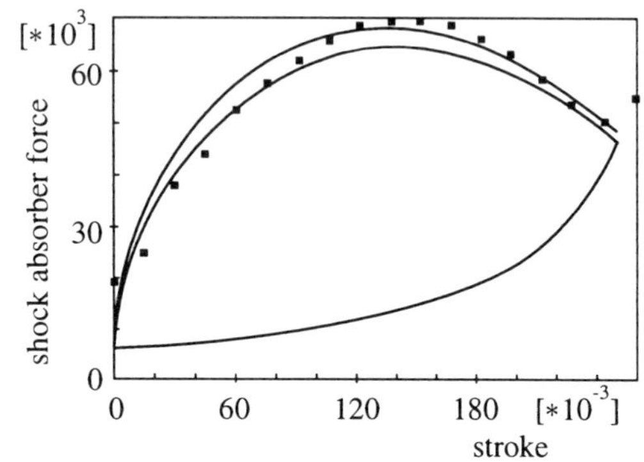

Figure 14.34: Drop Test: Simulation and Measurement

SHOCK ABSORBER OPERATION

According to eqs. (14.86) the shock absorber forces depend on the gas work, on the oil flow dissipation and on dry friction effects. Figure 14.33 portrays some typical force relationships. For the landing procedure the course of the force F_1 over the displacement Δy is of value to determine the work absorbed by the landing gear. The bottom curve is the gas spring force, the next higher curve also contains the damping force, and the upper curve shows the complete force, including F_C. The point of change in the stage of the shock absorber is clearly visible in the bottom curve. The effect of the wheel spin-up on the friction force is indirectly seen at the beginning of the compression. The accuracy of the shock absorber model has been validated in the comparison between simulation and measurement of a drop test as shown in Fig. 14.34. The drop test was performed by a German airplane manufacturer.

14.5 Assembly Processes

14.5.1 Introduction

Typical assembly tasks require the manipulator to operate close to or in contact with the environment. Small deviations in the initial conditions can therefore determine whether contact occurs they can lead to contact configurations completely different from those under ideal conditions. As a result, the uncertainties related to real-world assembly systems may produce excessive mating forces which affect the quality of the surfaces of the mated parts or can even cause the assembly operation to fail [89, 90, 91, 99, 100].

There has already been a great deal of research in the field of automated assembly which addressed those problems predominantly in an experimental context. Various forms of compliance mechanisms, based on the remote center compliance (RCC), have been developed, and the use of active force feedback and part mating strategies have been presented in order to compensate for external disturbances and to reduce mating forces (for an overview see, e.g., [83]). The interaction of the manipulator dynamics and the involved contact mechanics has been studied only from the theoretical point of view in the past few years, mainly in the context of contour tracking problems where essentially constraints with nonchanging structures have been dealt with (e.g., [84]).

This section deals with the dynamic formulation of the manipulator's motion in the presence of time-varying constraints during assembly tasks. The equations of motion for such systems are formally derived by adding the contact forces in the form of Lagrange multipliers to the open-loop dynamics and by considering an additional set of equations describing the acting geometric constraints (see Chapter 5). The current contact configuration and the respective system of differential equations are controlled by so-called kinematic and kinetic indicator magnitudes which are characteristic for such systems with time-varying topologies. Contact is controlled by the relative distance between two bodies and the constraint force perpendicular to the contact plane, while the states of sliding and stiction are indicated by the relative velocity and the difference between maximum static friction force and constraint friction force. These magnitudes indicate all changes in the acting constraints and hence in the manipulator's mobility.

The dynamics of assembly processes can then be investigated in two stages. On the one hand, we consider the assembly process under ideal conditions, which yields the necessary mating forces and the required extra actuator input torques. On the other hand, we analyze the complete system behavior in the presence of tolerances which allows us to determine the sensitivity of an assembly process toward various parameter uncertainties. As illustrating examples the mating of a flexible snap joint and the peg-in-hole problem are considered. In addition we shall investigate the behavior of flexible rubber-like rings when assembled in a corresponding hole. A combination of impacts, friction and nonlinear deformation behavior will be observed [57].

14.5.2 Mechanical and Mathematical Models

During research the example of assembly processes significantly influenced the methodical development of the complementarity aspects for multibody systems with multiple nondecoupled contacts. It is a typical example where all the methods and equations of Chapters 6 and 7 apply. Therefore we shall discuss the matrices and vectors for the examples mentioned.

Peg-in-Hole Problem

The combination of robot dynamics control and contact dynamical processes has been studied for the planar peg-in-hole insertion task. For verification of our theoretical approach, experiments with a five-degree-of-freedom laboratory robot with a force-torque sensor were carried out. The fixture, housing the complementary part, is equipped with six distance sensors that measure the gripper's position and orientation and dispose of two translational and three rotational ajustments which allow production of definite positioning errors with respect to the parts to be mated. In experiments we investigated the insertion process when there is a lateral error between the peg and the hole.

For establishing first the theoretical model, we consider the constraints of the system. The geometric constraints of an assembly operation basically require the parts to not penetrate each other. In order to establish the underlying constraint inequalities we assume that the geometry of the bodies can be described by piecewise plane surfaces characterized by a series of straight lines in a given cross section. The contour of the parts is defined in body-fixed coordinate frames ${}_BK$ and ${}_LK$ whose mutual position is specified by a translation $\boldsymbol{r}$ and rotation φ in task space (Fig. 14.35).

Each corner point that might come into contact with a surface of the complementary part represents a possible contact point or constraint, respectively, and can be collected in a set $\mathcal{M}_L$ of all latent constraints. A point of contact is defined by the condition that the distance between two bodies becomes zero:

$$g_N(\boldsymbol{r}, \varphi) = \boldsymbol{n}^T(\boldsymbol{x} - \boldsymbol{p}) = 0 \ , \quad g_N \in \mathcal{M}_L \tag{14.93}$$

$\boldsymbol{n}$: normal vector of the contact surface
$\boldsymbol{x}$: coordinates of any point within the contact surface
$\boldsymbol{p}$: coordinates of the corner point under consideration of the complementary body

whereas for all passive constraints $g_N(\boldsymbol{r}, \varphi) > 0$ must be satisfied. Constraints of type (14.93) implicitly describe the restriction of the possible gripper motion and $\dim(\boldsymbol{r}, \varphi)$. Since the gripper position depends on the manipulator's generalized coordinates $\boldsymbol{q}$, (14.93) can also be expressed as a holonomic constraint in the form of $g_N(\boldsymbol{q}) = 0$.

Thus, the set of latent constraints $\mathcal{M}_L$ is solely defined by the geometry of the components to be assembled, which generally holds for all insertion tasks of arbitrarily shaped parts. On the other hand, their evaluation is based in

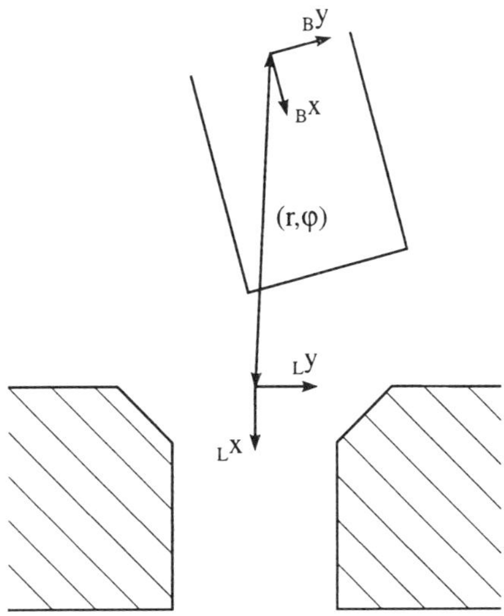

Figure 14.35: Coordinate System of the Mating Parts [89]

a straightforward manner on the manipulator's kinematics so that it can be carried out automatically.

The geometrical constraints for the mating parts are derived by the methods of Chapter 4, and the relevant equations of motion follow from Chapters 6 and 7, particularly from eqs. (6.1), (6.36), (7.11), (7.12) in combination with some equations of Chapter 8. For more detail see, for example, [89, 99]. In considering assembly by manipulators, we fuse the nonlinear contact-driven dynamics of the assembly process itself with the dynamics of the robot, which, for this specific case, usually can be linearized around the "point" of assembly. This set of equations underlies the structure-variant behavior of the contact events during assembly, which is controlled by indicators. As discussed in Part 1 the event that causes a change in the acting constraints and hence in the system structure can be traced by introducing a set of kinematic and kinetic indicator magnitudes which corresponds to an investigation in two steps. First we select from the entirety of latent constraints those which unilaterally constrain the manipulator's motion for the present state $(\boldsymbol{q}^T \; \dot{\boldsymbol{q}}^T)^T$. This subset is determined with the aid of kinematic test functions:

(a) relative distance between two bodies

the relative position of two parts is tested by examining the distance g_N of potential contact configurations; for $g_N > 0$ the constraint under consideration is not active, $g_N = 0$ means geometric contact

(b) relative position s of the contact point with respect to the surface edges in direction of insertion

if the objects under consideration are defined by a series of bounded surface fragments the constraints change when a contact point passes over an edge separating two surface fragments; for that reason we specify an additional geometric magnitude s that controls the distance between a contact point and the constraint surface edge in direction of insertion

(c) relative velocity $\dot{\boldsymbol{g}}_T$ in the tangential contact plane

as long as $\dot{\boldsymbol{g}}_T$ is nonzero sliding friction is present, otherwise stiction may be possible

These unilateral constraints then define a set of conditions that must be satisfied with respect to kinetics. If there is geometric contact $g_{Ni} = 0$ the permissible motion of the contact point perpendicular to the constraint surface is limited to $g_{Ni}(\boldsymbol{q}) \geq 0$ or, equivalently, $\ddot{g}_{Ni} \geq 0$, while the constraint force is simultaneously restricted by $\lambda_{Ni} \geq 0$ with ($\ddot{g}_{Ni} = 0$, $\lambda_{Ni} \geq 0$) or ($\ddot{g}_{Ni} \geq 0$, $\lambda_{Ni} = 0$). On the other hand, if the relative velocity becomes zero in some contact point i the maximum friction force according to Coulomb's law is limited by the friction saturation $\lambda_{H0i} = \mu_0 \lambda_{Ni} - |\lambda_{Ti}| \geq 0$. If stiction is present $\ddot{g}_{Ti} = 0$ and $\lambda_{H0i} \geq 0$ must be satisfied, whereas at a transition to sliding the friction force λ_{Ti} opposes the starting relative motion characterized by $\ddot{g}_{Ti} \neq 0$ and, hence, $\ddot{g}_{Ti}\lambda_{Ti} \leq 0$. Assuming a steady characteristic relating the coefficient of friction and the relative velocity the magnitude of the sliding friction force is furthermore specified by $\lambda_{H0i} = 0$ with $\mu(v_{rel} = 0) = \mu_0$.

Figure 14.36 illustrates some typical contact configurations for the plane peg-in-hole insertion and the most important coordinate frames applied for the robot (I inertial, G gripper, B mating body).

Assembly with Compliant Parts

Assembling compliant parts we leave the assumptions of rigid body contacts and therefore a large part of the theory presented in Part 1. But on the other hand we still have to deal with indicators to prove contact events, and we still have to consider friction influences. Nevertheless the complementarity idea does not apply here due to the fact that by elasticity of the mating parts the contacts are decoupled. In the following we shall investigate particularly a peg-in-hole problem where an elastic rubber-like ring will be inserted into a rigid hole. The robot is a PUMA robot modeled with rigid links but with elastic joints with dry friction and backlashes [57].

The robot model is straightforward and will not be discussed here [57]. The model of the nonlinear elastic ring consists of a series of elementary

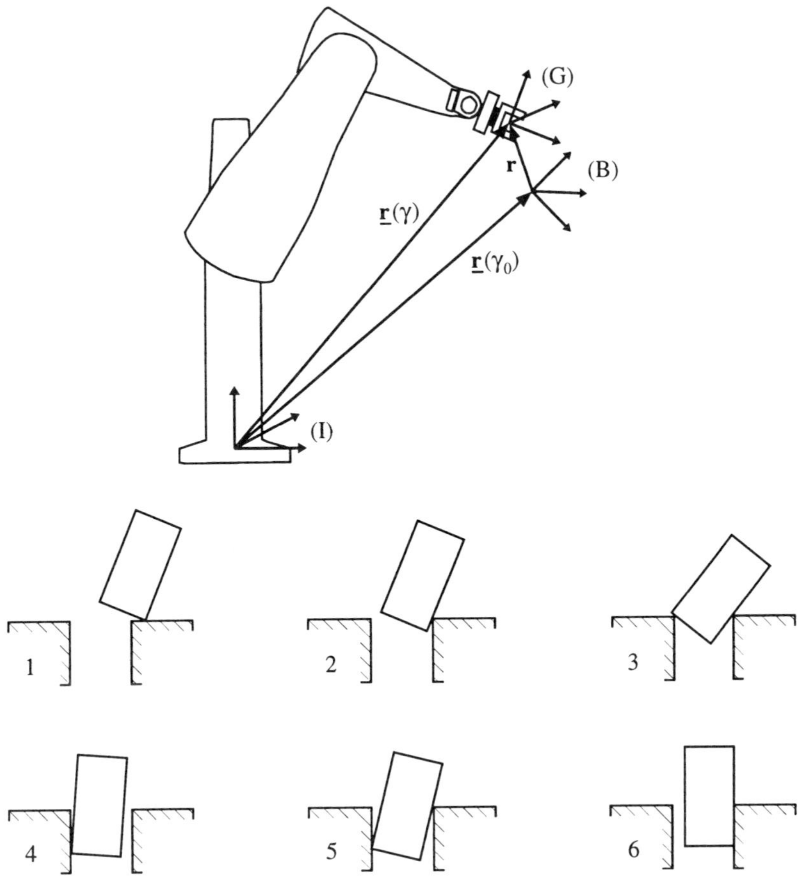

Figure 14.36: Coordinate Systems for the Manipulator and Some Typical Contact Configurations [89]

infinitesimal springs in the circumferential direction, where the spring characteristics are determined from FEM calculations. We regard the elasticity of an infinitesimal sector of the torus, when $d\varphi$ becomes zero ($d\varphi \to 0$, Fig. 14.37), and thus it is possible to evaluate the resulting forces and torques acting on the piston by integrating over the circumference. Friction between the ring and the wall of the hole is taken into account. Notches in the wall are permissible. The outer contour of the ring is assumed to be circular and rigid so that the contact point with the environment may be easily determined. The problem of obtaining the stiffness of the spring is solved by a

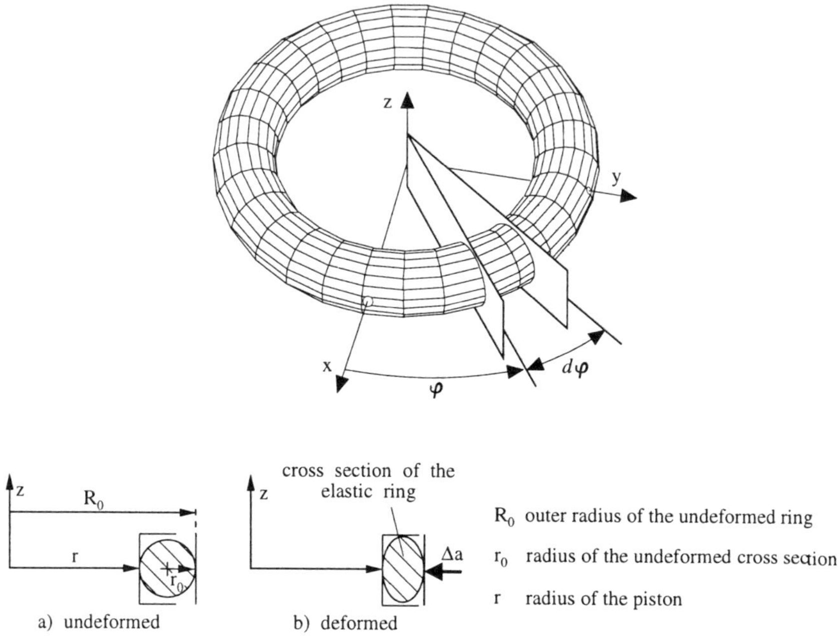

Figure 14.37: Model of the Elastic Ring [57] (© 1994 IEEE)

simplified two-dimensional FEM computation with axisymmetric elements, as suggested in Fig. 14.37. The torus is thus concentrically deformed from all sides. The radial reaction force onto the inner wall of the FEM model (Fig. 14.37) is divided by the circumference to get a line load. From this load characteristic the stiffness parameters can be determined. They are also dependent on the specific material properties, which are taken into account in the FEM computations.

We must combine the model of the ring with a gripper model and with the hole, where the ring must be inserted. For this purpose we define the coordinate frames depicted in Fig. 14.38. Starting from the inertial system $\boldsymbol{I}$ the position and orientation of the gripper system $\boldsymbol{G}$ and the (inertially fixed) hole system $\boldsymbol{C}$ are described by the vectors ${}_I\boldsymbol{r}_{IG}$ and ${}_I\boldsymbol{r}_{IC}$ and the roational transformation matrices $\boldsymbol{A}_{IG}$ and $\boldsymbol{A}_{IC}$, respectively. The reference system $\boldsymbol{R}$ on the piston is defined in the $\boldsymbol{G}$ system of the gripper of the robot. The vector ${}_G\boldsymbol{r}_{GR}$ then represents the position of the piston relative to the gripper in the $\boldsymbol{G}$ frame. For each ring an additional local system $\boldsymbol{L}$ is introduced in which the following calculations are made. Each $\boldsymbol{L}$ system has a parallel z-axis with the $\boldsymbol{C}$ system because we further assume that the ring does not tilt in the hole. This is a feasible assumption since the orientational

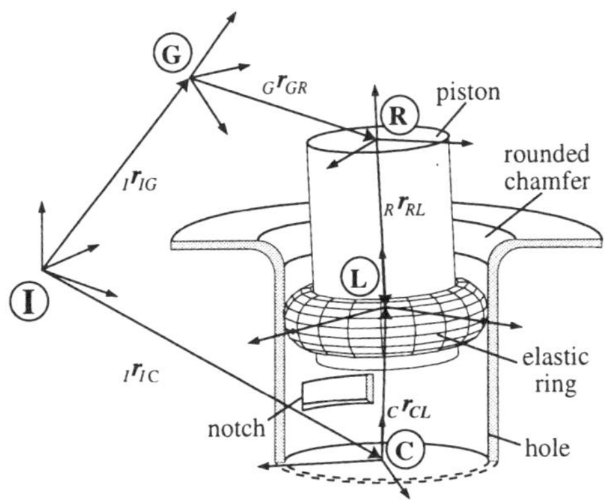

Figure 14.38: Coordinate Frames [57] (© 1994 IEEE)

errors of industrial robots are generally small. The vector ${}_R\boldsymbol{r}_{RL}$ has only a nonzero z-component and defines the position of the ring on the piston. The $\boldsymbol{L}$ system has a radial displacement Δr and a rotation φ_{CL} with respect to the $\boldsymbol{C}$ system (Fig. 14.39). The x_L axis coincides with the shortest stretch between piston and hole. Δr and φ_{LC} are evaluated from the previously defined vectors:

$$ {}_C\boldsymbol{r}_{CL} = \boldsymbol{A}_{CI} \left[{}_I\boldsymbol{r}_{IG} - {}_I\boldsymbol{r}_{IC} + \boldsymbol{A}_{IG} \left({}_G\boldsymbol{r}_{GR} + \boldsymbol{A}_{GR} \, {}_R\boldsymbol{r}_{RL} \right) \right] \tag{14.94} $$

$$ \Delta r = \sqrt{ {}_C r_{CL,x}^2 + {}_C r_{CL,y}^2 } \tag{14.95} $$

$$ \varphi_{CL} = \arctan \left(\frac{ {}_C r_{CL,x} }{ {}_C r_{CL,y} } \right) . \tag{14.96} $$

The slit function in the $\boldsymbol{L}$ system (Fig. 14.39) is approximated by a cosine function:

$$ a(\varphi_L) \approx R - r - \Delta r \cos(\varphi_L) . \tag{14.97} $$

The characteristic of the elastic ring is evaluated with the FEM program MARC. The rubber material of the ring is described by the MOONEY-RIVLIN material law. The resulting force in radial direction is taken from the FEM calculation divided by the circumference of the ring $2r\pi$ and plotted over the radial displacement Δa. The characteristic is nonlinear, as shown in Fig. 14.40, and is approximated by a quadratic polynomial function:

$$ p_N(\Delta a) = c_1 \Delta a + c_2 \Delta a^2 . \tag{14.98} $$

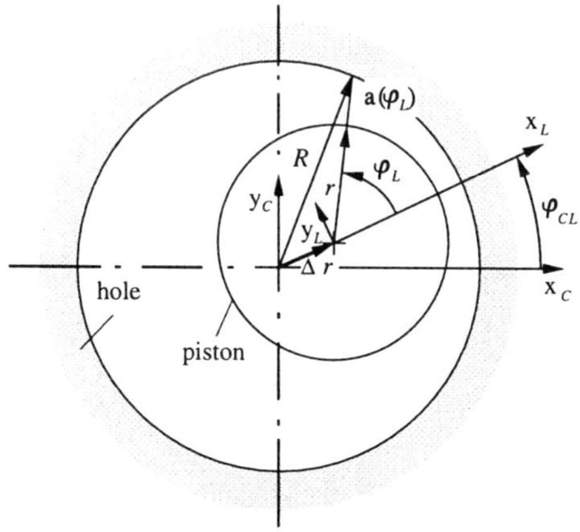

Figure 14.39: Displacement between Piston and Hole [57](© 1994 IEEE)

With the use of eqs. (14.97) and (14.98) the line load over the circumference resulting from the radial displacement of the ring Δr is determined:

$$\begin{aligned} \Delta a(\varphi_L) &= R_0 - r - a(\varphi_L) = R_0 - R + \Delta r \cos(\varphi_L) \\ p_N(\varphi_L) &= k_0 + k_1 \cos(\varphi_L) + k_2 \cos^2(\varphi_L) \end{aligned} \tag{14.99}$$

where

$$\begin{aligned} k_0 &= c_1(R_0 - R) + c_2(R_0 - R)^2 \\ k_1 &= \Delta r(2c_2(R_0 - R) + c_1) \\ k_2 &= \Delta r^2 . \end{aligned}$$

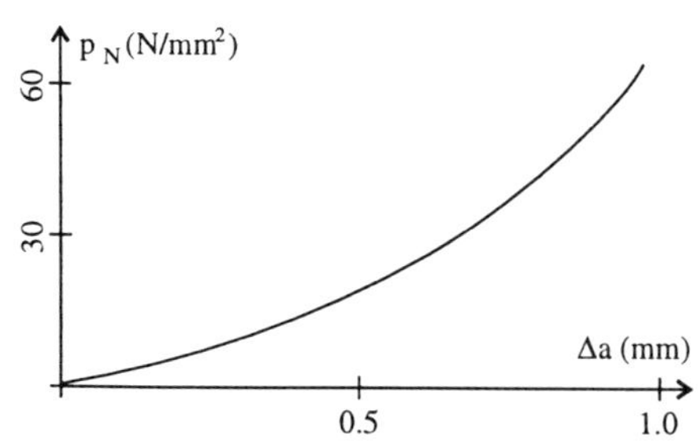

Figure 14.40: Nonlinear Characteristic for the Elastic Ring [57] (© 1994 IEEE)

This line load may be split into the cartesian directions of the $\boldsymbol{L}$ system:

$$\begin{aligned} p_x(\varphi_L) &= p_N(\varphi_L)\cos(\varphi_L) \\ p_y(\varphi_L) &= p_N(\varphi_L)\sin(\varphi_L) \\ p_z(\varphi_L) &= \mu p_N(\varphi_L) \end{aligned} \tag{14.100}$$

where μ denotes the coefficient of friction. The resulting forces and torques are then found via an integration along the circumference of the ring. If the hole contains notches, the integration is performed over a bounded region.

The above model is coupled with the robot model via the forces and torques on the gripper and resulting from the mating process. We get the following equations of motion for the mounting task by supplementing the manipulator's equations by the generalized mating forces:

$$\boldsymbol{M}\ddot{\boldsymbol{q}} + \boldsymbol{h}(\dot{\boldsymbol{q}}, \boldsymbol{q}, t) = {}_R\left(\boldsymbol{J}_T^T, \boldsymbol{J}_R^T\right)_R \begin{pmatrix} \boldsymbol{f}_R \\ \boldsymbol{\tau}_R \end{pmatrix} . \tag{14.101}$$

The PUMA robot is modeled with $f = 9$ degrees of freedom ($\boldsymbol{q} \in \mathbb{R}^9$). Its motion is contained in the left-hand side of eqs (14.101). The coupling between the assembly process and the robot's motion in eq. (14.101) arises from the mating forces, which depend on the kinematics ${}_I\boldsymbol{r}_{IG}(\boldsymbol{q})$ and $\boldsymbol{A}_{IG}(\boldsymbol{q})$ of the process model and the generalized coordinates $\boldsymbol{q}$ of the dynamic model, respectively.

14.5.3 Results

PEG-IN-HOLE PROBLEM

With respect to applications in manufacturing processes we have two goals. First we want to map an insertion as detailed as possible to establish a basis for manufacturing planning. Some typical problems are selection of robot type and the tools, the arrangement of robots in a manufacturing cell, and the configuration of robot and environment for certain assembly tasks. Second we are interested in the question, how much error tolerance can be permitted for a given robot-assembly combination? This directly influences the accuracy required for the manipulator type to be selected.

To verify the theory some experimental tests have been performed. Figure 14.41 shows the principle of this experiment and the laboratory test setup. In the first experiment we investigated the insertion process when there is a lateral error between the peg and the hole. The respective numerical and experimental results are depicted in Fig. 14.42. The peaks in the force history at $t = 0.9$ s result from the impact when the peg hits the chamfer. When two-point contact is reached at $t = 1.2$ s the manipulator's motion decreases quickly until the jamming condition is reached. Only when the controller

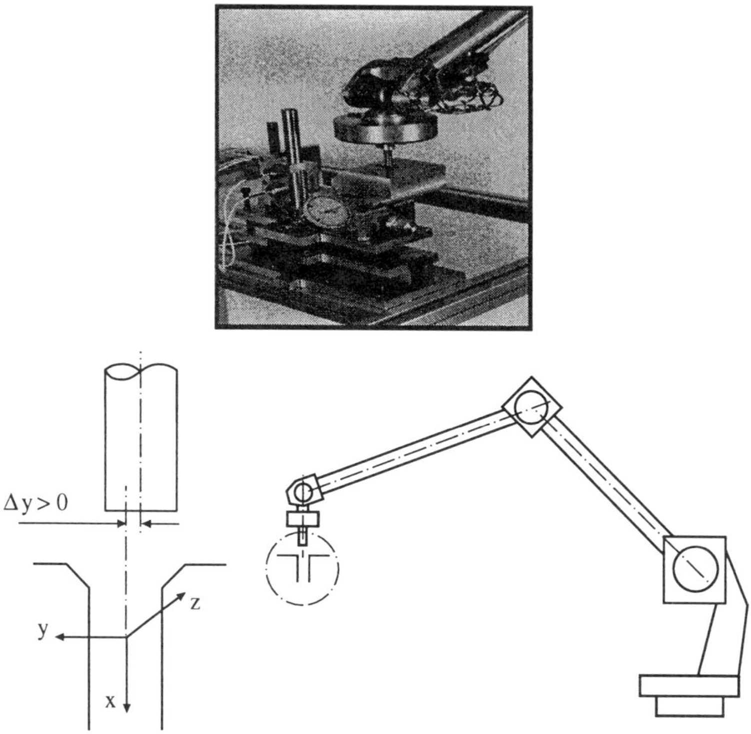

Figure 14.41: Peg-in-Hole Experiment

torques become large enough is the motion continued with sliding in two, and finally one, contact points. This cycle represented by the oscillations in the force history repeats until the end of the trajectory and demonstrates the phenomenon of changing constraints during an assembly operation.

In the second experiment we investigated the permissible tolerance range of lateral and angular errors of the peg so that a maximum force of 5 N is not exceeded during the insertion phase. In this context we did not take into consideration the forces that develop while the peg is guided along the chamfer, since this automatically involves chamfer design [104]. Figure 14.43 contrasts the numerical results with some data obtained by experiment.

The respective lines identified by theory correspond to the existence of different contact configurations in which the maximum insertion force of 5 N is reached and the range of permissible positioning tolerances is limited. The difference between numerical and experimental data mainly results from backlash in the last joint of the robot which was not modeled. Note that in both

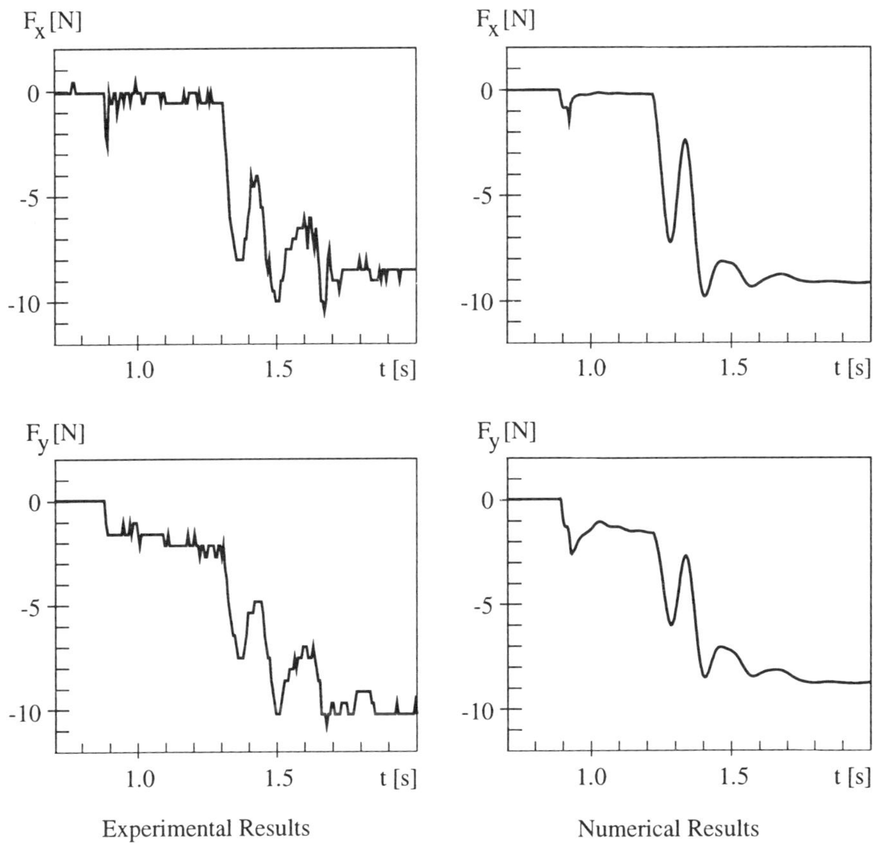

Figure 14.42: Comparison of Simulation and Measurements for a Peg-in-Hole Insertion with Lateral Error (F_x Insertion Force, F_y Lateral Force)

experiments the sensor served only as a force measuring device for data processing, since a simple PD joint controller was used.

Compliant Mating Parts

We consider an elastic rubber-like ring on a cylinder which is inserted into a rigid hole (Fig. 14.38) by a PUMA 562 robot. For verification of the theory an experiment with a rubber ring on a piston being mated in a hole under perfect conditions is performed. The ring possesses the parameters $r = 10.2$ mm, $r_0 =$ 1.5 mm (Fig. 14.37), and the hole has a radius $R = 13$ mm. Only forces in the z-direction of the insertion are generated. The comparison in Fig. 14.44

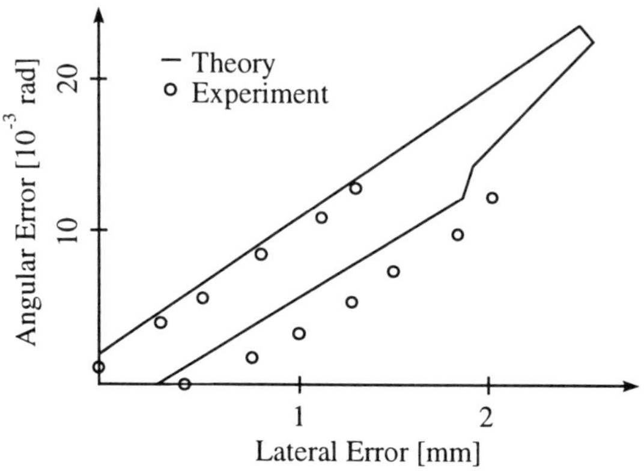

Figure 14.43: Permissible Positioning Errors for a Maximum Insertion Force of 5 N

shows good agreement.

After verification of the theoretical model an insertion task with two elastic rings is investigated. They are fabricated from the same material as the ring in the experiments, but they are larger. The ring distance amounts to 10 mm. Figure 14.45 portrays a variation of the parameter Δr, which is the radial

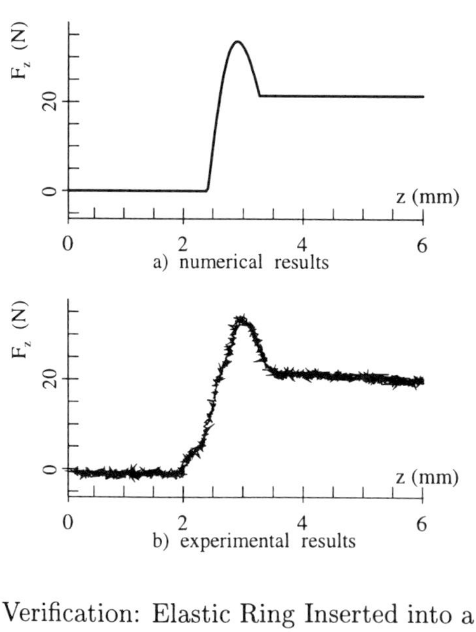

Figure 14.44: Verification: Elastic Ring Inserted into a Rigid Hole [57] (© 1994 IEEE)

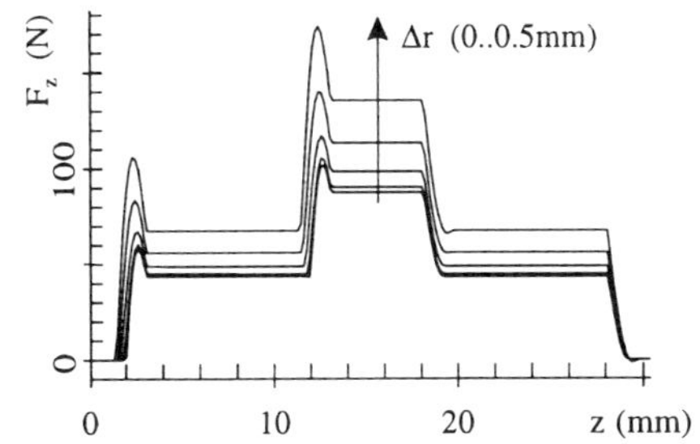

Figure 14.45: Insertion with Two Elastic Rings and Var. Lateral Errors [57] (© 1994 IEEE)

displacement error at the beginning of the insertion process (parameters $r = 10.45$ mm, $r_0 = 2.5$ mm, $R = 13$ mm). As could be expected, the mating force F_z increases significantly with growing lateral displacement. From this, conclusions can be drawn for the necessary control accuracy of the robot dependent on its force budget. With regard to the ring we thus can find the necessary accuracy and from this the maximum applied forces that will not destroy the ring.

In considering the combined dynamics of the robot together with the double-o-ring insertion, the model of the o-rings is coupled in the described way with the dynamic model of the robot. For the insertion task a point-to-

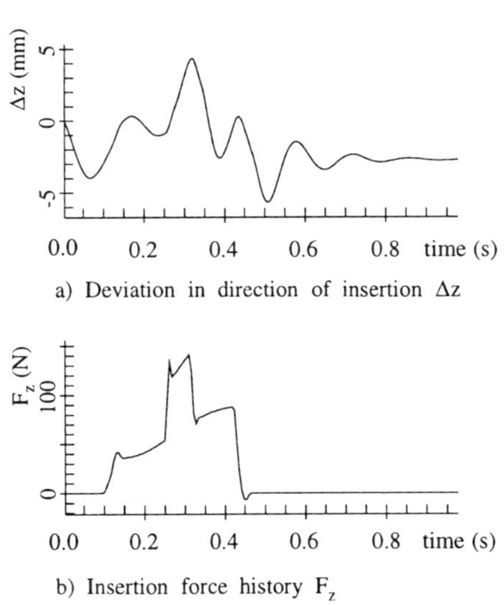

Figure 14.46: Results from the Insertion Task

point (PTP) path is used. The robot moves 3 cm in the negative z-direction (see Fig. 14.38) in 0.5 s.

In the other two cartesian directions no motion is desired. In Fig. 14.46(a) we plot the deviation of the gripper from the planned path in the z-direction. It can be recognized that the greatest positional error of 5 mm occurs when the force on the piston (Fig. 14.46b) reaches a maximum. The second ring then has just passed the rounded chamfer (see Fig. 14.38). After 0.5 s the last o-ring has left the hole and the resulting jump in the mating force excites an oscillation, which fully decays after 0.9 s. The end position of the gripper is about 3 mm lower than desired because of the elasticity of the gears and the controller. Through the weight of the arms the spring in the gear model is tightened. In Fig. 14.46(b) we see the actual active force in the z-direction resulting from the rings on the piston. Differences in the force under ideal conditions (Fig. 14.45) arise from the radial position errors of the gripper with respect to the hole.

REFERENCES

[1] T. Alishenas, "Zur numerischen Behandlung, Stabilisierung durch Projektion und Modellierung mechanischer Systeme mit Nebenbedingungen und Invarianten," TRITA-NA-9202, Dept. of Numerical Analysis and Computing Science, The Royal Institute of Technology, Sweden, 1992.

[2] L. S. An, "The painleve paradoxes and the law of motion of mechanical systems with Coulomb friction," in *PMM U.S.S.R.,* Vol. 54, No. 4, 1990, pp. 430–438.

[3] W. Blajer, "A projection method approach to constrained dynamic analysis," in *J. Appl. Mech.,* Vol. 59, 1992, pp. 643–649.

[4] R. M. Brach, "Friction, restitution and energy loss in planar collisions," in *Trans. ASME,* Vol. 51, 1984, pp. 164–170.

[5] R. M. Brach, "Rigid body collisions," in *J. App. Mech.,* Vol. 56, 1989, pp. 133–138.

[6] H. Brandl and M. Hajek, "Mechanische Systeme mit Trockenreibung," in *ZAMM,* Vol. 68, No. 4, 1988, T59–T61.

[7] H. Brandl, R. Johanni, and M. Otter, "A very efficient algorithm for the simulation of robots and similar multibody systems without inversion of the mass-matrix," in *Proc. IFAC/IMACS Int. Symp. Theory of Robots,* Wien, 1987.

[8] H. Brandl, R. Johanni, and M. Otter, "An algorithm for the simulation of multibody systems with kinematic loops," in *Proc. 7th World Congress on the Theory of Machines and Mechanisms (IFToMM),* Sevilla, 1987.

[9] J. Braun, "Dynamik und Regelung elastischer Flugzeugfahrwerke," Thesis at the Dept. of Mech. Eng., Inst. B f. Mech., Tech. Univ. of Munich, Oct. 1989.

[10] H. Bremer, "Kinetik starr-elastischer Mehrkörpersysteme," Fortschr.-Ber. VDI, Reihe 11, No. 53, 1983.

[11] H. Bremer, *Dynamik und Regelung mechanischer Systeme.* Teubner Studienbücher, 1988.

[12] H. Bremer, "Zwang und Zwangskräfte," in *Arch. Appl. Mech.,* Vol. 62, 1992, pp. 158–171.

[13] H. Bremer and F. Pfeiffer, *Elastische Mehrkörpersysteme.* Teubner-Verlag, Stuttgart, 1992.

[14] O. Buxbaum, *Betriebsfestigkeit.* Verlag Stahleisen mbH, Düsseldorf, 1986.

[15] O. Buxbaum and O. Svenson, "Zur Beschreibung von Betriebsbeanspruchungen mit Hilfe statistischer Kenngrößen," in *ATZ Automobiltechnische Zeitschrift,* Vol. 75, No. 6, 1973.

[16] R. W. Cottle and G. B. Dantzig, "Complementary pivot theory of mathematical programming," in *Linear Algebra Appl.,* Vol. 1, 1968, pp. 103–125.

[17] J. le R. d'Alembert, *Traité de dynamique.* Paris, 1743.

[18] E. Eich, *Projizierende Mehrschrittverfahren zur numerischen Lösung von Bewegungsgleichungen technischer Mehrkörpersysteme mit Zwangsbedingungen und Unstetigkeiten.* VDI-Fortschrittberichte Mechanik / Bruchmechanik, Reihe 18, Nr. 109, VDI-Verlag, Düsseldorf, 1992.

[19] A. Fritzer, *Nichtlineare Dynamik von Steuertrieben.* VDI-Fortschrittberichte Schwingungstechnik, Reihe 11, Nr. 176, VDI-Verlag, Düsseldorf, 1992.

[20] C. Führer and B. Leimkuhler, "Formulation and numerical solution of the equations of constrained mechanical motion," DFVLR-Forschungsbericht DFVLR-FB 8908, März 1989.

[21] Ch. Glocker, *Dynamik von Starrkörpersystemen mit Reibung und Stößen.* Fortschrittberichte VDI, Reihe 18, Nr. 182, VDI-Verlag, Düsseldorf, 1995.

[22] Ch. Glocker and F. Pfeiffer, "Dynamical systems with unilateral contacts," in *Nonlinear Dynam.,* Vol. 3, 1992, pp. 245–259.

[23] Ch. Glocker and F. Pfeiffer, "An LCP-approach for multibody systems with planar friction," in *Proc. Contact Mechanics Int. Symp.,* Lausanne, 1992, pp. 13–30.

[24] Ch. Glocker and F. Pfeiffer, "Complementarity problems in multibody systems with planar friction," in *Arch. Appl. Mech.*, Vol. 63, 1993, pp. 452–463.

[25] Ch. Glocker and F. Pfeiffer, "Multiple Impacts with Friction in Rigid Multibody Systems," in *Nonlinear Dynam.*, Vol. 7, 1995, pp. 471–497.

[26] Ch. Glocker and F. Pfeiffer, "Stick-slip phenomena and application," in *Nonlinearity and Chaos in Engineering Dynamics,* IUTAM Symposium, UCL, July 1993, John Wiley & Sons, 1994, pp. 103–113.

[27] J. Guckenheimer and P. Holmes, *Nonlinear oscillations, dynamical systems and bifurcations of vector fields.* Springer-Verlag, New York, Berlin, Heidelberg, Tokyo 1983.

[28] M. Hajek, "Numerische Behandlung mechanischer Systeme mit Trockenreibung," in *ZAMM,* Vol. 69, No. 5, 1989, T 297–T 299.

[29] M. Hajek, *Reibungsdämpfer für Turbinenschaufeln.* Fortschrittberichte VDI, Reihe 11, Schwingungstechnik, Nr. 128, VDI-Verlag, Düsseldorf, 1990.

[30] G. Hamel, *Theoretische Mechanik.* Springer-Verlag 1967.

[31] J. Holland, "Die instationäre Elastohydrodynamik," in *Konstruktion,* Vol. 30, No. 9, 1978, pp. 363–369.

[32] P. J. Holmes, "The dynamics of repeated impact with a sinusoidally vibrating table," in *J. Sound Vib.*, Vol. 84, 1982, pp. 173–189.

[33] M. Jean and J. J. Moreau, "Dynamics in the presence of unilateral contacts and dry friction; a numerical approach," Note Technique 85-5, Lab. de Mécanique Générale des Milieux Continues, Univ. des Sci. et Techniques du Languedoc, Montpellier, 1985.

[34] M. Jean and J. J. Moreau, "Unilaterality and dry friction in the dynamics of rigid body collections," in *Proc. Contact Mechanics Int. Symp.*, Lausanne, 1992, pp. 31–48.

[35] K. L. Johnson, *Contact mechanics.* Cambridge University Press, Cambridge, 1985.

[36] N. Juranek, "Erweiterte Simulation eines Turbogenerator-Antriebsstrangs mit elektrischen Zustandsgrößen," Thesis at the Dept. of Mech. Eng., Inst. B f. Mech., Tech. Univ. of Munich, 1992.

[37] K. Karagiannis, *Analyse stoßbehafteter Schwingungssysteme mit Anwendung auf Rasselschwingungen in Getrieben.* VDI-Fortschrittberichte Schwingungstechnik, Reihe 11, Nr. 125, Düsseldorf, 1989.

[38] K. Karagiannis and F. Pfeiffer, "Theoretical and experimental investigation of gear-rattling," in *Nonlinear Dynam.*, Vol. 2, 1991, pp. 367–387.

[39] G. Kirchhoff, *Vorlesung über mathematische Physik, Mechanik.* Teubner-Verlag, Leipzig, 1883.

[40] A. Klarbring, "Quadratic programs in frictionless contact problems," in *Int. J. Eng. Sci.*, Vol. 24, No. 7, 1986, pp. 1207–1217.

[41] A. Klarbring, "A mathematical programming approach to three-dimensional contact problems with friction," in *Computer Meth. Appl. Mech. Eng.*, Vol. 58, 1986, pp. 175–200.

[42] A. Klarbring and G. Björkman, "A mathematical programming approach to contact problems with friction and varying contact surface," in *Computers & Structures*, Vol. 30, No. 5, 1988, pp. 1185–1198.

[43] K. Klotter, *Technische Schwingungslehre.* Springer-Verlag, Berlin, Heidelberg, New York, 1981 (reprint).

[44] F. Küçükay, *Dynamik der Zahnradgetriebe.* Springer-Verlag, Berlin, Heidelberg, New York, 1987.

[45] F. Küçükay and F. Pfeiffer, "Über Rasselschwingungen in Kfz-Schaltgetrieben," in *Ing.-Arch.*, Vol. 56, 1986, pp. 25–37.

[46] A. Kunert, *Dynamik spielbehafteter Maschinenteile.* VDI Fortschrittberichte, Reihe 11, Schwingungstechnik, Nr. 175, VDI Verlag, Düsseldorf, 1992.

[47] J. L. Lagrange, *Méchanique Analytique Reprint.* Librairie Scientifique et Technique, Albert Blanchard, Paris, 1965.

[48] C. E. Lemke, "Recent results in complementarity problems," in J. B. Rosen et al., Eds., *Nonlinear programming,* Academic Press, New York, 1970, pp. 349–384.

[49] A. J. Lichtenberg and M. A. Lieberman, *Regular and stochastic motion.* Springer-Verlag, New York, Heidelberg, Berlin, 1983.

[50] P. Lötstedt, "A numerical method for the simulation of mechnical systems with unilateral constraints," Technical Report TRITA-NA-7920, Dept. of Numerical Analysis and Computing Science, The Royal Institute of Technologie, Sweden, 1979.

[51] P. Lötstedt, "Mechanical systems of rigid bodies subject to unilateral constraints," in *SIAM J. Appl. Math.*, Vol. 42, No. 2, 1982, pp. 281–296.

[52] P. Lötstedt, "Numerical simulation of time-dependent contact and friction problems in rigid body mechanics," Technical Report TRITA-NA-8214, Dept. of Numerical Analysis and Computing Science, The Royal Institute of Technologie, Sweden, 1982.

[53] K. Magnus, *Schwingungen.* Teubner-Verlag, Stuttgart, 1976.

[54] K. Magnus, *Grundlagen der Technischen Mechanik.* Teubner-Verlag, Stuttgart, 6. Auflage, 1990.

[55] K. Magnus, *Kreisel, Theorie und Anwendung.* Springer-Verlag, 1971.

[56] M. T. Mason and Y. Wang, "On the inconsistency of rigid-body frictional planar mechanics," in *IEEE CH2555-1*, 1988.

[57] Th. Meitinger and F. Pfeiffer, "Automated assembly with compliant mating parts," in *Proc. IEEE Conf. on Robotics and Automation,* San Diego, California, 1994, pp. 1462–1467.

[58] G. Meltzer, R. Melzig-Thiel, and M. Schatte, "Ein mathematisches Schwingungsmodell für das menschliche Hand-Arm-System," in *Machinenbautechnik,* Vol. 29, No. 2, 1980, pp. 54–58.

[59] F. C. Moon, *Chaotic vibrations.* John Wiley & Sons, 1987.

[60] J. J. Moreau, "Application of convex analysis to some problems of dry friction," in *Trends in applications of pure mathematics to mechanics,* Vol. 2, London, 1979, pp. 263–280.

[61] J. J. Moreau, "Dynamique de systèmes a liaisons unilatérales avec frottement séc éventuel: essais numériques," Technical Report Note Technique 85-1, Lab. de Mécanique Générale des Milieux Continues, Univ. des Sci. et Techniques du Languedoc, Montpellier, 1985.

[62] J. J. Moreau, "Standard inelastic shocks and the dynamics of unilateral constraints," in CISM courses and lectures, *Unilateral problems in structural analysis,* Springer-Verlag, Wien, New York, 1985, pp. 173–221.

[63] J. J. Moreau, "Une formulation de la dynamique classique," Technical Report 5, C.R. Acad. Sci. Paris, Série II, 1987.

[64] J. J. Moreau, "Unilateral contact and dry friction in finite freedom dynamics," in CISM courses and lectures, *Nonsmooth mechanics and applications,* Springer-Verlag, Wien, New York, 1988.

[65] Sh. Murato et al., "Synchronizer and shift system optimization for improved manual transmission shiftability," SAE Technical Paper Series, ISSN 0148-7191, Society of Automotive Engineers, Inc., 1989.

[66] K. G. Murty, *Linear complementarity, linear and nonlinear programming.* Heidermann-Verlag, Berlin, 1988.

[67] P. D. Panagiotopoulos, *Ungleichungsprobleme in der Mechanik.* Habilitation, Aachen, 1977.

[68] P. D. Panagiotopoulos, *Inequality problems in mechanics and applications.* Birkhäuser-Verlag, Basel, Boston, Stuttgart, 1985.

[69] P. D. Panagiotopoulos, *Hemivariational inequalities—applications in mechanics and engineering.* Springer-Verlag, Berlin, Heidelberg, New York, 1993.

[70] L. A. Pars, *A treatise on analytical dynamics.* Ox Bow Press, Woodbridge, 1979.

[71] F. Pfeiffer, "Mechanische Systeme mit unstetigen Übergängen," in *Ing.-Arch.,* Vol. 54, 1984, pp. 232–240.

[72] F. Pfeiffer, "Über unstetige, insbesondere stoßerregte Schwingungen," in *Flugwiss. Weltraumforsch.,* Vol. 12, 1988, pp. 358–367.

[73] F. Pfeiffer, "Seltsame Attraktoren in Zahnradgetrieben," in *Ing.-Arch.,* Vol. 58, 1988, pp. 113–125.

[74] F. Pfeiffer, "Theorie des Getrieberasselns," in *VDI-Berichte,* Nr. 697, 1988, pp. 45–65.

[75] F. Pfeiffer, *Einführung in die Dynamik.* Leitfäden der angewandten Mathematik und Mechanik Bd. 65, Teubner Studienbücher Mechanik, Teubner-Verlag, Stuttgart, 1989.

[76] F. Pfeiffer, "Dynamische Systeme mit unstetig auftretenden Bindungen," in *ZAMM,* Vol. 70, No. 4, 1990, T37–T38.

[77] F. Pfeiffer, "Dynamical systems with time-varying or unsteady structure," in *ZAMM,* Vol. 71, 1991, T6–T22.

[78] F. Pfeiffer and A. Fritzer, "Resonanz und Tilgung bei spielbehafteten Systemen," in *ZAMM,* Vol. 72, No. 4, 1992, T38–T40.

[79] F. Pfeiffer, Ch. Glocker, and M. Hajek, "Turbine blade damper—an example for stick-slip-motion," in *Unilateral contact and dry friction,* Abstracts of Communications of Euromech 273, La Grand Motte, 1990.

[80] F. Pfeiffer and M. Hajek, "Stick-slip motion of turbine blade dampers," in *Phil. Trans. R. Soc. London,* A 338, 1992, pp. 503-517.

[81] F. Pfeiffer and F. Küçükay, "Eine erweiterte mechanische Stoßtheorie und ihre Anwendung in der Getriebedynamik," in *VDI-Z.*, Vol. 127, No. 9, 1985, pp. 341–349.

[82] F. Pfeiffer and W. Prestl, "Hammering in diesel-engine driveline systems," in *Nonlinear Dynam.*, Vol. 5, 1994, pp. 477–492.

[83] F. Pfeiffer and E. Reithmeier, *Roboterdynamik.* Teubner Studienbücher Mechanik, Teubner, 1987.

[84] F. Pfeiffer, W. Seyfferth, and H. Wapenhans, "Dynamics and control of automated assembly processes," in *Dynamics and control of mechanical systems in 1993 and beyond,* Kluwer Academic Publishers, Dordrecht, The Netherlands, 1995.

[85] F. Pfeiffer and H. Wapenhans, "Simulation of the nonlinear dynamics of airplane landing gear," in *Proc. 4th Japanese-German Sem. on Nonlinear Probl. in Dyn. Sys.*, Kobe, Japan, Oct. 1989.

[86] F. Pfeiffer and H. Wapenhans, "Simulation des Synchronisationsvorgangs an der ZF-D-Synchronisierung," Lehrstuhl B für Mechanik, TU-München, Dezember 1992.

[87] W. Prestl, *Zahnhämmern in Rädertrieben von Dieselmotoren.* VDI-Fortschrittberichte Schwingungstechnik, Reihe 11, Nr. 145, VDI-Verlag, Düsseldorf, 1991.

[88] R. T. Rockafellar, *Convex analysis.* Princeton University Press, Princeton, New Jersey, 1972.

[89] W. Seyfferth, *Modellierung unstetiger Montageprozesse mit Robotern.* Fortschrittberichte VDI, Reihe 11, Nr. 199, VDI Verlag, Düsseldorf, 1993.

[90] W. Seyfferth and F. Pfeiffer, "Dynamics of rigid and flexible part mating with a manipulator," in *Proc. IMACS Symp. MCTS,* Lille, 1991.

[91] W. Seyfferth and F. Pfeiffer, "Dynamics of assembly processes with a manipulator," in *Proc. 1992 IEEE/RSJ Int. Conf. on Int. Robots and Systems,* Vol. 2, 1992, pp. 1303–1310.

[92] W. Seyfferth and F. Pfeiffer, "Modelling of time-varying contact problems in multibody systems," in *Proc. of 12th Symp. on Eng. Appl. of Mechs,* Montral, Canada, 1994, pp. 579–588.

[93] S. W. Shaw, "The dynamics of a harmonically excited system having rigid amplitude constraints. Part I: Subharmonic and local bifurcations," in *J. Appl. Mech.,* Vol. 52, June 1985, pp. 453–458.

[94] S. W. Shaw, "The dynamics of a harmonically excited system having rigid amplitude constraints. Part II: Chaotic motion and global bifurcations," in *J. Appl. Mech.*, Vol. 52, June 1985, pp. 459–464.

[95] S. W. Shaw and P. J. Holmes, "A periodically forced piecewise linear oscillator," in *J. Sound Vib.*, Vol. 90, No. 1, 1983, pp. 129–155.

[96] Y. Wang and M. T. Mason, "Modeling impact dynamics for robotic operations," in *Proc. IEEE Int. Conf. on Robotics and Automation,* 1987, pp. 678–685.

[97] Y. Wang and M. T. Mason, "Two-dimensional rigid-body collision with friction," in *J. Appl. Mech.*, Vol. 59, 1992, pp. 635–642.

[98] H. Wapenhans, "Dynamik und Regelung von Flugzeugfahrwerken," Thesis at the Dept. of Mech. Eng., Inst. B f. Mech., Tech. Univ. of Munich, April 1989.

[99] H. Wapenhans, *Optimierung von Roboterbewegungen bei Manipulationsvorgängen.* Fortschrittberichte VDI, Reihe 2, Nr. 304, VDI Verlag, Düsseldorf, 1994.

[100] H. Wapenhans, W. Seyfferth, and F. Pfeiffer, "Hybrid position and force control with unsteady assembly dynamics," in *IMACS/SICE Int'l Symp. RMS*^{2}, 1992, pp. 1279–1284.

[101] A. Waterstraat, "Das instationär belastete Gleitlager endlicher Breite," Dissertation, Universität Hannover, 1981.

[102] M. Weck and P. Fritsch, "Zahnflankenhämmern," in *VDI-Z.*, Vol. 127, No. 7, April(I), 1985.

[103] M. Weck, S. Lachenmaier, and H. Salje, "Numerische Simulation des dynamischen Leerlaufverhaltens von Pkw-Getrieben," in *VDI-Z.*, Vol. 126, No. 18, Sept.(II), 1984.

[104] D. E. Whitney, R. E. Gustavson, and M. P. Hennessey, "Designing chamfers," in *Int. J. Robotics Res.*, Vol. 2, No. 4, 1983, pp. 3-18.

INDEX